21st Century Manufacturing

The Role of the Manufacturing Extension Partnership Program

Committee on 21st Century Manufacturing:
The Role of the Manufacturing Extension Partnership Program
of the National Institute of Standards and Technology

Board on Science, Technology, and Economic Policy

Policy and Global Affairs

Philip P. Shapira
Committee Chairman

Charles W. Wessner
Editor

NATIONAL RESEARCH COUNCIL
OF THE NATIONAL ACADEMIES

THE NATIONAL ACADEMIES PRESS
Washington, D.C.
www.nap.edu

THE NATIONAL ACADEMIES PRESS 500 Fifth Street NW Washington DC 20001

NOTICE: The project that is the subject of this report was approved by the Governing Board of the National Research Council, whose members are drawn from the councils of the National Academy of Sciences, the National Academy of Engineering, and the Institute of Medicine. The members of the committee responsible for the report were chosen for their special competences and with regard for appropriate balance.

This study was supported by: Contract/Grant No. SB134106Z0011, Task Order #9, between the National Academy of Sciences and the National Institute of Standards and Technology. Any opinions, findings, conclusions, or recommendations expressed in this publication are those of the author and do not necessarily reflect the views of the organizations or agencies that provided support for the project.

International Standard Book Number 13: 978-0-309-29117-0
International Standard Book Number 10: 0-309-29117-8

Additional copies of this report are available for sale from the National Academies Press, 500 Fifth Street, NW, Keck 360, Washington, DC 20001; (800) 624-6242 or (202) 334-3313; http://www.nap.edu/ .

Printed in the United States of America

THE NATIONAL ACADEMIES

Advisers to the Nation on Science, Engineering, and Medicine

The **National Academy of Sciences** is a private, nonprofit, self-perpetuating society of distinguished scholars engaged in scientific and engineering research, dedicated to the furtherance of science and technology and to their use for the general welfare. Upon the authority of the charter granted to it by the Congress in 1863, the Academy has a mandate that requires it to advise the federal government on scientific and technical matters. Dr. Ralph J. Cicerone is president of the National Academy of Sciences.

The **National Academy of Engineering** was established in 1964, under the charter of the National Academy of Sciences, as a parallel organization of outstanding engineers. It is autonomous in its administration and in the selection of its members, sharing with the National Academy of Sciences the responsibility for advising the federal government. The National Academy of Engineering also sponsors engineering programs aimed at meeting national needs, encourages education and research, and recognizes the superior achievements of engineers. Dr. C. D. Mote, Jr., is president of the National Academy of Engineering.

The **Institute of Medicine** was established in 1970 by the National Academy of Sciences to secure the services of eminent members of appropriate professions in the examination of policy matters pertaining to the health of the public. The Institute acts under the responsibility given to the National Academy of Sciences by its congressional charter to be an adviser to the federal government and, upon its own initiative, to identify issues of medical care, research, and education. Dr. Harvey V. Fineberg is president of the Institute of Medicine.

The **National Research Council** was organized by the National Academy of Sciences in 1916 to associate the broad community of science and technology with the Academy's purposes of furthering knowledge and advising the federal government. Functioning in accordance with general policies determined by the Academy, the Council has become the principal operating agency of both the National Academy of Sciences and the National Academy of Engineering in providing services to the government, the public, and the scientific and engineering communities. The Council is administered jointly by both Academies and the Institute of Medicine. Dr. Ralph J. Cicerone and Dr. C. D. Mote, Jr., are chair and vice chair, respectively, of the National Research Council.

www.national-academies.org

Committee on 21st Century Manufacturing: The Role of the Manufacturing Extension Partnership Program of the National Institute of Standards and Technology*

Philip P. Shapira, *Chair*
Professor of Management, Innovation and Policy
Director, Manchester Institute of Innovation Research
Manchester Business School
University of Manchester
and
Professor, School of Public Policy
Director, Georgia Tech Program in Science, Technology and Innovation Policy
Georgia Institute of Technology

Edward Breiner
President & CEO
Schramm, Inc.

Mary L. Good (NAE)
Dean Emeritus, Donaghey College of Engineering and Information Technology
Special Advisor to the Chancellor for Economic Development
University of Arkansas at Little Rock

James Griffith
President & CEO
The Timken Company

Robert James
President
Stoneleigh Strategies Inc.

Ginger Lew
Managing Director
Enduring Hydro

Deborah J. Nightingale (NAE)
Professor of the Practice of Aeronautics and Astronautics and Engineering Systems
Massachusetts Institute of Technology

Luis M. Proenza
President
University of Akron

Paul K. Wright (NAE)
Director
Center for Information Research in the Interest of Society
A. Martin Berlin Chair in Mechanical Engineering
University of California, Berkeley

*As of September 2013

PROJECT STAFF

Charles W. Wessner
Study Director

McAlister T. Clabaugh
Program Officer

Sujai J. Shivakumar
Senior Program Officer

David E. Dierksheide
Program Officer

CONSULTANTS

Robin Gaster
Innovation Competititons LLC

Thomas R. Howell
Consultant

David B. Watters
Global Advantage Consulting Group, Inc.

Jan Youtie
Georgia Institute of Technology

For the National Research Council (NRC), this project was overseen by the Board on Science, Technology and Economic Policy (STEP), a standing board of the NRC established by the National Academies of Sciences and Engineering and the Institute of Medicine in 1991. The mandate of the Board on Science, Technology, and Economic Policy is to advise federal, state, and local governments and inform the public about economic and related public policies to promote the creation, diffusion, and application of new scientific and technical knowledge to enhance the productivity and competitiveness of the U.S. economy and foster economic prosperity for all Americans. The STEP Board and its committees marshal research and the expertise of scholars, industrial managers, investors, and former public officials in a wide range of policy areas that affect the speed and direction of scientific and technological change and their contributions to the growth of the U.S. and global economies. Results are communicated through reports, conferences, workshops, briefings, and electronic media subject to the procedures of the National Academies to ensure their authoritativeness, independence, and objectivity. The members of the STEP Board* and the NRC staff are listed below:

*As of August 2013.

continued

Contents

Preface

Manufacturing strength remains tightly linked to the innovative potential and competitiveness of nations. "In many sectors, innovative methods and ideas are generated and perfected through the process of making things. In the recent Report to the President on Ensuring American Leadership in Advanced Manufacturing, the President's Council of Advisors on Science and Technology (PCAST) and the President's Innovation and Technology Advisory Committee (PITAC) emphasized the critical importance of advanced manufacturing in driving knowledge production and innovation in the United States.[1] Manufacturing companies in the United States are responsible for over two-thirds of the industrial research and development (R&D), employing the majority of domestic scientists and engineers. Furthermore, manufacturing R&D is a primary source of innovative new service-sector technologies, so that its benefits reach beyond the manufacturing arena."[2]

The Manufacturing Extension Partnership (MEP)—a program of the U.S. Department of Commerce's National Institute of Standards and Technology (NIST)—has sought for more than two decades to strengthen American manufacturing. It is a national network of affiliated manufacturing extension centers and field offices located throughout all fifty states and Puerto Rico. Funding for MEP centers comes from a combination of federal, state, local, and private resources. Centers work directly with manufacturing firms in their state or sub-state region. MEP centers provide expertise, services, and assistance directed toward improving growth, supply chain positioning, leveraging emerging technologies, improving manufacturing processes, work

[1]President's Council of Advisors on Science and Technology, 2011, "Report to the President on Ensuring American Leadership in Advanced Manufacturing." Access at *<http://www.whitehouse.gov/sites/default/files/microsites/ostp/pcastadvanced-manufacturing-june2011.pdf>*.

[2]The current status of U.S. manufacturing is discussed in detail by a new report by the Department of Commerce, written in consultation with the National Economic Council. This report argues that, despite recent declines, manufacturing remains a vital part of the U.S. economy. U.S. Department of Commerce, "The Competitiveness and Innovative Capacity of the United States," Washington, DC, January 2012.

force training, and the application and implementation of information in client companies through direct assistance provided by center staff and from partner organizations and third-party consultants.

Given the importance of innovation to economic growth and competitiveness, MEP today is seeking to evolve beyond its traditional "technology push" mission to increase the innovative capacity of the nation's manufacturers.

THE STEP BOARD'S RESEARCH ON INNOVATION AND COMPETITIVENESS

The National Research Council, under the auspices of its Board on Science, Technology, and Economic Policy (STEP), has since 1991 undertaken a program of activities to improve policymakers' understandings of the interconnections of science, technology, and economic policy and their importance for the American economy and its international competitive position. The Board's activities have contributed to increased policy recognition of the importance of technology, innovation, and entrepreneurship to economic growth. This work is in many ways congruent with economic growth theory, which emphasizes the role of technology creation in the generation of significant growth externalities.[3] In addition, many economists have recognized the limitations of traditional trade theory, particularly with respect to the reality of imperfect international competition. Public-private partnerships are increasingly recognized for their contributions to the commercialization of state and national investments in research and development. Such partnerships help address the challenges associated with the transition of research into products ready for the marketplace.[4]

One important element of STEP analysis has concerned the growth and impact of foreign technology programs. [5] U.S. competitors have launched substantial programs to support new technologies, small firm development, innovative production at large companies, and consortia among large and small firms to strengthen national and regional positions in strategic sectors. Some governments overseas have chosen to provide public support to innovation to overcome the market imperfections apparent in their national innovation systems. They believe that the rising costs and risks associated with new

---

[3]National Research Council, *Enhancing Productivity Growth in the Information Age*, D. W. Jorgenson and C. Wessner, eds., Washington, DC: The National Academies Press, 2007.

[4]National Research Council, *Government-Industry Partnerships for the Development of New Technologies: Summary Report*, C. Wessner, ed., Washington, DC: The National Academies Press, 2003.

[5]For a review of the challenges and opportunities faced by the United States in the face of unprecedented global competition for developing, commercializing, and manufacturing the next generation of technologies, see National Research Council, *Rising to the Challenge: U.S. Innovation Policy for the Global Economy*, C. Wessner and A. Wm. Wolff, eds., Washington, DC: The National Academies Press, 2012.

Statement of Task

An ad hoc committee will carry out an evaluation of the operation, achievements, and challenges of the Manufacturing Extension Partnership (MEP) program at the National Institute of Standards and Technology. The committee will hold a series of fact-finding workshops and commission research papers and case studies to review and document the program's current achievements, challenges, and new opportunities; identify and review similar national programs from abroad in order to draw on foreign practices, funding levels, and accomplishments as a point of reference; and discuss current needs and initiatives in light of the global focus on advanced manufacturing.

One workshop summary will be prepared in the course of the study. The committee will develop findings and recommendations to improve program operations and impact for inclusion in the committee's final consensus report.

potentially high-payoff technologies, and the growing global dispersal of technical expertise, underscore the need for national R&D programs to support new and existing high-technology firms within their borders.[6]

THE MEP STUDY

In 2011, MEP requested the National Academies' Board on Science, Technology, and Economy Policy to undertake a review of MEP. As noted in the Statement of Task, this study seeks to generate a better understanding of the operation, achievements, and challenges of the Manufacturing Extension Partnership program in its mission to support, strengthen, and grow U.S. manufacturing.

ACKNOWLEDGMENTS

On behalf of the National Academies, we would like to express our appreciation for the many contributions to the study. We would particularly like to express our thanks to Robin Gaster for his contributions to the Committee's analysis of the MEP and to Thomas Howell for his contributions to our understanding of a broad range of foreign programs and to David Watters for his advancement of the Committee's understanding of the Canadian Industrial Research Assistance Program. The Committee would also like to acknowledge the many contributions of our foreign colleagues who provided invaluable insights into the goals, structure, operations, and achievements of the manufacturing support programs reviewed in the course of the study. Without

[6]For a review of global initiatives in this regard, see National Research Council, *Rising to the Challenge: U.S. Innovation Policy for the Global Economy*, op. cit.

their cooperation and enthusiastic support we would not have been able to carry out the study. Similarly, we would like to acknowledge the many contributions of the leadership and staff of the MEP centers who spoke at the conference, participated in our workshops, hosted our site visits, and enabled us to draw on their insights and experience. We would also like to express our thanks to the project staff, notably Sujai Shivakumar, who devoted his energy and experience to refining the report and supporting the review process, to McAlister Clabaugh and David Dawson, who played key roles in organizing, and in the case of Mr. Clabaugh, participating in the multiple missions abroad and the many off-site meetings around the country, while contributing directly to the preparation of the final report.

ACKNOWLEDGMENT OF REVIEWERS

This report has been reviewed in draft form by individuals chosen for their diverse perspectives and technical expertise, in accordance with procedures approved by the National Academies' Report Review Committee. The purpose of this independent review is to provide candid and critical comments that will assist the institution in making its published report as sound as possible and to ensure that the report meets institutional standards for objectivity, evidence, and responsiveness to the study charge. The review comments and draft manuscript remain confidential to protect the integrity of the process.

We wish to thank the following individuals for their review of this report: Timothy Bartik, W.E. Upjohn Institute for Employment Research; David Bodde, Clemson University; Gary Cowger, GLC Ventures, LLC; George Dieter, University of Maryland; Dietmar Harhoff, Institute for Innovation Research, Technology Management and Entrepreneurship; Christopher Hill, George Mason University; Jennie Hwang, H-Technologies Group; Ron Jarmin, U.S. Census Bureau; Robert Kill, Enterprise Minnesota; Sridhar Kota, University of Michigan; Jean-Michel LeRoux, Institut Carnot Centrale Supélec; Thomas Mahoney, West Virginia University; Elizabeth Reynolds, Massachusetts Institute of Technology; and Gavriel Salvendy, Purdue University.

Although the reviewers listed above have provided many constructive comments and suggestions, they were not asked to endorse the conclusions or recommendations, nor did they see the final draft of the report before its release. The review of this report was overseen by Irwin Feller, Pennsylvania State University and Robert Frosch, Harvard University. Appointed by the National Academies, they were responsible for making certain that an independent examination of this report was carried out in accordance with institutional procedures and that all review comments were carefully considered. Responsibility for the final content of this report rests entirely with the authoring committee and the institution.

Philip Shapira Charles W. Wessner

Summary

Through its national network of affiliated manufacturing extension centers and field offices located throughout all fifty states and Puerto Rico, the Manufacturing Extension Partnership (MEP) provides small and medium-sized manufacturers access to expertise and offers services and assistance directed toward improving growth, supply chain positioning, leveraging emerging technologies, improving manufacturing processes, work force training, and the application and implementation of information in client companies. These services are provided in some cases through direct assistance provided by MEP center staff and, in other cases, from partner organizations and third-party consultants.

After more than 20 years of operation, the MEP program faces new challenges as the nation's manufacturing firms adapt to a changing competitive environment. As summarized in Box S-1, these challenges include assisting firms that vary widely in terms of size, technological focus, and business needs.

Given the growing recognition of innovation and manufacturing to nation's economic growth and competitiveness, MEP today is seeking to evolve beyond its traditional "technology push" mission to increase the innovative capacity of the nation's manufacturers.

In 2011, NIST requested the National Academies' Board on Science, Technology, and Economy Policy (STEP) to undertake a review of MEP. In response, a committee of the STEP Board held a series of fact-finding workshops and commissioned research papers and case studies to review and document the program's current achievements, challenges, and new opportunities. The Committee also identified and reviewed similar national programs from abroad in order to draw on foreign practices, funding levels, and accomplishments as a point of reference and to discuss current U.S. needs and initiatives in light of the global focus on advanced manufacturing.

Box S-1
MEP's New Environment

Changing economic environment

- **U.S. manufacturing** has faced increasing challenges and recent developments have accelerated what many see as the need for rapid change and adjustment in this sector.
- **Globalization**, and especially the globalization of supply chains in manufacturing, means that American third, fourth, and fifth-tier suppliers can now face direct challenges from foreign manufacturers with different cost structures.
- **Innovation,** and especially the accelerating pace of technological change in manufacturing, means that firms need to continually and rapidly adapt to—and develop—new products and processes.
- **Regulation,** for example environmental regulation, is changing the way that manufacturing operates in the United States, and firms must adapt here too.

The population of manufacturing companies and MEP's market targets

- **The actual number of potential clients**—i.e., manufacturing firms—in any given center's region varies substantially, from almost 30,000 firms in southern California to less than 3,000 in some smaller states
- **Manufacturers range in size** from the large majority with fewer than 20 employees, to a handful with many thousands.
- **The actual market for MEP services** varies by region, but it likely does NOT include:
    - The very large number of firms that are in the main too small to take advantage of MEP's services.
    - Firms larger than 250 employees who have other alternatives and tend to prefer commercial services providers.
    - The substantial numbers of smaller and mid-sized firms that have established customers, established products, and see limited value in change. They may not all be correct in this assessment, but their self-exclusion from MEP services is nonetheless a reality.

- Given very limited resources for outreach, centers must therefore decide how to target their efforts to maximize impact, meet program objectives, and at the same time generate sufficient fee revenues to meet federal match requirements.

MEP structure and funding

- **Diversity.** NIST MEP must lead a system that includes 60 centers that are quite diverse in terms of size, capabilities, market penetration, strategic orientation, functional organization, and openness to innovation.
- **Balance of authority**. Centers have their own local supporters and funders, which translate into considerable independence. NIST MEP can influence center choices and encourage new initiatives, e.g., the Next Generation Strategy, but insofar as NIST MEP provides a maximum of one third of center funding, this tends to limit leverage on day-to-day operations. Moreover, in practice, center funding formulas are largely fixed, with very limited discretion for MEP management to shift resources across centers.

Varying center capabilities

- Centers have to a considerable degree focused on their existing core services for more than a decade. Staff are tuned to providing these services—and introduction of new services will require substantial readjustment of personnel.
- Some centers have developed significant in-house capabilities. Others rely almost exclusively on external consultants to provide services.
- Some centers have developed training and support programs focused on linking to C-level staff; others have not begun this process.
- Some centers are already well along the road toward more innovation-oriented practice, while many others have not yet engaged.

Based on this review, the National Academies Committee for the Study of the Manufacturing Extension Partnership developed a series of findings and recommendations, which are detailed in Chapter 8 of this report. The core findings and recommendations of are as follows:

CORE FINDINGS

- **The Value of MEP:**
  - Spread across 50 states and 60 centers, MEP is the leading U.S. Government program designed explicitly to provide support services to the nation's small and medium manufacturers.[1]

[1]See Finding 2.

- Overall, MEP's support for lean manufacturing shows evidence of success. Evidence of this overall success is found in academic reviews, as well as in the analysis, case studies and interviews of company staff conducted for this study.[2]
- MEP's annual budget is relatively modest, given the importance of SME manufacturing in the U.S. and in comparison to similar foreign programs.[3]

- **Focus on Innovation:**

  - MEP efforts to add support for innovation and growth through its national network are commendable. This strategy is in line with recent academic and policy analyses that call for U.S. manufacturers to become more innovative and more focused on growth and international competitiveness.[4]
  - NIST-MEP evaluations and programs are evolving from a focus on quantified metrics derived from a survey of clients and center reporting to a new emphasis on qualitative metrics.[5]
  - Implementation of the Next Generation Strategy (NGS) poses risks and challenges for MEP centers. MEP centers encounter significant risks as they seek to transition from a tight focus on lean production to a much wider range of services that require new clients, new contacts, new kinds of client conversations, new services, new toolsets and capabilities.[6]

- **U.S. Support for Manufacturing:**

  - While the United States has numerous public organizations engaged in applied research, notably in fields such as medicine, agriculture, energy, and defense, large segments of the U.S. manufacturing sector are often underserved with respect to technological support from the national research base.[7]
  - Major U.S. trading partners have initiated substantial programs to support manufacturing. Drawing on a portfolio of programs and policies, they provide shared use facilities, expertise garnered from working on multiple projects over long periods of time, and connections to other parts of the innovation ecosystem, be it suppliers,

---

[2]See Finding 3.
[3]See Finding 4.
[4]See Finding 7.
[5]See Finding 12.
[6]See Finding 8.
[7]See Finding 15.

customers, or sources of finance, as well as a ready and low cost source of technical support.[8]

- **Best Practices**

  - The most successful foreign applied research programs around the world share elements of the following characteristics:[9] The institutions benefit from substantial and sustained funding over a long time horizon, providing first class equipment and infrastructure and retaining qualified staff to maintain focus on core missions and client base. They often have a high degree of autonomy in establishing technology strategies, deploying resources, and developing relationships with the private sector, and rely on funding from public and private contract research. They also build strong links to local clusters, partner with universities, and provide a source of training. Many of these institutions also benefit from periodic assessments by independent reviewers on a longer-term basis.

CORE RECOMMENDATIONS

While the committee finds that the MEP program provides valuable help to small manufacturers, with the enhancements recommended here, the program will be an increasingly important element in the nation's portfolio of programs to support manufacturing and the jobs it brings.

- **Funding for MEP should be commensurate with the importance of manufacturing:**

  - Funding for MEP should be commensurate with the importance of manufacturing to the growth of the economy and the program's proven ability to contribute to improved firm performance and adapt to the changing needs of the manufacturing sector. The current level of funding is not adequate to maintain the program's focus on small firms, build new services around the Next General Strategy, and provide the resources required to drive the improvements recommended by this assessment.[10]

---

[8]See Finding 13.
[9]See Finding 16.
[10]See Recommendation 6.

Box S-2
Challenges NIST Needs to Address for MEP

To adapt the MEP program to the new global environment, NIST needs to:

- Continue efforts to enhance growth, increase innovation, and improve sustainability though its Next Generation Strategy."[a]
- Continue to encourage lean manufacturing.[b]
- Significantly improve its collection and analysis of performance data.[c]
- Take into account lessons from U.S. and international best practice.[d]
- Identify best practices, and measure the relative accomplishments of individual initiatives and center activities.[e]
- Use its resources to leverage maximum beneficial outcomes for the U.S. manufacturing sector rather than focus on reaching the maximum number of manufacturers.[f]
- In order to refine the MEP to better serve its mission:
    - Single provider contracts should be used sparingly, should require a detailed justification, and should have clear deliverables and metrics, and be closely monitored.
    - NIST MEP should be more flexible in the management of the funding of MEP centers.[g]
    - The fixed federal match of one-to-two from the states and centers should be changed to a match approach with more flexibility for NIST management and the centers.[h]

The Findings and Recommendations outlined in Chapter 8 spell out these issues in greater detail. However, these are actions that NIST should take to enhance the effectiveness of MEP.

[a]See Recommendation 4.
[b]See Recommendation 3.
[c]See Recommendation 5.
[d]See Recommendation 8.
[e]See Recommendation 5.
[f]See Recommendation 2.
[g]See Recommendation 7.
[h]See Recommendation 7.

  - NIST-MEP should be more flexible in the management of the funding of MEP centers. The fixed federal match of one-to-two from the states and centers should be changed to a match approach with more flexibility for NIST management and the centers.[11]

- **NIST-MEP should significantly improve its collection and analysis of performance data[12]**

  - MEP should develop both in-house and external data and the analytic capacity to provide ongoing and independent evaluations of its new initiatives.
  - MEP data and analyses should identify best practices, and measure the relative accomplishments of individual initiatives and center activities. The centers should be given considerable latitude and be viewed as laboratories for new initiatives, where rapid and effective assessment in turn leads to the adoption and transfer of best practices.

- **Take into account lessons from U.S. and international best practice.**

  - Any effort to establish programs to further support manufacturing should thoroughly assess existing U.S. resources, organizations and institutions already engaged in applied research and should take into account lessons from U.S. and international best practice.[13]

[11]See Recommendation 7.
[12]See Recommendation 5.
[13]See Recommendation 8.

Chapter 1

The Structure and Role of MEP

The Hollings Manufacturing Extension Partnership (MEP) traces its origins to the establishment of the Manufacturing Technology Centers Program in 1989.[1] This program was developed as a part of the nation's response to the perceived decline in position of the United States vis-à-vis Japan as a leading manufacturer of high-technology goods. Located within the National Institute of Standards and Technology (NIST), MEP has offered technical and business support primarily to the nation's small- and medium-sized manufacturers.

Two decades later, the rapid rise of China as a global locus of manufacturing is once again raising concerns about U.S. competitiveness.[2] To address these concerns, MEP is seeking to refine and adapt its mission to encourage product innovation and commercial development among the nation's manufacturers. In its own words, it has begun a transition from "reactive" strategies to the "proactive pursuit of increased profits and overall growth."[3]

This study, initiated at the request of MEP, seeks to illuminate the operation, achievements, and challenges of the program in its mission to support, strengthen, and grow U.S. manufacturing.[4] Given that MEP intends to stimulate greater productivity and innovativeness in the United States among

---

[1]The Manufacturing Technology Centers (MTC) Program at the National Institute of Standards and Technology (NIST) began in 1989 with regional centers in three states—South Carolina, Ohio, and New York. The mission of these regional centers was to support the transfer of manufacturing technology to improve the productivity and technological capabilities of America's small manufacturers. The number of centers grew rapidly to provide services to all 50 states and Puerto Rico, and in 1998, the program was renamed the Manufacturing Extension Partnership (MEP). In 2004, the program was redesignated the Hollings Manufacturing Extension Partnership in honor of Senator Fritz Hollings. Source SC-MEP at <*http://www.scmep.org/history/*>.

[2]For a review of the current competitive challenges from China and other nations facing the United States, see National Research Council, *Rising to the Challenge: U.S. Innovation Policy for the Global Economy*, C. Wessner and A. Wm. Wolff, eds., Washington, DC: The National Academies Press, 2012.

[3]NIST, "The Future of the Hollings Manufacturing Extension Partnership." Accessed on April 17, 2013, at <*http://www.nist.gov/mep/upload/MEP_ExecSummary72dpi.pdf*>.

[4]See the Preface to this volume for the Committee's Statement of Task.

Box 1-1
The Hollings Manufacturing Extension Partnership

The Manufacturing Extension Partnership (MEP), administered by the National Institute of Standards and Technology (NIST) within the Department of Commerce, has sought for more than two decades to strengthen American manufacturing.

Mission. MEP's mission is to "act as a strategic advisor to promote business growth and connect manufacturers to public and private resources essential for increased competitiveness and profitability."[a]

Program Scale: The NIST-MEP federal budget for 2013 is $123 million. The total NIST MEP headquarters staff numbers some 45 people who focus on setting strategy, evaluating the needs and demands of clients, helping facilitate the development of tools, and bringing together the centers into a shared network. Currently, NIST funding is matched 1:2 by individual state centers, using funding primarily from state governments and client fees. The nationwide network includes some 1,300 staff supported by over 2,300 third-party service providers, and the overall budget for the MEP system was about $300 million in 2012.

Decentralized Structure: NIST-MEP works cooperatively with accredited organizations that include nonprofits, state government agencies, and universities to complete the MEP mission. In all, some 60 MEP centers are located across the country, with centers in every state. They vary widely in structure and operating strategy. Pennsylvania, for example, has seven centers; many states have only a single MEP center. California, which accounted for 13 percent of the nation's manufacturing gross domestic product (GDP) in 2011, has two MEP centers serving the state. The work of these centers is further dispersed among some 300 field offices. The centers rely heavily on local partners to design and deliver services that are tailored to the needs of the manufacturing clients.

Evolving Focus: According to then MEP Director Roger Kilmer, "Part of our evolution was to change from offering a technology push, where we knew about which technologies work in a federal lab, to looking at what manufacturers really needed. It also meant learning to look at the entire manufacturing enterprise—not just the tech piece of it, but everything else: the financing, workforce development, marketing, and sales."[b] From an early focus on off-the-shelf manufacturing technologies, basic technical assistance, and plant layout, MEP evolved towards "lean production" in response to demand from companies. The program continues to adapt with a new emphasis on growth and on innovation, reflecting the need for firms to be more proactive in an increasingly competitive world economy.

---

[a]NIST, "The Future of the Hollings Manufacturing Extension Partnership." Accessed on April 17, 2013, at <*http://www.nist.gov/mep/upload/MEP_ExecSummary72dpi.pdf*>

[b]Roger Kilmer, "MEP's Place in the Innovation Chain," in National Research Council, *Strengthening American Manufacturing: The Role of the Manufacturing Extension Partnership—Summary of a Symposium*, C. Wessner, Rapporteur, Washington, DC: The National Academies Press, 2013.

small- and medium-sized manufacturers, it is important to ensure that MEP is configured in such a way that it can meet these challenges.

MEP SUPPORT FOR SMALL MANUFACTURERS

Small and medium manufacturers represent 98 percent of all manufacturing enterprises in the United States. They account for two-thirds of all manufacturing employment and employ nearly 11.3 million people, and contribute over half of the total value added by all U.S. manufacturers.[5] But these small firms face unique challenges.

- **Market Failures:** Small manufacturers often confront failures in information markets due to difficulty in evaluating what information they need and the quality of the available consultants.[6] Many small and medium manufacturers interviewed for this report said that they cannot attract the interest or afford the fees of private consultancies given their small scale, limited resources, and remote locations. In many cases, they observed that private consultancies do not adequately provide the type of services required for their scale of operation.[7]

- **Distinctive Focus:** While the United States has numerous public organizations engaged in applied research, notably in fields such as medicine, agriculture, energy, and defense, large segments of the U.S. manufacturing sector are often underserved with respect to technological support from the national research base.[8] These federal

[5]Susan Helper, Timothy Krueger, and Howard Wial, *Locating American Manufacturing: Trends in the Geography of Production*, Washington, DC: Brookings Institution, April 2012.

[6]For a review of market failures facing small manufacturers and the role for public policy, see Susan Helper and Howard Wial, *Strengthening American Manufacturing: A New Federal Approach*, Washington, DC: The Brookings Institution, September 2010. For a review of the role of MEP in addressing market failures, see Russell M. Frazier, "The Imperatives of Successful Policy Implementation: A Case Study of the Hollings National Institute of Standards and Technology-Manufacturing Extension Partnership (NIST-MEP) Program's Implementation in Arkansas," *International Journal of Learning & Development*, 2012, Vol. 2, No. 4; and Philip Shapira, "Manufacturing Extension: Performance, Challenges and Policy Issues," in L. M. Branscomb and J. Keller, eds., *Investing in Innovation*, Cambridge, MA: MIT Press, 1998, pp. 250-275.

[7]These views were reinforced by MEP Center directors, including those in Pennsylvania, California, and Ohio, interviewed for this study in October-November 2012.

[8]A partial list of federal programs to assist manufacturing includes the Department of Defense's ManTech Program, the National Science Foundation's Engineering Research Centers and Industry/University Cooperative Research Centers, the Department of Energy's Energy Assessment

programs do not focus as directly on providing technical and management advice to these small firms. MEP's focus on small firms represents an important segment of U.S. industry, one that can contribute to enhancing the performance of the small companies that add to the nation's productivity and employment, and one that can serve as a valuable component to the nation's manufacturing strategy.

A series of studies have examined the characteristics of market failures affecting manufacturers, especially small- and medium-sized enterprises (SMEs).[9] These studies highlight asymmetrical information availability and the presence of negative externalities which particularly affect small firms. As a class, small firms underinvest in research and development, innovation, and training not only because of limited resources but also because it is difficult for them to capture the full set of returns from such investments. Additionally, while small manufacturing firms have expertise in their current technologies, they may lack knowledge about new and improved technologies and methods. This may limit investment or lead to unproductive investments. Investment in new and improved technologies and methods also may be hampered by lack of knowledge about new markets, limited or weak networking connections, by limitations or the expense of private consulting expertise, or the short-term nature of demand. Where there is suboptimal investment by small firms in new and improved technologies and methods, this is likely to result in negative externalities for supply chains, regional economies, manufacturing jobs and wages, and overall economic performance. Expert and unbiased technical advice, as offered by the MEP, can help manufacturers diagnose their needs and make optimal choices. MEP services also address transaction costs which limit private market provision of information and guidance by undertaking front-end diagnosis, then referring companies to qualified consultants and private vendors.

In addition to market failure arguments, there are three other complementary rationales for manufacturing extension. First, "public" failure, whereby governmental institutions place disproportionate burdens (e.g., burdens of regulatory compliance) on small firms with limited managerial capability to address or where publicly funded federal research institutions or universities have expertise which is not oriented to or easily accessible by small firms. While

---

Centers, and the Department of Commerce's Export Assistance Centers. At the state level, a number of technology-based economic development organizations also provide support for small manufacturing companies.

[9]Dan Luria, "Why Markets Tolerate Mediocre Manufacturing," *Challenge*, 1996, 39, 4, 11-16; National Academy of Public Administration, The National Institute of Standards and Technology's Manufacturing Extension Program, Report 1, "Reexamining the Core Premise of the MEP Program," 2003; Susan Helper and Marcus Stanley, "Creating Innovation Networks Among Manufacturing Firms: How Effective Extension Programs Work," in Scott Shane, ed., *Economic Development through Entrepreneurship: Government, University and Business Linkages*; Edward Elgar, 2006; Stephen J. Ezell and Robert D. Atkinson, *International Benchmarking of Countries' Policies and Programs Supporting SME Manufacturers*, Washington, DC: Information Technology and Innovation Foundation, 2011.

ideally such government and public failures should be remedied at source, as a practical matter, given the inevitable complexity of government and its institutions, there is an ongoing need to find ways to help small firms directly navigate through these systems. In such cases, manufacturing extension agents can advise or serve as brokers to level the field for small firms. Among the array of publicly sponsored business services, the MEP is the sole federal program primarily focused on strategically assisting small and mid-sized manufacturers.[10] For example, extension agents are able to advise manufacturers on more efficient ways to meet environmental regulatory requirements not just through "end of pipe" solutions but through reorganizing manufacturing operations, reducing materials use, and eliminating waste.

Second, there are strategic rationales for manufacturing extension, to ensure the retention of supply chains and capacity in certain key sectors (including defense) and to help firms address competition from firms in other countries which benefit from well-established public systems of technology and manufacturing support.[11]

Third, there is a cooperative technology rational for manufacturing extension and technology services where such services bring companies, universities, associations, and other organizations together in innovation networks, collaborations, and value chains to reduce transaction costs and coordinate and leverage resources.[12]

MEP STRUCTURE

MEP can be divided into the headquarters, regional offices, and partners, including the MEP centers. There is also a national advisory board that provides strategic advice but is not operationally integrated into MEP. [13] The headquarters staff includes a director and deputy director, as well as senior staff responsible for system operations, program development, center operations, communications, strategic partnerships, and manufacturing policy and research. They are supported by an administration unit.

---

[10]Stone and Associates and the Center for Regional Economic Competitiveness, "Re-examining the Manufacturing Extension Partnership Business Model," report prepared for the NIST Manufacturing Extension Partnership, 2010.

[11]Stephen J. Ezell and Robert D. Atkinson, *International Benchmarking of Countries' Policies and Programs Supporting SME Manufacturers*, Washington, DC: Information Technology and Innovation Foundation, 2011.

[12]Henry Etzkowitz, *The Triple Helix: University-Industry-Government Innovation in Action*, Routledge, 2008; A. Schrank and J. Whitford, "Industrial Policy in the United States: A Neo-Polanyian Interpretation, *Politics and Society*, 2009, 37, 521.

[13]More detail on the Advisory Board as well as its annual reports can be found on the MEP website.

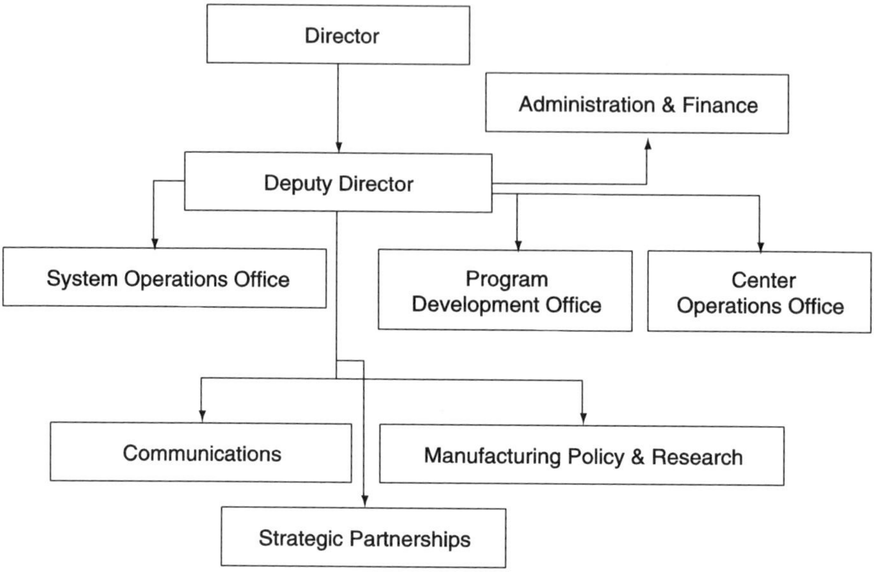

FIGURE 1-1 The MEP organization chart (June 2013).
SOURCE: NIST MEP.

MEP Mission

According to MEP, its mission is "To act as a strategic advisor to promote business growth and connect manufacturers to public and private resources essential for increased competitiveness and profitability."[14]

Because MEP is construed as a partnership between the federal government and local centers operated independently and funded in part both by federal and state dollars, NIST MEP has always sought to maintain a delicate strategic balance between encouraging centers to adopt what it sees as advanced services, and recognizing that its leverage is limited.

NIST MEP influences individual centers by evaluating them on terms designed by NIST MEP, with the implicit sanction of the withdrawal of funding. Yet, centers can only be influenced up to a point. They have other stakeholders locally—usually including the state in which they operate—and they have an overriding imperative to maintain a business that is sustainable over the long run. (As we shall see, this latter concern is of particular relevance in the context of NIST MEP's Next Generation Strategy that seeks to encourage major changes in center operations and strategy.) While NIST MEP does sometimes decertify a center leading to a recompetition for the region, in fact,

[14]NIST MEP, "About the next generation strategy," <*http://www.nist.gov/mep/ngs.cfm*>. Accessed July 11, 2013.

recompetition as a sanction is rarely applied. Most recompetitions occur as a result of changes in state policy or an institution's inability to meet matching requirements.[15] While, in some cases, informal pressure is sometimes applied through direct contact by NIST MEP management, this appears to be neither systematic nor transparent.

The Role of the MEP Centers

NIST MEP partners with local organizations to provide services to local manufacturers. It currently accredits and funds 60 centers with about 370 field locations, with a total of about 1,450 nonfederal staff. The centers, in turn, contract with more than 2,400 third-party service providers.

The centers are operated by independent entities, and are not under the direction of MEP—hence the term *Partnership* in the program title. Centers operate at the hub of a network of relationships. As we will see in more detail below, these relationships can be a major strength for an effective center, leveraging resources on behalf of SMEs and providing a conduit for other organizations into the local population of SMEs. Currently, 17 centers are affiliated with universities and 4 are operated by state governments. The remaining centers are nonprofit organizations.

Most centers use external consultants to provide some services, but few rely on them heavily. Typically centers utilize consultants for 30 percent or less of work for clients, although some—like those in Oklahoma and Southern California—deliberately utilize consultants to the maximum degree possible.

Under growing pressure to enhance center revenue, centers have increasingly moved toward in-house delivery. This trend goes back to the early 2000s, but has accelerated: In 2011, third-party providers overall accounted for only about 20 percent of total hours (see Figure 1-3).

MEP CUSTOMERS

Most MEP customers are small- and medium-sized enterprises, ranging from small job shops focused on meeting the specific demands of existing customers to firms that seek to develop new products and new markets. In some

---

[15] According to the 2011 GAO Report, "Arizona MEP was awarded to a new operator because the center wanted to be a state-based center (it had previously been run out of the Maine MEP center). The Illinois center was awarded to a new operator because the previous operator could not draw down the full federal funds because it could not raise enough nonfederal funds to meet cost share requirements. Additionally, the center was not achieving required metrics after being notified during annual performance reviews conducted by NIST. Annual reviews of MEP centers are required. 15 C.F.R. § 290.8 outlines performance standards and other related requirements for MEP centers. The Pennsylvania MEP in Central Pennsylvania was awarded to a new operator in order to split an existing MEP service area into two distinct service areas." See GAO, "Factors for evaluating the cost share of Manufacturing Extension Program to assist small and medium-sized manufacturers," GAO-11-437R, April 4, 2011.

TABLE 1-1 List of MEP Centers (July 2013)

MEP Centers
Alabama: Alabama Technology Network
Alaska: Alaska Manufacturing Extension Partnership
Arkansas: Arkansas Manufacturing Solutions
California: California Manufacturing Technology Consulting
California: Corporation for Manufacturing Excellence (Manex)
Colorado: Colorado Association for Manufacturing and Technology
Connecticut: Connecticut State Technology Extension Program
Delaware: Delaware Manufacturing Extension Partnership
Florida: Florida Manufacturing Extension Partnership
Georgia: Georgia Manufacturing Extension Partnership
Hawaii: INNOVATE Hawaii
Idaho: Idaho TechHelp
Illinois: Illinois Manufacturing Excellence Center - Downstate
Indiana: Indiana MEP Purdue Technical Assistance Program
Iowa: Iowa Center for Industrial Research and Service
Kansas: Mid-America Manufacturing Technology Center
Kentucky: Advantage Kentucky Alliance
Louisiana: Manufacturing Extension Partnership of Louisiana
Maine: Maine Manufacturing Extension Partnership
Massachusetts: Massachusetts Manufacturing Extension Partnership
Michigan: Michigan Manufacturing Technology Center
Minnesota: Enterprise Minnesota
Mississippi: InnovateMEP Mississippi
Missouri: Missouri Enterprise
Montana: Montana Manufacturing Extension Center
Nebraska: Nebraska Manufacturing Extension Partnership
Nevada: Nevada Industry Excellence
New Hampshire: New Hampshire Manufacturing Extension Partnership
New Jersey: New Jersey Manufacturing Extension Program
New Mexico: New Mexico Manufacturing Extension Partnership
New York: New York Manufacturing Extension Partnership
North Carolina: North Carolina Manufacturing Extension Partnership
North Dakota: North Dakota Manufacturing Extension Partnership
Ohio: Ohio Manufacturing Extension Partnership
Oklahoma: Oklahoma Manufacturing Alliance
Oregon: Oregon Manufacturing Extension Partnership
Pennsylvania: Catalyst Connection
Pennsylvania: Delaware Valley Industrial Resource Center
Pennsylvania: IMC-PA
Pennsylvania: MANTEC
Pennsylvania: Manufacturers Resource Center
Pennsylvania: Northeastern Pennsylvania Industrial Resource Center
Pennsylvania: Northwest Pennsylvania Industrial Resource Center
Puerto Rico: Puerto Rico Manufacturing Extension Inc.
Rhode Island: Rhode Island Manufacturing Extension Services
South Carolina: South Carolina Manufacturing Extension Partnership
South Dakota: South Dakota Manufacturing and Technology Solutions
Tennessee: Tennessee Manufacturing Extension Partnership

Texas: TMAC
Utah: Utah Manufacturing Extension Partnership
Vermont: Vermont Manufacturing Extension Center
Virginia: GENEDGE ALLIANCE
Washington: Impact Washington
West Virginia: West Virginia Manufacturing Extension Partnership
Wisconsin: Northwest Wisconsin Manufacturing Outreach Center
Wisconsin: Wisconsin Manufacturing Extension Partnership
Wyoming: Manufacturing-Works

SOURCE: MEP. <*http://patapsco.nist.gov/mep/centers-near-you/index.htm*>. Accessed July 12, 2013.

Box 1-2
Examples of MEP Center Models

The MEP program has evolved to include several differing center business models. Current MEP cooperative models include:

Recipient is a nonprofit organization: The nonprofit organization partners with other organizations, including the state government; academia, including community colleges; economic development organizations; and consultants in the region. MEP center staff could be employees of the nonprofit or employees of the partner organization, where they are managed under a specific contractual arrangement.

Single recipient located at a host organization. As employees of the host organization, typically a university, staff may work on MEP and other projects. The center may also partner with other organizations in the state to secure additional staff resources, specialized services, and technical capabilities.

Single recipient located in a state government agency or foundation. This model applies to state recipients in Ohio, New York, and Kansas. In Ohio, the state recipient manages the performance of several state-funded nonprofit organizations. In New York, the state periodically makes competitive awards to 10 nonprofit organizations to serve the state's manufacturers. In Kansas, the state Department of Commerce manages the federal and state funding associated with the MEP program.

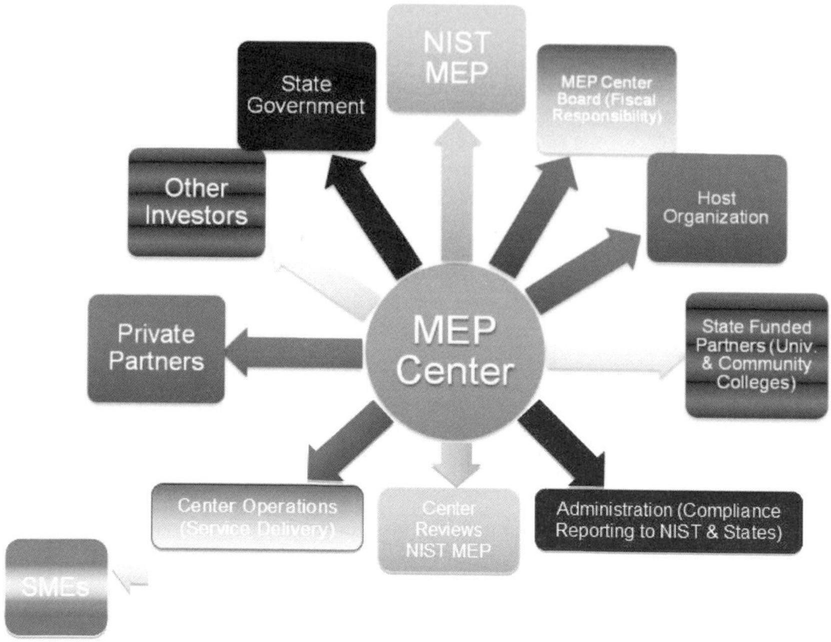

FIGURE 1-2 Schematic of center partnerships and relationships.
SOURCE: MEP.

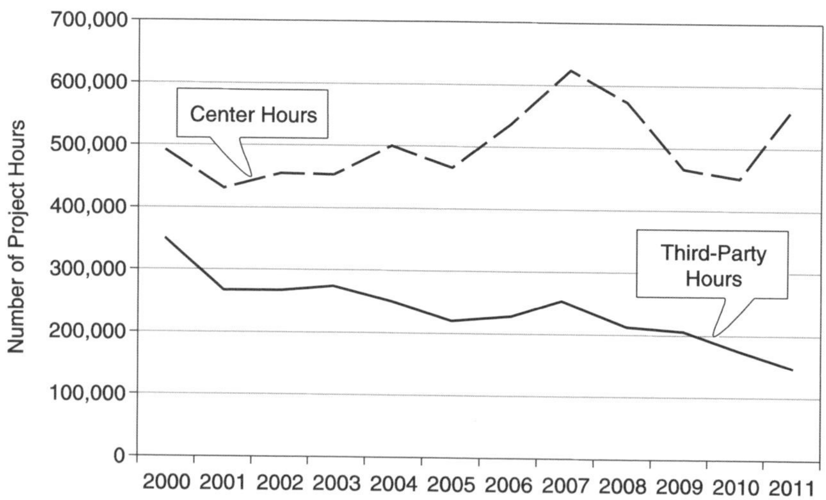

FIGURE 1-3 Distribution of project hours by provider location.
SOURCE: MEP.

cases, MEP centers work also with large companies such as Honda and Navistar to develop integrated supply chain initiatives that foster improvements for lower-tier small suppliers.

Discussions with center staff indicate that there are wide differences in attitude and in capability across the range of customers, and that one of the more difficult tasks for each center is to balance the need to serve the entire population with the need to generate sufficient client revenue to stay in business: The smallest firms typically produce minimal revenues and at the same time require (in aggregate) substantial expenditure of time and resources. More than three-quarters of U.S. small- and mid-sized manufacturing enterprises employ fewer than 20 workers, while under one-tenth are in the prime MEP target range of 50-299 workers. [16]MEP customers range from those eager to grow and willing to adopt new approaches to those who are simply seeking to stay afloat in what have been hard times for manufacturing.

MEP BUDGET

The total MEP budget is about $300 million. One-third of this is provided by the federal government, with the remainder coming from state and industry sources.[17] The federal contribution in FY 2013 was $123 million, with more than three-quarters going directly to the centers.

TABLE 1-2 MEP Federal Budget

Item	Millions of Dollars	Percent of Total
Center Renewals	93.5	76.0
Strategic Competitions	3.9	3.2
Support for Centers	11.4	9.3
Programmatic Requirement/Administration	3.2	2.6
Centralized MEP System Support	8.2	6.7
MEP Labor + Benefits	7.8	6.3
Other Objects	0.9	0.7
NIST Overhead	5.5	4.5
Total	123.0	

SOURCE: NIST Presentation of June 2013 to the MEP Advisory Board.

[16]Calculated from U.S. Census Bureau, Statistics of U.S. Business, 2010 Annual Data.
[17]MEP. Private communication July 21, 2012.

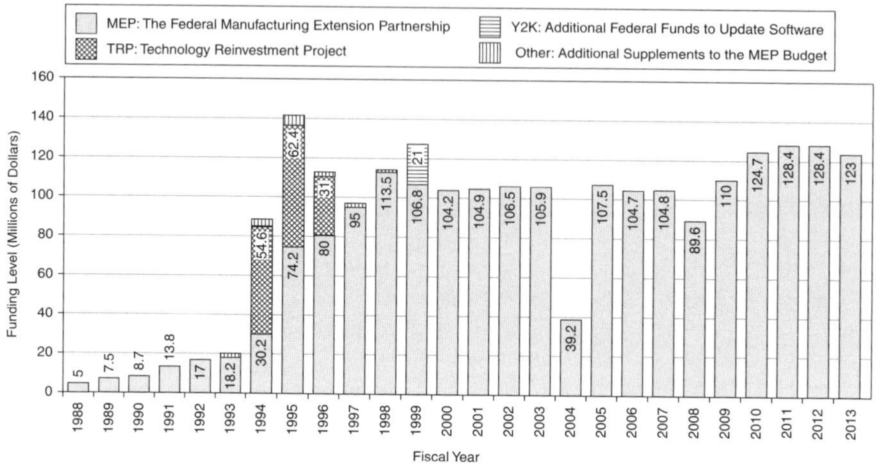

FIGURE 1-4 MEP Federal budget, FY 1988-2011 (current dollars).
SOURCE: MEP.
NOTE: Data is presented in current dollars.

The federal contribution has varied over time, although funding has increased from the recent low of $89.6 million in 2008, reaching $128.4 million in 2011, as shown in Figure 1-4.

THE IMPACT OF THE MEP COST-SHARE

Based on the congressionally defined MEP program cost-share formula, the MEP centers receive one-third of their operating expenses from federal funds. Centers are expected to match federal funds with funding from other sources, notably fees for services to clients and funding from state governments. To meet the nonfederal cost-share, centers draw on a variety of sources. These include:

- Fees for services provided by MEP centers;
- Cash or in-kind contributions from state and local governments;
- Cash or in-kind contributions from institutions such as trade associations, community colleges, and economic development organizations;
- Interest and dividends earned on nonfederal funds.[18]

[18]GAO, GAO-11-437R, "NIST Manufacturing Extension Partnership Program Cost Share," p.6, April 2011.

According to MEP staff, Federal funding for MEP centers was $90 million in FY 2011. Centers generated $102 million from client fees, and raised a further $51 million in cash (and $3 million in in-kind contributions) from the states.[19] (See summary table.) We note that there are discrepancies between these figures and data reported in the 2011 Government Accountability Office (GAO) report referenced below, although it is also apparent that the GAO figures themselves have been challenged directly by MEP staff.

FY 2011-2012 Funding breakout by source of funding (Millions of Dollars)	
Federal (MEP)	90
Income from client fees	102
State funding (cash)	51
State funding (in kind)	3
Total	246

SOURCE: MEP staff, private communication.

State funding for the MEP has been under stress in recent years. In some large states—for example, California—state funding has been eliminated altogether. According to GAO, 7 centers reported no state funding in FY 2010, although 15 also reported that they received more than half of their funding from the states.[20] Some MEP centers have also reported that the recent downturn has made it more challenging to meet the required cost-share. According to the GAO, the number of centers reporting difficulties increased from 11 to 32 after 2008.[21] The increasing need to rely on fees from clients is changing the dynamics of the program: Centers must ensure that their consulting business generates substantial and sustainable revenue. Pressure to switch to the Next Generation Strategy (NGS) with public funds limited and without clear *prior* evidence of market demand adds substantial risk (see Chapter 5 for detailed analysis of NGS).

The MEP cost-share has attracted increasing policy attention, including a recent if inconclusive analysis by the GAO.[22] That report reviewed the origins of the cost-share and its evolution from a 1:1 basis to a 2:1 share that was passed in 1998 and identified a number of considerations to be taken into account. It did not make any formal recommendation with respect to the appropriate cost-share. The report did identify both positive and negative effects of the current cost-share.

Positive Effects of the Cost-Share. In response to open-ended questions about the positive effects of the current MEP cost-share, a significant number of

[19]MEP staff. Private communication July 6, 2013.

centers (14) reported that there were none. Other MEP centers did identify positive effects of the current 2:1 cost-share, suggesting that:

- Centers are encouraged to leverage resources and improve partnerships with other organizations, "ensuring that more investors are involved in manufacturing extension other than just the federal government."[23]
- Some centers suggested that the need to collect client fees gives manufacturers a stake in the program, increasing buy-in and a commitment to making the partnership with the center more productive. Others pointed out that companies recognize value when they have to pay for services, with free services perceived as having low or no value.[24]
- The fees result in centers emphasizing services that are relevant to manufacturers, with an emphasis on results. A number of centers also argue that the cost-share helps them to avoid duplication of efforts with other partners within the economic development community.

Negative Effects of the Cost-Share .Notwithstanding these affirmative responses, the GAO notes that MEP centers cited more negative than positive effects in the current MEP cost-share. For example, about a quarter of the MEP centers reported that the current cost-share mechanism requires them to spend more time and effort seeking funds, establishing and maintaining partnerships, and accounting for in-kind contributions. Some centers have argued these administrative efforts are detracting from serving clients. A significant percentage of MEP centers reported that they are seeking projects outside their mission in order to obtain revenue-generating projects.

- The current match also encourages MEP centers to shift their focus to larger clients who can pay higher fees. Some 80 percent of the MEP centers reported that they are likely to shift their focus towards larger clients, e.g., over 150 employees, that can afford to pay the increased rates, with less attention to smaller companies that most need the services.
- A large majority of the MEP centers reported that they are focusing more on multiple projects with repeat clients. This runs directly

---

[20]GAO, GAO-11-437R, "NIST Manufacturing Extension Partnership Program Cost Share," p.10, April 2011.

[21]GAO, op. cit. p.7.

[22]GAO, op. cit. The objectives of the GAO report were to (1) provide information on the various cost-share structures in the MEP program, (2) identify the effect of such cost-share structures on individual centers and the overall program, and (3) provide recommendations on how to best structure the cost-share requirement. At its conclusion, the study did not provide recommendations on how to best structure the cost-share but identified a number of factors to be taken into account.

[23]GAO, op. cit. p. 8.

[24]Ibid.

counter to NIST's emphasis on the need for centers to reach out to clients not currently served. Indeed, "almost all the MEP centers express concern about limiting how long they can serve existing clients."[25]

- The current cost-share also pushes MEP centers to focus less on rural clients. Centers reported to the GAO that the high level of cost-share burdens their partners with paperwork and makes it more difficult to serve small, rural manufacturers.

Reflecting these concerns, the GAO reported that the majority of the MEP centers supported a reduced cost-share, with half supporting a 1:1 cost-share. The GAO notes that many MEP centers reported that "seeking out and accounting for in-kind contributions is burdensome and diverts resources from supporting small manufacturers."[26]

A separate study, commissioned by NIST, also recommended that the cost-share requirements be changed to a 1:1 ratio.[27] That review of the MEP business model noted that the MEP program was the only program at the Department of Commerce with a cost-share that exceeded a 1:1 match. Indeed, many of the Department of Commerce programs have no nonfederal cost-share requirement. It appears that few federal programs have a matching requirement that is as rigorous and as high as the MEP's current ratio. The matching requirement also pushes centers to seek in-kind contributions, which involve considerable complexity in terms of definitions, monitoring, and management.[28]

While the benefits of the cost-share noted in the GAO report cited above are real, they would arguably be unaffected by the reduction in the ratio of the cost-share. Centers would still need to seek partnerships, revenue, and state support. There is an understandable concern that easing the cost-share could result in a decline in incentives for state support. At the same time, centers that are able to adjust their products and services to meet new or emerging state priorities may well be able to maintain continued state support. The annual federal contribution to each center has remained fixed for several years, and even with that predictability, it is worth noting that in some cases states have reduced or eliminated their contribution.[29] In short, there may well be no automatic relationship between state support for the centers and the cost-

[25]Ibid. p. 9.

[26]Ibid. p. 11.

[27]Stone & Associates and the Center for Regional Economic Competitiveness, "Re-examining the Manufacturing Partnership Business Model," Washington, DC, October 2010.

[28]For discussion of in-kind contributions see GAO, op. cit. p. 12, and Stone & Associates, op. cit.

[29]GAO, op. cit. p. 10. For example, Pennsylvania made dramatic reductions in its support for the MEP program during the recent economic downturn. See the presentation by Eric Esoda, Executive Director, Northeastern Pennsylvania Industrial Resource Center from the National Academies Symposium, "Diversity and Achievements: The Role of the Manufacturing Extension Partnership in the Midwest," Akron, OH, March 26, 2012.

share, particularly given the growing emphasis on manufacturing and mechanisms to support it both at the federal and state levels.[30]

The GAO report and related analyses would therefore seem to suggest that the disadvantages of the current 2:1 cost-share outweigh the advantages of a system that appears to be too rigid, based on an outmoded formula, and one which inhibits the type of flexibility and responsiveness that would enhance the value of the program.

Changing the cost-share will require additional funds from NIST, but there is no reason to believe that these additional federal resources will lead states to reduce their level of support. Funding support from states for MEP centers is typically based on policy, except when fiscal considerations make this impossible. In addition, given the limited leverage that NIST has over the individual centers, we have no reason to believe that changing the match will tilt more control to NIST.

MEP'S EVOLVING ROLE

Since its establishment in 1989, the MEP system has continued to grow to serve U.S. small and medium manufacturers. It now reaches out to some 7,000 manufacturers a year, providing a variety of services, from short-term cost reduction through lean manufacturing to longer-term growth initiatives.

Reflecting the changing face of global competition, MEP's role is evolving from an emphasis on lean production to a focus on enhancing the innovative capacity of small manufacturers. The MEP centers were created initially to transfer federally sponsored, state-of-the-art technology to firms. Later they delivered pragmatic assistance, appropriate to state and local conditions, with business services, quality systems, manufacturing systems, information technology, human resources, engineering, and product development—"the 'soft' business practices."[31] Today, the goal of the partnership is to increase the competitiveness and productivity of U.S. manufacturing by helping small manufacturers improve their production performance and grow their business through innovation in product development and production.[32]

Indeed, there is a need and opportunity for NIST MEP and the centers to play a bigger role in helping a variety of related organizations appreciate the value of manufacturing to communities, states, and country. The case of the

---

[30]See, for example, the discussion of Ohio's support for innovation and manufacturing. National Research Council, *Building the Ohio Innovation Economy: Summary of a Symposium,* C. Wessner, Rapporteur, Washington, DC: The National Academies Press, 2013.

[31]Philip Shapira, Jan Youtie, and Luciano Kay, "Building Capabilities for Innovation in SMEs: A Cross Country Comparison of Technology Extension Policies and Programmes," *International Journal of Innovation and Regional Development*, 2011, Vol. 3 (3-4).

[32] See Roger Kilmer, "MEP's Place in the Innovation Chain," in National Research Council, *Strengthening American Manufacturing: The Role of the Manufacturing Extension Partnership—Summary of a Symposium*, op. cit.

Northeastern Ohio MAGNET program is illustrative of the role of this type of public-private partnership in bringing together manufacturers, elected officials, industry and academic leaders, and philanthropic foundations and regional development organizations.[33]

To address these partnership goals, over the past few years, NIST MEP has made a concerted effort to encourage the MEP centers to develop a wider range of services focused on growth, in particular through innovation. NIST MEP is also seeking to realign its metrics to accommodate and reflect the introduction of these new services. Figure 1-5 illustrates some of the changes.

The changing orientation illustrated above is of course a considerable simplification. In practice, MEP activities have been less neatly divided, with overlap and accumulation of functions over time.

The red lines indicate changes in the tracking and evaluation system, with the adoption of a set of standardized metrics in the early 2000s, followed in FY 2011-2012 by the transition to the new CORE metrics now under way.

MEP Within the NIST Portfolio

In the context of U.S. manufacturing, the MEP program is relatively small at around $300 million, with a federal contribution in 2013 of $123 million. Even so, the program can serve as a valuable element in an overall

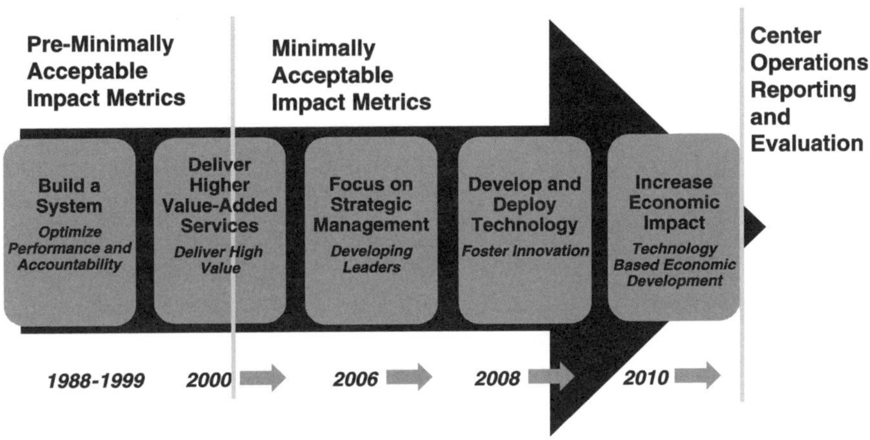

FIGURE 1-5 Evolution of the MEP program.
SOURCE: MEP.

---

[33]See National Research Council, *Building the Ohio Innovation Economy: Summary of a Symposium*, op. cit.

national manufacturing strategy. As Figure 1-6 shows, NIST sees MEP as a part of a portfolio of programs to support manufacturing from basic research through development, demonstration, and deployment. Many of these foreign programs, described in Chapter 7 (and Appendix A) of this study, focus on Technology Readiness Levels 3 to 7 (with some variations across programs.) They are also substantially larger than MEP's combined budget.

KEY CHALLENGES FOR MEP

As it evolves from an emphasis on lean production to a focus on enhancing the innovative capacity of small manufacturers, MEP faces a variety of challenges related to the nature of its mission and operations:

- **Diversity of Clientele:** The manufacturing sector is not homogenous. Small firms have widely varying needs and capacities. They generally do not invest in anticipation of emerging (i.e., not yet proven) technologies.[34] There is also significant dynamism in the market with business failures as well as new firms being created.[35]
- **Operational Challenges:** These include improving the efficiency of the program operations, expanding its scale and reach, measuring the program's impacts, and its improving its transition to a new innovation-oriented strategy.[36]

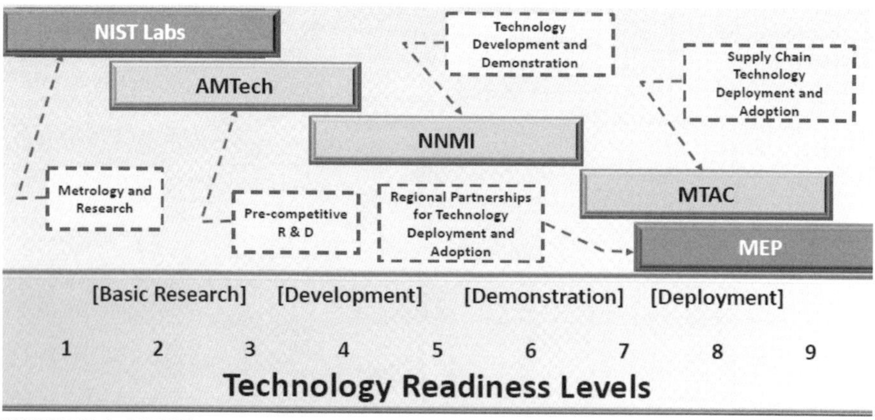

FIGURE 1-6 NIST programs to support manufacturing.
SOURCE: NIST, June 2013.

[34]See Chapter 2 for a description of the diversity of manufacturing firms.
[35]In recent years roughly 7 percent of small businesses exit each year and about 5 percent are newly formed. (Census: Business Dynamic Statistics.)
[36]See Chapters 3 through 6 for a discussion of these operational challenges.

- **Mission and Funding Challenges:** Especially at a time of shrinking support by state governments, MEP centers need to derive revenue from SMEs in order to attract matching federal funds and remain financially viable. They need to address market demand while, at the same time, perform a public service in assisting companies to adopt new and innovative practices. This creates a tension between NIST MEP's efforts to drive centers to adopt particular services and the demand for services by manufacturing companies served by MEP centers.[37]

THIS REPORT

At the request of NIST, an ad hoc committee of the National Academies has prepared a report that addresses two complementary tasks: an evaluation of the operation, achievements, and challenges of MEP and a review of a selected number of similar national programs from abroad. This consensus report brings together this work, with recommendations to improve the MEP program's operations and impact.

Scope of the Study

As called for in the Statement of Task, the committee convened a major conference, held a series of fact-finding workshops, and commissioned research to review and document MEP's "current achievements, challenges, and new opportunities."[38] Reflecting its second mandate "to identify and review similar national programs from abroad in order to draw on foreign practices, funding levels, and accomplishments as a point of reference; and discuss current needs and initiatives in light of the global focus on innovation," the committee also visited and met with representatives of leading national programs from abroad that support applied research and manufacturing. The committee's goal is to inform a wide array of stakeholders, from federal and state policymakers and NIST and other federal agencies to small and large manufacturers, academic researchers, and others concerned about the manufacturing challenge and the sustained effort of other countries, and identify steps MEP can take to enhance its ability to address this challenge.

Plan of the Report

This final consensus report underscores the value, limitations, and potential of the MEP program to support the development and manufacture of innovative products in the United States. This chapter reviewed the founding rationale and described the organization and funding sources of the program.

---

[37]See Chapter 2 for a discussion of the MEP cost-share and its impacts.
[38]Please see the Statement of Task in the Preface of this volume.

Chapter 2 explains why a strong manufacturing sector is important for the nation's growth, employment, and national security. It describes the recent declines in U.S. manufacturing and its impacts, previews potential opportunities for "re-shoring" U.S. manufacturing, and discusses how MEP can be a valuable part of a larger national strategy to support manufacturing.

Chapters 3 through 6 of this report describe and assess the MEP program. Chapter 3 describes MEP's focus on lean manufacturing services. Chapter 4 describes how MEP measures its performance, while Chapter 5 provides an analysis of outcomes. Chapter 6 describes MEP's new innovation-oriented strategy that is based on an expanded view of its role in supporting manufacturing and also analyzes the challenges MEP faces in implementing this strategy.

Addressing the second facet of the Statement of Task, which is to "identify and review similar national programs from abroad in order to draw on foreign practices, funding levels, and accomplishments as a point of reference; and discuss current needs and initiatives in light of the global focus on advanced manufacturing." Chapter 7 provides an overview of leading foreign programs to support applied research and manufacturing, revealing the significant scope and commitment of efforts around the world. This analysis is further deepened by a review, found in Appendix A, of five leading national programs to support applied research and manufacturing: Canada's Industrial Research Assistance Program (IRAP), Germany's Fraunhofer Institutes, Taiwan's Industrial Technology Research Institute (ITRI), the United Kingdom's Catapult program, and France's Carnot initiative.

In light of the evaluation of the MEP program and the review of foreign programs, Chapter 8 of this report draws together the committee's findings and sets out its consensus recommendations to improve MEP operations and impact and outline potential lessons for U.S. policy.

Chapter 2

U.S. Manufacturing in Global Context

The National Institute of Standards and Technology (NIST) Manufacturing Extension Partnership (MEP) program was founded in 1989 as a means of supporting the manufacturing sector and in particular small manufacturers. At the time, the primary challenge was seen as coming from Japan, with its high-quality and very efficient manufacturing processes. Consequently the MEP program focused much of its effort until quite recently on helping manufacturers to implement lean production processes.

This focus is evolving as rapid changes in both the global and U.S. economies are driving manufacturing firms to adapt to new circumstances. Even if their markets are strictly domestic, they face the globalization of supply chains and competition from abroad. In particular, they face productivity and marketing challenges.

The MEP focuses most of its services on small- and medium-sized manufacturers (SMEs[1]). SMEs are a critical component of the U.S. economy. Recent data from the Census Bureau indicates that firms with fewer than 500 employees account for 99 percent of firms engaged in manufacturing.[2] However, while small firms with fewer than 20 employees account for almost 75 percent of all manufacturing firms, they have only 9.1 percent of total manufacturing employment, compared with almost three-quarters of manufacturing employment generated by larger firms with at least 100 employees (see Figure 2-1). This has implications for MEP strategy (discussed below).

[1]Here we have adopted the European terminology for small- and medium-sized businesses. MEP itself does not appear to utilize the associated size bands for any analytic or program purposes.
[2]U.S. Bureau of the Census, Statistics of U.S. Businesses, *<http://www.census.gov/epcd/susb/2008/us/US--.HTM>*. Accessed July 24, 2012.

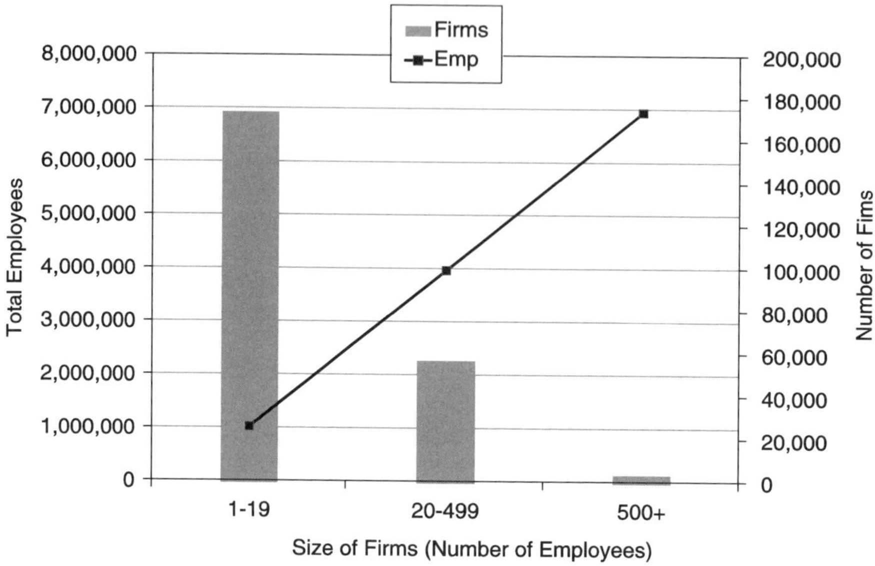

FIGURE 2-1 Total employment in U.S. manufacturing by size of firm, 2011.
SOURCE: U.S. Census: Longitudinal Business Database.

Over the past 10 years, the productivity gap between small and large manufacturers[3] has been expanding. During that period, large firm productivity grew by about 300 percent (see Figure 2-2), while that of SME manufacturers has grown by 200 percent. Moreover, as large firms started from a much higher productivity baseline in 1997, the absolute gap had widened sharply and was by 2007 about $250,000 per employee.

Over the long run, as NIST MEP points out in Figure 2-2, this growing gap means that SMEs will certainly face increasing competitive pressures.[4] Other changes have occurred as well. The globalization of the world economy means that even small manufacturers have the need to become more agile and better marketers as they seek a niche in lengthening supply chains. The impacts of globalization have in part driven MEP's new strategic thrust, discussed in Chapter 6, which focuses on helping companies innovate and grow rather than just cut costs (see below).

---

[3]In most industries, small businesses are defined by the Small Business Administration as those with fewer than 500 total employees. *<http://www.sba.gov/content/what-sbas-definition-small-business-concern>*. Accessed July 27, 2012.

[4]For a discussion of this divergence in manufacturing productivity, see Susan Helper, "The High Road for U.S. Manufacturing, *Issues in Science and Technology*, Winter 2009. She notes that higher productivity is associated with higher worker training, use of information technologies, and greater integration into high-value supply chains.

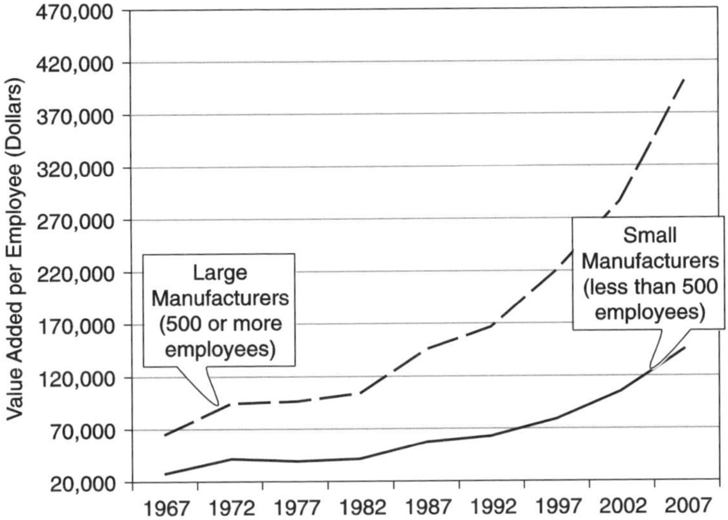

FIGURE 2-2 Value added per employee small and large manufacturers, 1967-2007. SOURCE: MEP "Delivering Measurable Results to Manufacturing Clients," 2009 p.3; data from U.S. Census Bureau, 2007 Economic Census: Manufacturing (November 2010), calculations by MEP. Data presented in constant dollars.

THE IMPORTANCE OF A STRONG U.S. MANUFACTURING INDUSTRY

There are a series of interrelated reasons which prompt ongoing attention to manufacturing and which set the context for understanding and probing MEP's relevance and operation.

Manufacturing remains an important component of the U.S. economy. It contributes $1.6 trillion to gross domestic product (GDP), and employs 11 million workers, with many of the manufacturing jobs providing above average pay and benefits.[5] The manufacturing sector also has powerful indirect employment effects on other sectors of the U.S. economy, supporting millions of additional supply chain jobs across the economy.

A Major Source of R&D

Manufacturing companies in the United States represent 12 percent of U.S. GDP, manufacturing and account for nearly 70 percent of private-sector

[5]"Total hourly compensation in the manufacturing sector is, on average, 22 percent higher than that in the services sector. About 91 percent of factory workers have employer-provided benefits, compared to about 71 percent of workers across all private sector firms." See Executive Office of the President of the United States, 2009, *A Framework for Revitalizing American Manufacturing*, *<http://www.whitehouse.gov/sites/default/files/microsites/20091216-manufacturing-framework.pdf>*
.

research and development (R&D) and 60 percent of all U.S. R&D employees.[6] Reflecting this, the sector employs 57 percent of the nation's industrial scientists and engineers.[7] While some service industries do a moderate amount of R&D internally, that amount "pales in comparison with the amount done by the manufacturing sector."[8]

The Largest Contributor to U.S. Exports

As an economic sector, manufacturing is the largest contributor to U.S. exports.[9] In 2010, the United States exported over $1.1 trillion worth of manufactured goods, accounting for 86 percent of all U.S. goods exports and 60 percent of U.S. total exports. These exports have a direct and positive impact on job creation. "Exports contributed 46 percent to the growth of the U.S. economy between 2009 and 2011. This export surge helped create 600,000 jobs nationally in 2010, even while the rest of the economy was shedding them."[10]

Linkages to Innovation

A strong manufacturing sector is also of central economic importance because of its strong linkage to innovation. Susan Helper and Howard Wial report that 22 percent of manufacturers introduce new processes to increase productivity compared to just 8 percent of non-manufacturers.[11] U.S.-based manufacturing also helps to sustain an "industrial commons," a term that describes the complex and enduring partnerships among manufacturers, universities, technical colleges, firms, research institutes, financing entities, and other links in the supply chain.[12] Recognizing the importance of U.S.-based manufacturing for sustaining a robust innovation ecosystem, recent reports by the President's Council of Advisors on Science and Technology have

---

[6]See the Council of Economic Advisors, 2013 Economic Report of the President, Chapter 7, March 2013.

[7]National Center for Science and Engineering Statistics, *Research and Development in Industry: 2006-07*, Detailed Statistical Tables NSF 11–301, Arlington, VA: National Science Foundation, 2011. Available at *<http://www.nsf.gov/statistics/nsf11301/>*.

[8]See Gregory Tassey, "The Manufacturing Imperative," in National Research Council, *Strengthening American Manufacturing: The Role of the Manufacturing Extension Partnership*, C. Wessner, Rapporteur, Washington, DC: The National Academies Press, 2013.

[9]In order to stimulate the creation of additional jobs, President Obama's National Export Initiative has set the ambitious goal of doubling U.S. exports by the end of 2014.

[10]Emilia Istrate, "Making It Here, Selling It There, and Creating U.S. Jobs," Washington, DC: Brookings, March 8, 2012.

[11]Susan Helper and Howard Wial, op. cit., 2010. In general, small businesses have been found to renew the U.S. economy by introducing new products and new lower-cost ways of doing things, sometimes with substantial economic benefits. For review of the empirical evidence supporting the finding of high-innovation performance of small firms, see Zoltan J. Acs and David B. Audretsch, "Innovation in Large and Small Firms, An Empirical Analysis," *The American Economic Review*, Vol. 78, No. 4, 1988, pp. 678-690.

[12]See Gary Pisano and Wily Shih, "Restoring American Competitiveness," *Harvard Business Review*, July 2009.

Box 2-1
The Link Between Manufacturing and Innovation

Manufacturing is integral to new product development. Production lines are links in an iterative innovation chain that includes precompetitive R&D, prototyping, product refinement, early production, and full-scale production. U.S. corporations still dominate a number of industries, such as personal computers and certain semiconductors, even though end products are produced offshore. America's logic chip-design industry, which includes companies like Qualcomm, Nvidia, and Broadcom, relies almost entirely on silicon wafers fabricated in Asian foundries, while Apple iPods, iPhones, and iPads are assembled in China by the Taiwanese firm Hon Hai Precision Industry. In such products, the greatest economic value is in software, microprocessors, and proprietary designs, while the hardware is generally comprised of standardized parts and assembled with standard production processes.

In many high-technology industries, however, design is not so easily separated from manufacturing. Production processes for advanced solar cells, lithium-ion vehicle batteries, and next-generation solid-state lighting devices are highly proprietary to the producing company and often constitute a competitive advantage. If new U.S. companies lack the domestic capability to scale up, Intel founder Andy Grove warns, "we don't just lose jobs—we lose our hold on new technologies. Losing the ability to scale will ultimately damage our capacity to innovate."[a]

[a]National Research Council, *Rising to the Challenge; U.S. Innovation Policy for the Global Economy*, C. Wessner and A. Wolff, eds., Washington, DC: National Academies Press, 2012, p. 84.

emphasized the critical importance of advanced manufacturing in driving knowledge production and innovation in the United States.[13]

Linkages to Advanced Services

A robust manufacturing sector does not operate in isolation; it consumes a variety of services and relies heavily on them to operate. As a recent McKinsey report notes, "service inputs (everything from logistics to advertising) make up an increasing amount of manufacturing activity. In the United States, every dollar of manufacturing output requires 19 cents of

[13]President's Council of Advisors on Science and Technology, 2011, "Report to the President on Ensuring American Leadership in Advanced Manufacturing," *<http://www.whitehouse.gov/sites/default/files/microsites/ostp/pcastadvanced-manufacturing-june2011.pdf>*.

Box 2-2
The Debate on Manufacturing Productivity Growth

Data from the Bureau of Labor Statistics show manufacturing productivity doubled between 1990 and 2010.[a] Over the same period, manufacturing employment fell, dropping rapidly between 2000 and 2010. Looking at these trends, many analysts concluded that U.S. manufacturing has become more efficient.

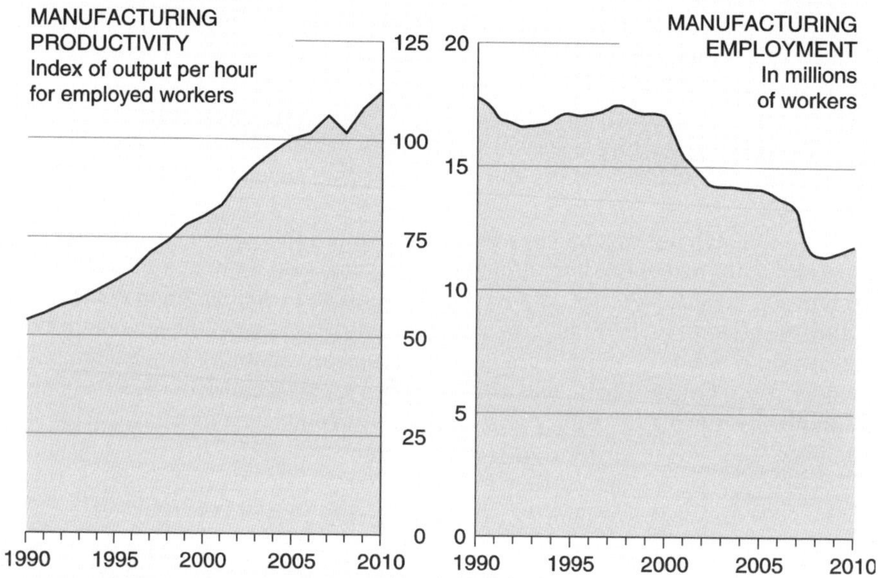

FIGURE B-2-2 Manufacturing productivity and employment 1990-2010.
SOURCE: Bureau of Labor Statistics.

Recently, some economists have disputed this view. They have sought to show that U.S. statistics may be overstating the gains in manufacturing productivity because they fail to reflect the value of imported inputs adequately in manufactured products and because they do not adequately account for the growing use of temporary factory workers.[b] They also note that gains in manufacturing productivity are unevenly distributed, with the significantly higher productivity in computer and electronics manufacturing masking the trends in other sectors. Foreign manufacturing and trade practices, particularly those of China in the first decade of this century, have also negatively affected U.S. manufacturing employment. Susan Houseman estimates that these measurement biases may have accounted for as much as half of the measured growth of U.S. manufacturing output from 1997 to 2007, excluding computers and electronics manufacturing.[c]

[a]Manufacturing productivity is calculated by dividing the value of U.S. manufacturing output (as adjusted for inflation and quality changes) by the number of manufacturing-worker hours.
[b]According to Susan Houseman, the integration of a variety of outsourced components in the production process makes it difficult to count U.S. manufacturing output separately. This means that price savings that U.S. factories have realized from outsourcing can show up as gains in U.S. output and productivity. Economists also note that "since the late 1980s manufacturers have increasingly used workers employed by temporary help services to work in their factories. Although they work in factories alongside manufacturers' employees, these workers do not count as manufacturing workers in the official statistics. Yet the goods that they help produce count as manufacturing output. For this reason, some economists believe that manufacturers' productivity is overstated when they use temporary help services." See Susan Houseman et al., "Offshoring Bias in U.S. Manufacturing," *Journal of Economic Perspectives*, 25: 111-132.
[c]Ibid.

services. And in some manufacturing industries, more than half of all employees work in service roles, such as R&D engineers and office-support staff."[14]

Manufacturing and National Security

A strong manufacturing sector is also essential for the nation's security; global leadership in development and production of advanced manufacturing technologies provide the basis for military superiority and preparedness.[15] However, there is growing concern that the broader deterioration of the nation's industrial base over the last three decades (described below) has also left the United States dependent on production facilities in other nations for a variety of strategic components and systems and high-tech materials.[16] Recognizing this vulnerability, the director of national intelligence has announced the preparation of a National Intelligence Estimate (NIE) on the state of American manufacturing.[17]

Citing these considerations, a growing number of analysts argue that maintaining a competitive onshore manufacturing sector and the associated skilled labor and technical institutions are linked and essential for long-term

[14]McKinsey Global Institute, "Manufacturing the Future; The Next Era of Growth and Innovation," November 2012. Access at <*http://www.mckinsey.com/insights/manufacturing/the_future_of_manufacturing*>.
[15]See Erica R. H. Fuchs, "Innovation, Production, and Sustainable Job Creation: Reviving U.S. Prosperity," Department of Engineering and Public Policy, Carnegie Mellon University, February 2012.
[16]See, for example, Richard Silberglitt, James T. Bartis, Brian G. Chow, David L. An, and Kyle Brady, *Critical Materials: Present Danger to U.S. Manufacturing,* Santa Monica: Rand Corporation, 2013.
[17]Richard McCormack, "Intelligence Director Will Look at National Security Implications of U.S. Manufacturing Decline," *Manufacturing and Technology News*, February 3, 2011, Vol. 18, No. 2.

national competitiveness.[18] They note that once manufacturing activity moves overseas, so do the required skills, networks, and supply chains; and once offshore they, and the learning they engender, are difficult to recover. These analysts therefore argue that it is important for U.S. policymakers to be concerned with the capabilities and composition of the economy, just as policymakers are elsewhere.[19]

RECENT DECLINES IN U.S. MANUFACTURING

While the United States is still the world's largest manufacturer with a global share of about 22 percent of global output, and industrial output has risen in 2013, U.S. producers continue to face intense and growing competition from around the world. There is a growing awareness in this country that thriving manufacturers are critical to America's economic recovery. Policymakers are increasingly emphasizing that the United States cannot completely move into a knowledge-based and services-based economy; it also has to produce tangible assets. There concerns are buttressed by a growing and authoritative concern that the erosion of America's manufacturing and high-technology base threatens to undermine U.S. leadership in next-generation technologies.[20]

Sharp Declines in Manufacturing Employment

Jobs in the U.S. manufacturing sector have declined by about 8 million in the last 26 years.[21] Particularly in the last decade, employment levels in manufacturing have declined steeply by about one-third. This decline, at least initially, was the result of greater competition from low-wage countries, leading to the offshoring of low-skilled jobs to lower-cost locations.[22] Manufacturing

---

[18]For an analysis of the linkages between manufacturing and R&D, see E. Fuchs and R. Kirchain, "Design for Location? The Impact of Manufacturing Offshore on Technology Competitiveness in the Optoelectronics Industry", *Management Science*, 56(12): 2323-2349, 2010. See also Greg Tassey, "Rationales and Mechanisms for Revitalizing U.S. Manufacturing R&D Strategies," *The Journal of Technology Transfer*, June 2010, Vol. 35, Issue 3, pp. 283-333; Gary P. Pisano and Willy C. Shih, "Restoring American Competitiveness," *Harvard Business Review*, July-August 2009; and Ro Khanna, *Entrepreneurial Nation: Why Manufacturing is Still Key to America's Future,* New York: McGraw Hill, 2012.

[19]For a review of current programs and policy focus around the world to support innovation-led competitiveness, see National Research Council, *Rising to the Challenge: U.S. Innovation Policy for the Global Economy*, C. Wessner and A. Wm. Wolff, eds., Washington, DC: The National Academies Press, 2012.

[20]See, for example, E. Fuchs and R. Kirchain, "Design for Location? The Impact of Manufacturing Offshore on Technology Competitiveness in the Optoelectronics Industry," *Management Science*, 56(12): 2323-2349, 2010.

[21]This information is based on data prepared by the U.S. Census. Access at *<http://www.ces.census.gov/index.php/bds/sector_line_charts>*. For additional information, see Robert D. Atkinson, 2011, "Explaining Anemic U.S. Job Growth: The Role of Faltering U.S. Competitiveness," Information Technology and Innovation Foundation, December 2011.

[22]For example, one study has shown that "between one-quarter and more than one-half of the lost manufacturing jobs in the 2000s are the result of import competition from China". See David Autor,

employment fell by 16.1 percent from 2003 to 2009, before recovering by 4.6 percent to end 2012.[23]

The Growing Trade Deficit

These employment and wage trends also roughly coincide with trade deficit in the U.S. manufacturing sector, which was -$585 billion in 2011.[24] Americans are consuming more manufactured goods than ever, but U.S. share in production of manufactured goods is declining. Worryingly, the United States continues to lose ground in key manufacturing sectors, including those sectors that are likely to drive our economy in the future. Until 2001, the United States ran a trade surplus in "advanced technology products," which includes biotechnology products, computers, semiconductors, and robotics. By 2010, however, the United States ran an $81 billion trade deficit in this important sector (see Figure 2-3).[25] This represents a very significant shift.

The Impact of Offshoring Manufacturing

Much of the trade deficit in advanced technology is attributable to the phenomenon of progressive offshoring over the last few decades. First, U.S. manufacturers began by setting up manufacturing facilities abroad, either to be near growing markets, to make use of skilled, low-cost labor, or both. The offshore facility did a small amount of R&D in order to move products into the market. As host countries became able to provide more skilled labor, they often pressed companies to provide R&D training and expand their internal R&D infrastructures to capture synergies at the "entry" tier of the high-tech supply chain. For example, Taiwan and Korea became skilled at producing electronic components, while China excelled at assembly. In this way, those countries gradually became competitive in their own submarkets.[26]

David Dorn, and Gordon Hansen, 2011, "The China Syndrome: The Local Labor Market Effects of Import Competition," MIT Working Paper; econ-<*http://www.mit.edu/files/6613*>. Some of this decline is conventionally described as due primarily to increased efficiencies and productivity gains, though the basis for this view has been questioned by Susan Helper and Susan Houseman, among others.

[23]BEA. Access at <*http://data.bls.gov/timeseries/CES3000000001?data_tool=XGtable*>.

[24]Bureau of Economic Analysis, "International Economic Accounts, Trade in Goods and Services 1992-Present," <*http://www.bea.gov/international/index.htm*>. Accessed December 2011.

[25]U.S. Census, Trade in Advanced Technology Products 2010, <*http://www.census.gov/foreign-trade/balance/c0007.html*>.

[26]See Gregory Tassey, "The Manufacturing Imperative," in National Research Council, *Strengthening American Manufacturing: The Role of the Manufacturing Extension Partnership*, op cit. See also Gregory Tassey, *The Technology Imperative*, Edward Elgar, 2009.

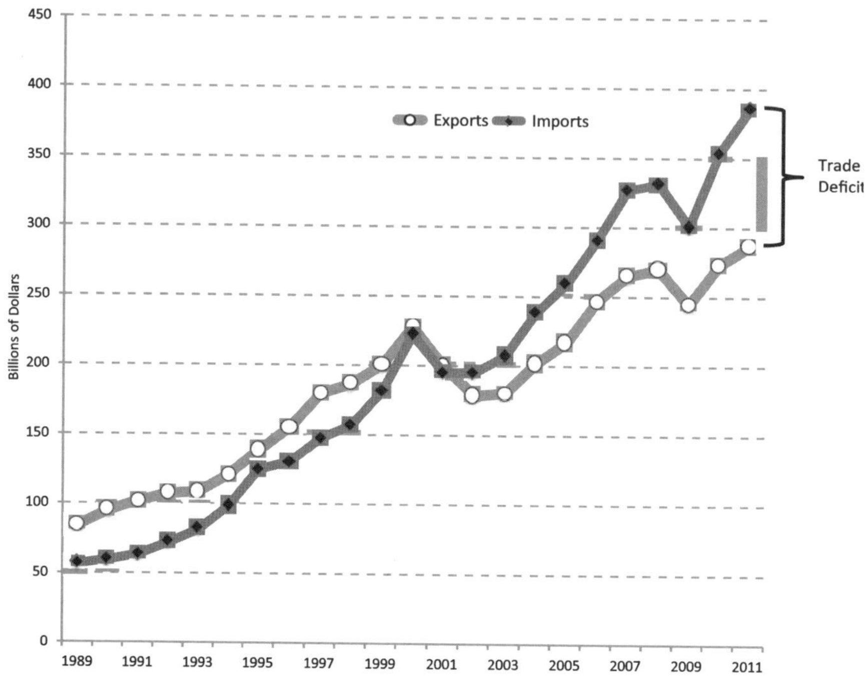

FIGURE 2-3 U.S. exports and imports of high-technology products in constant dollars.
SOURCE: U.S. Census Bureau.

The Role of Foreign Applied Research Programs

Traditionally, U.S. firms have excelled in being first to acquire knowledge, thanks in large part to the steady production of good ideas through substantial and sustained federal investments in basic research. However, U.S. policy has traditionally not supported major national programs to accelerate the application of new ideas through engineering and the commercialization of new products in the market.[27]

As documented in Appendix A of this report, many other countries do intervene with substantial resources and well-equipped facilities early in the product development cycle. They provide shared-use facilities, expertise garnered from working on multiple projects over long periods of time, and connections to other parts of the innovation ecosystem, be it suppliers, customers, or sources of finance, as well as a ready and low-cost source of technical support. They believe that a robust domestic industrial base, which

[27]For a review of leading national programs to advance innovation in manufacturing in Canada, Germany, France, the United Kingdom, and Taiwan, see Appendix A of this report. For a review of foreign programs to accelerate innovation, see National Research Council, *Rising to the Challenge: U.S. Innovation Policy for the Global Economy*, op. cit., Ch. 6.

can produce advanced products in high volumes and the high-skilled jobs that this productive activity generates, is integral to maintaining their global competitiveness.[28]

The Hollowing Out of U.S. Supply Chains

As economies specialize in a particular tier of the high-tech supply chain, they begin to integrate backward along the supply chain, taking more value added from the Western economies, including the United States. As Gregory Tassey and others maintain, this "hollowing out" of supply chains has cost the United States dearly in terms of wealth creation, high-value jobs, and technology sales. Although the United States had been the "first mover" in developing many commercial technologies, "poor technology life-cycle management," has led to a gradual loss of market share in products such as oxide ceramics, semiconductor memory devices, semiconductor production equipment, lithium ion batteries, flat-panel displays, robotics, and advanced lighting.[29]

The Importance of Co-Location

There is growing recognition, based on empirical research, concerning the importance of colocation in manufacturing activity. Manufacturing in itself is desirable, playing an important role in the economy, as noted above. As noted above, while representing 12 percent of U.S. GDP, manufacturing accounts for nearly 70 percent of private-sector research and development and 60 percent of all U.S. R&D employees.[30] A recent MIT institute-wide research effort reaffirms the importance of proximity to manufacturing and innovation. From extensive interviews with managers at small- and medium-sized U.S. manufacturers, the MIT researchers found that these companies "often repurpose existing technologies or techniques and apply them to make new products. And they often bundle products together with services—thus blurring the boundary between the manufacturing and service industries." They conclude that proximity and collaboration matter in this sphere: "A key to innovation for these firms is being located in a diverse industrial ecosystem that offers many complementary resources, such as training and opportunities for collaborative research."[31]

---

[28]See David Cameron, Speech at the Lord Mayor's Banquet, November 12, 2012. In his speech, Prime Minister Cameron promoted investments in innovation and manufacturing, citing their major contributions to advancing Britain's foreign and economic policies.
[29]Gregory Tassey, "Rationales and Mechanisms for Revitalizing U.S. Manufacturing R&D Strategies,
The Journal of Technology Transfer, January 29, 2010.
[30]See the CEA 2013 Economic Report of the President, Ch. 7, March 2013.
[31]Suzanne Berger et al., "A Preview of the Production in the Innovation Economy Report," Cambridge: MIT Press, 2013.

For the fast-growing high-tech services sector, it is particularly important to have close ties to its manufacturing base to fuel innovation. There are co-location synergies between services and the sources of their technology. Those working in a manufacturing supply chain find increasingly important interactions with workers in related activities. "These co-location synergies flow between the tiers of the supply chain and ultimately the hardware and software that are used by the service industries.[32]

Moreover, as the military comes to rely more heavily on complex and advanced technology systems, retaining the capacity and knowledge necessary to manufacture these goods in the United States becomes more important.[33] The ability to source critical infrastructure components, from communications equipment to power generation also affects our ability to protect against disruptions in the supply chain.[34]

Recent empirical research also suggests the importance of spillover benefits from manufacturing to the location and to the country in which the activity occurs. For example, a 2010 study by Greenstone, Hornbeck, and Moretti showed significant productivity gains for firms due to the development of a new manufacturing plant in the area.[35] A study by Lee Branstetter in 2001 showed that spillovers tend to be intranational, captured by the country in which the activity occurs.[36]

The MIT *Production in the Innovation Economy* study cited above focuses on the interdependence between production and innovation. Their initial task force paper states that "learning takes place as engineers and technicians on the factory floor come back with their problems to the design engineers and struggle with them to find better resolutions; learning takes place as users come back with problems." This is perhaps best illustrated by the decision of companies such as Intel or GLOBALFOUNDRIES to keep their design and some production facilities closely linked through co-location.[37]

---

[32]See Gregory Tassey, "The Manufacturing Imperative," in National Research Council, *Strengthening American Manufacturing: The Role of the Manufacturing Extension Partnership*, op cit. In its 2011 manufacturing report (op. cit. p. 11) the PCAST states: "Proximity is important in fostering innovation. When different aspects of manufacturing—from R&D to production to customer delivery—are located in the same region, they breed efficiencies in knowledge transfer that allow new technologies to develop and businesses to innovate."

[33]U.S. Department of Commerce, Revitalizing Manufacturing, January 2012.

[34]A key goal of the Obama administration's Advanced Manufacturing Partnership is to "jumpstart domestic manufacturing capability essential to our national security." See Sridhar Kota, "Revitalizing American Manufacturing," in National Research Council, *Strengthening American Manufacturing: The Role of the Manufacturing Extension Partnership*, op cit.

[35]Michael Greenstone, Richard Hornbeck, and Enrico Moretti, "Identifying Agglomeration Spillovers: Evidence from Winners and Losers of Large Plant Openings," April 2010. Access at *<http://emlab.berkeley.edu/~moretti/mdp2.pdf>*.

[36]Lee G. Branstetter, "Are Knowledge Spillovers International or Intranational in Scope? Microeconometric Evidence from the U.S. and Japan," *Journal of International Economics*, 53, (2001), 53-79.

[37]For example, GLOBALFOUNDRIES $2.2 billion Technology Development Center in Malta, New York, which will be located "right next to the fab," offers real-time advantages. This means that the same engineers that operate the fab can participate in R&D, and this proximity enables them to

Box 2-3
Losing Manufacturing *and* Innovation?

"The danger is that as U.S. companies shift the commercialization of their technologies abroad, their capacity for initiating future rounds of innovation will be progressively enfeebled. …The loss of companies that can make things will end up in the loss of research that can invent them."

Suzanne Berger, *Report on the MIT Taskforce on Innovation and Production*, Cambridge, MA: MIT Press, 2013.

NEW OPPORTUNITIES FOR "RE-SHORING" MANUFACTURING

According to a 2012 MIT survey on re-shoring, "Manufacturing is now going through a genuine transformational period," although the scale and permanence of this change is not certain.[38] Currently, this "trend" towards re-shoring is being led by a group of relatively capital-intensive manufacturing industries including computers and electronics, machinery, fabricated metals, electrical equipment, and plastics and rubber. The MIT survey found that manufacturing firms increasingly see the benefits of offshore production being eroded by rising offshore labor costs, and shifting currency exchange rates.

At the same time, the surveyed firms cited a number of factors including lower energy costs and advances in manufacturing techniques that favor the re-shoring of manufacturing to the United States.[39] Recent years have yielded advances in hydraulic fracturing or "fracking" technology resulting in a significant increase in the supply of proven natural gas reserves and declines in natural gas prices.[40] Advances in automation, additive manufacturing and nanotechnology also have the potential to shift the nature of manufacturing from

---

"collaboratively discuss challenges—there is no substitute for right next door." Remarks by Mike Russo, "Breaking New Ground: The New York Advantage." Presentation at the National Academies Conference on New York's Nanotechnology Model: Building the Innovation Economy, Troy, New York, April 4, 2013.

[38]MIT Forum for Supply Chain Innovation and Supply Chain Digest, "U.S. Re-Shoring: A Turning Point," Cambridge, MA: MIT, 2012. Three hundred and forty participants completed the MIT survey, of which 198 were manufacturing-only companies. Out of those 198 companies, 156 were U.S. companies, defined as having their headquarters in the United States. "Specifically, 33.6 percent of respondents stated that they are 'considering' bringing manufacturing back to the United States, while only 15.3 percent of U.S. companies stated that they are 'definitively' planning to re-shore activities. Time-to-market and controlling costs were two main reasons for re-shoring, according to the survey."

[39]TD Economics, "Onshoring, and the Rebirth of American Manufacturing," October 15, 2012. Access at *<http://www.td.com/document/PDF/economics/special/md1012_onshoring.pdf>*.

[40]See PWC, "Shale Gas: A Renaissance in U.S. Manufacturing?" 2011. Access at *<http://www.pwc.com/en_US/us/industrial-products/assets/pwc-shale-gas-us-manufacturing-renaissance.pdf>*.

mass production of generic products towards efficient and localized production of personalized products.[41]

New Policy Opportunities

The MIT report on onshoring concludes that "there exists a huge opportunity for U.S. companies and policy makers to accelerate this trend and return the country to an era of manufacturing growth."[42] The federal government can help sustain this trend to re-shore through its trade, taxation, and manufacturing policies and through its investments in the nation's research and development infrastructure.[43] Major initiatives, such as the announced development of a National Network of Manufacturing Institutes (NNMI), can accelerate the adoption, refinement, and application of a variety of emerging technologies. As described below, a manufacturing strategy that includes sustained support for an improved MEP program can further contribute to the strengthening of U.S. manufacturing.

The Call for a National Manufacturing Strategy

As we have seen, a growing number of experts have called for a national strategy to support U.S.-based manufacturing. They highlight the importance of manufacturing to innovation, economic growth, and job creation, and draw attention to new opportunities to re-shore production.[44] These calls are being backed by the development of new institutes to support advanced manufacturing.[45] There are bipartisan bills before Congress to promote growth of the manufacturing sector, support the development of a skilled manufacturing workforce, enable innovation and investment in domestic manufacturing, and support national security.[46]

---

[41]See the *Economist*, "Additive Manufacturing, Print Me a Jet Engine," November 22, 2012.
[42]MIT, "U.S. Re-Shoring: A Turning Point," Cambridge, MA: MIT, 2012.
[43]U.S. Department of Commerce, "Revitalizing Manufacturing," January 2012.
[44]For a comprehensive review of the rationale and potential of such a program, see the prepared remarks by Daniel Eugene Sperling of the National Economic Council, "Case for a Manufacturing Renaissance," at the Brookings Institution, July 25, 2013.
[45]In addition to the pilot institute for additive manufacturing headquartered in Youngstown, Ohio, competitions have been launched to create three additional manufacturing innovation institutes with a federal commitment of $200 million across five federal agencies—Defense, Energy, Commerce, NASA, and the National Science Foundation.
[46]On June 20, 2013, U.S. Rep. Dan Lipinski (D-IL) introduced H.R. 2447, "The American Manufacturing Competitiveness Act of 2013," a bill that would bring together the private and public sectors to develop recommendations to revitalize American manufacturing and create good-paying, middle-class jobs here at home. U.S. Rep. Adam Kinzinger (IL-16) is the lead Republican cosponsor.

Box 2-4
The Five Pillars of the Proposed National Manufacturing Strategy

In prepared remarks at the Brookings Institution, Gene Sperling, the director of the National Economic Council described the five key thrusts of the Obama administration's National Manufacturing Strategy.[a]

1. **Making the United States more cost competitive for production:** To make production in the United States more cost competitive, the administration has proposed tax reforms that lower the overall rates to 28 percent and 25 percent for manufacturers. Also proposed is the modernization of the nation's infrastructure to reduce supply chain and transportation costs. The strategy also foresees seizing cost advantages from new technologies in accessing natural gas and energy resources.
2. **Spurring innovation based on next-generation technologies:** This includes creating a network of 15 advanced manufacturing innovation institutes to bring companies and universities, supported by federal agencies, to coinvest in world-leading technologies and capabilities. The administration also proposes to increase federal investment in advanced manufacturing R&D, as well as invest in U.S. leadership in key technologies such as bio-based products, clean energy, advanced vehicles, materials, robotics, and other platform technologies with broad benefits.
3. **Strengthening skills, communities, and supply chains to attract investment**: This includes investing in a skilled workforce through a proposal to create an $8 billion Community College Career Fund and initiatives to strengthen manufacturing communities through a $6 billion Manufacturing Communities Tax Credit to help regions that are suffering job losses to attract investments. In addition, the Investing in Manufacturing Communities Partnership will seek to coordinate economic development resources across the federal government to support local strategies to compete for manufacturers.
4. **Access to markets**: A new Interagency Trade Enforcement Center has been created to pursue trade cases more actively. The administration is also seeking new trade agreements to open foreign markets to U.S. manufactured goods.
5. **Promoting investment and insourcing in the United States:** This includes launching SelectUSA, a federal program to attract and promote investments to the United States.

[a]Gene Sperling, "The Case for a Manufacturing Renaissance," address at the Brookings Institution, July 25, 2013.

IN SUMMARY

There is growing and authoritative concern that the erosion of America's manufacturing and high technology base threatens to undermine U.S. leadership in next-generation technologies and the high value-added employment gains that would follow expanded U.S. high technology production and exports. Moreover, some analysts argue that maintaining a competitive on-shore manufacturing sector and the associated skilled labor and technical institutions are linked and essential for long-term national competitiveness. They note that once manufacturing activity moves overseas, so do the required skills, networks and supply chains; and once offshore they, and the learning they engender, are difficult to recover. These analysts therefore argue that it is important for U.S. policymakers to be concerned with the capabilities and composition of the economy, just as policymakers are elsewhere. At the same time, the emergence of new technologies and other favorable developments, such as shifts in energy costs, open fresh windows of opportunity for manufacturing in the United States that can to be exploited by new policies.

Given that a strong domestic manufacturing base is integral to sustaining innovation and maintaining global competitiveness in advanced technologies and critical to national security, the United States needs to augment its efforts to support U.S.-based manufacturing. The next chapters describe how the MEP contributes to this mission, and how it could be improved and expanded, based on this assessment and as well as learning from best practice lessons from leading applied research and manufacturing programs from around the world.

Chapter 3

MEP and Lean Manufacturing

In the context of increasing global industrial competition and intensifying customer demands, there has been considerable pressure on U.S. small- and medium-sized (SME) manufacturers in recent years to upgrade efficiency and quality, raise productivity, lower unit costs, and improve delivery. Over the past two decades, most Manufacturing Extension Partnership (MEP) centers have deployed a suite of tools and services focused on lean manufacturing as a primary line of business.[1] There are many definitions of lean manufacturing, but MEP defines it as "the establishment of a systematic approach to eliminating waste and creating flow throughout the whole company."[2]

As an example of one of the major current thrusts within the MEP, this chapter examines how the MEP provides lean manufacturing services, including how individual MEP centers deliver these services to their clients. Variations in how MEP centers offer lean manufacturing services are reviewed.

ORIGINS OF LEAN MANUFACTURING CONCEPTS

According to James Womack, president of the Lean Enterprise Institute, much of the impetus for lean manufacturing came from the remarkable and unexpected success of Japanese auto companies in the U.S. market in the 1970s. Examining the sources of success, analysts focused on the Toyota Production System (TPS), which seemed to provide a critical competitive advantage not just for Toyota but for other Japanese firms that adopted variants on the TPS.

---

[1]The term "lean manufacturing" appears to have been coined by IT researcher John Krafcik in research published in James Womack, Daniel Jones, and Daniel Roos, *The Machine that Changed the World*, Harper Collins, 1990.

[2]Gary Yakimov, Lindsey Woolsey, et al., "Innovation and Product Development in the 21st Century," MEP Advisory Board white paper, February 2010.

The emergence of the TPS rests on the need for competitive improvement that the Japanese auto industry experienced after World War II. Womack notes that—

> *"This system in essence shifted the focus of the manufacturing engineer from individual machines and their utilization, to the flow of the product through the total process. Toyota concluded that by right-sizing machines for the actual volume needed, introducing self-monitoring machines to ensure quality, lining the machines up in process sequence, pioneering quick setups so each machine could make small volumes of many part numbers, and having each process step notify the previous step of its current needs for materials, it would be possible to obtain low cost, high variety, high quality, and very rapid throughput times to respond to changing customer desires. Also, information management could be made much simpler and more accurate."*[3]

The TPS was described in detail in Womack et al. 1990,[4] and again in Womack et al. 1996.[5] Variations on and components of the TPS approach now underpin many of the lean modalities in use at MEP centers, discussed below.

LEAN MANUFACTURING MODALITIES

Lean manufacturing encompasses a wide range of interrelated improvements in firm processes. Many of these, perhaps most, are found on or near the factory floor itself. However, other aspects include improvements in marketing, in hiring, and in intrafirm communication. This section provides an overview of some of the different kinds of lean manufacturing implemented with the help of MEP centers.

A good description of the multiple modalities that can be applied has been provided by TechHelp, the Idaho Manufacturing Extension Center, which uses the metaphor of multiple building blocks to capture the varied approaches that could be employed (see Figure 3-1).[6]

---

[3]Lean Enterprise Institute, "A Brief History of Lean," <*http://www.lean.org/whatslean/history.cfm*>. Accessed July 1, 2013.
[4]James Womack, Daniel Jones, and Daniel Roos, *The Machine that Changed the World*, op. cit.
[5]James Womack, Daniel Jones, and Daniel Roos, *Lean Thinking*, New York: Simon and Schuster.
[6]The House of Lean is a commonly used diagram. See David L. Goetsch and Stanley Davis, *Quality Management for Organizational Excellence*, 7th Edition, New York: Prentice Hall, 2012. For a description of the terms displayed in the diagram, see <*http://www.techhelp.org/index.cfm?fuseaction=services.lean_tools*>.

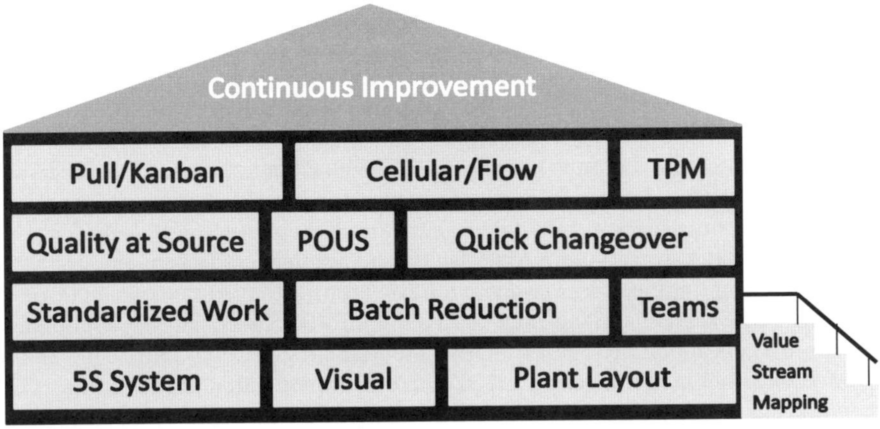

FIGURE 3-1 House of Lean diagram.
SOURCE: TechHelp MEP center, Idaho.

A given project may use some complement of these tools. The Idaho MEP website lists 14 separate offerings in this area, while the Purdue MEP center claims to offer more than 100 different services.[7] Still, implementation of selected tools had a substantial impact for some companies. TechHelp in Idaho for example claims that a typical company benefits from its lean manufacturing services in multiple ways:[8]

- Reduction in lead time—from several weeks to a few days.
- A 50 percent improvement in floor space utilization (eliminate costly plant expansion).
- A 70 percent reduction in travel distance.
- A 50 percent reduction in inventory levels (boost cash flow and profits).
- An increase in quality and productivity.

IMPLEMENTING LEAN PRACTICES IN MEP CENTERS

Many of the conceptual tools and related training programs used by MEP centers come from consulting companies that serve more than one center. It is important to see that a consulting intervention at a company is usually tightly interwoven with training for both management and workforce. Typically, the process involves discussions with management to set goals, discovery and assessment of the factory environment, introduction of specific tools for

[7]David R. McKinnis, The Purdue MEP Model, NAS Ohio workshop: "Diversity and Achievements: The Role of Manufacturing Extension Partnerships in the Midwest," March 26, 2012.
[8]TechHelp Idaho, Client Services, <*http://www.techhelp.org/index.cfm?fuseaction=services.lean_tools*>. Accessed July 17, 2010.

improvement, a blitz focus on effecting rapid change ("kaizan event"), and training both to implement the changes needed and to ensure that they are sustainable and indeed the basis for further rounds of improvement.[9]

Thus the overlap between lean manufacturing and the workforce development mission of MEP is central. Recently, this has been reflected in efforts by professional groups such as the Society of Manufacturing Engineers to develop curricula leading into specific professional certifications in various aspects of lean production.[10] In collaboration with the American Society for Quality and the Association for Manufacturing Excellence, the Society of Manufacturing Engineers has joined the Lean Certification Alliance in an effort to standardize curricula in this area. And these efforts may be gaining traction as local providers adopt them. For example, the Manchester Community College in New Hampshire is now offering a lean enterprise certification program (in conjunction with New Hampshire MEP).[11] Many other examples can be found; some are discussed in more detail in the workforce development section below. Consulting companies also provide systemic training programs. Six Sigma (a consulting company focused on quality issues) provides a number of different training programs, some of which have been adopted by various MEP centers (different centers adopt different programs). Among the main Six Sigma processes are:[12]

- 5S Programs.
- Theory of Constraints.
- The 7 Wastes.
- Toyota Production Systems (TPS).
- Demand Flow.
- Just in Time.
- Value Stream Mapping.

Some of these—notably "5S programs," the "7 Wastes," and "Value Stream Mapping"—appear in the offerings of most of the centers reviewed for this report. The more widely known offerings are described in more detail below.

---

[9]The impact of lean projects has not been assessed separately by MEP. However, there has until very recently been a close overlap between lean projects and MEP center projects overall. Hence the numerous assessments documented in Appendix A are to a considerable extent an effort to measure the outcomes of lean-oriented interventions.

[10]Society of Mechanical Engineers, Lean Certification, <*http://www.sme.org/lean-certification.aspx*>. Accessed July 17, 2012.

[11]New Hampshire MEP, Lean Enterprise Certification Program, <*http://www.nhmep.org/lean_enterprise_sep.html*>. Accessed July 17, 2012.

[12]Six Sigma, Lean Six Sigma and Lean Flow, <*http://www.6sigma.us/lean-manufacturing-flow.php?gclid=CImP8ajRgrECFYFo4Aod0CshKQ*>. Accessed July 5, 2012.

The 5S's

Most MEP centers offer a version of the 5S program as a core element of their lean production services.[13] 5S is based on Japanese words that begin with the letter S, as "5S" references the five elements of this system. The terminology of lean manufacturing is both varied and under continuous change (or perhaps improvement). The 5S methodology overlaps with Kaizen, which is a high-energy short-term effort to implement lean manufacturing principles (especially 5S), while there is substantial overlap between parts of the TPS, value stream mapping, and 5S, as well as other modalities.[14]

- ***Sort (Seiri)***—The first step focuses on removing all unnecessary items from the workplace. One approach is to red tag all potentially excludable items, which can then be addressed by management or a change team.
- ***Set in Order (Seiton)***—Step two focuses on efficient storage and location methods. Tools involved range from simple tools like marking tape, labeling systems, bins, and peg boards to complex electronic tracking via bar codes and radio-frequency identification.
- ***Shine (Seiso)***—Step three aims to introduce a spotless workplace and to maintain it, so that morale improves and machinery works better with lower downtime. Defects become more obvious, safety is improved, and daily inspections help identify problems before they become critical, as cleaning is itself a form of inspection.
- ***Standardize (Seiketsu)***—Step four is standardization, whereby activities—including both processes and responsibilities—within the factory become standardized, and can therefore be tracked more effectively.
- ***Sustain***—Step five requires that the new processes be sustained to the point of becoming habit, an important if difficult step.

Certifications created by Six Sigma and others now appear to be widely accepted by MEP client companies. A number of center directors have pointed out that their staff participate in "Black Belt" training.

Many centers report success stories related to 5S implementation. For example, in South Carolina, Beneteau Inc. is the American division of the world's largest sailboat manufacturer for models 30 feet and larger, employing more than 450 people. Using the 5S approach (modified here to 6S by adding "Safety") Beneteau rationalized its production lines to generate $225,000 in

---

[13]A search of the MEP success stories database for "5S" generated 131 references for 2009-2011, from 46 different MEP centers, *<http://ws680.nist.gov/mepmeis/SearchSS.aspx?ID=3143>*. Accessed July 22, 2012.

[14]For more details on 5S, see Edward Moulding, 5S, *A Visual Control System for the Workplace*, AuthorHouse UK, 2010.

annual cost savings and to reduce lead time from 10 weeks to 1 week. This led to further investment of $1.2 million in new equipment.[15]

In Rhode Island, Walco Electric found that a project focused on 5S with Rhode Island Manufacturing Extension Services generated substantial improvements. In this case, it took two rounds of process implementation with the center, as the first round of improvements were hard to sustain, but the net result was that on-time shipments improved 97 percent and productivity grew 15 percent, in turn generating ongoing cost savings.[16]

Value Stream Mapping

Value Stream Mapping (VSM) derives from another part of the Toyota Production System, known as Material and Information Flow Mapping.[17] Rother and Shook define the value stream as "all the actions (both value added and non-value added) currently required to bring a product through the main flows essential to every product: (1) the production flow from raw material into the arms of the customer, and (2) the design flow from concept to launch."[18]

As Rother and Shook observe, this is an ambitious and high-level vision. They recommend companies start by considering the value stream inside the factory, and once this has been addressed expand upstream into product development and downstream into sales and marketing.

The point however is that VSM provides companies with an overview of the production process that can help them to find and eliminate waste, and see opportunities for lower cycle times and lower costs. VSM is part of the training and consulting provided by Six Sigma and many other management consultants. It can include both training in VSM concepts and the provision of tools for generating visual maps of the production flow (e.g., Visio,[19] Smartdraw[20]).

VSM is the process of mapping the flow of materials and the processes needed to deliver goods or services to the customer. VSM is now quite widely adopted; in fact, a number of colleges and universities provide VSM training (e.g., University of Washington[21]). VSM can be extended to transactional

---

[15]MEP Success Stories, "Beneteau Inc.," <*http://ws680.nist.gov/mepmeis/SearchSS.aspx?ID=2848*>. Accessed July 22, 2012.

[16]MEP Success Stories, "Waco Inc.," <*http://ws680.nist.gov/mepmeis/SearchSS.aspx?ID=2857*>. Accessed July 22, 2012.

[17]Mike Rother and John Shook, *Learning to See: Value Stream Mapping to create Value and Eliminate Muda*, Lean Enterprise Institute, 2003, p. 3.

[18]Rother and Shook, op. cit. p. 4.

[19]See Visio overview, Creating A Value Stream Map, <*http://office.microsoft.com/en-us/visio-help/create-a-value-stream-map-HA010113024.aspx*>. Accessed July 5, 2012.

[20]See SmartDraw Inc., Easy Value Stream Mapping Software, <*http://www.smartdraw.com/specials/value-stream-mapping.htm?id=328924&gclid=COLxl8nkgrECFQff4AodSxZVhA*>. Accessed July 5, 2012.

[21]See University of Washington, Value Stream Management Course Overview, *courses.washington.edu/ie337/Value_Stream_Mapping.pdf*, n.d. Accessed July 5, 2012.

Box 3-1
Examples of Successful Value Stream Mapping Projects

There are dozens of examples of successful VSM projects with MEP centers. For example, Caldwell Tanks in Kentucky is a very well established company (founded 1887) that provides water and other storage systems to the water, wastewater, grain, coal, and process power and energy industries. The value stream mapping exercise undertaken by the Kentucky Manufacturing Assistance Center provided current state and future state value stream maps, and creating a prioritized action plan for moving from current state to future state. This was then implemented using a 5S program.

The result was a significant improvement in efficiency. Caldwell Tanks generated increased and retained sales of $4 million along with $180,000 in cost savings leading to $120,000 in additional investment, along with the intangible benefits of a better and leaner process. Barry Geswin, CFO, observed that "We are now more productive, better able to meet customers' demands and our employees are enthusiastic about the changes we've made."[a]

[a]MEP Success Stories, "Caldwell Tanks Inc.," <*http://ws680.nist.gov/mepmeis/SearchSS.aspx?ID=2755*>. Accessed July 23, 2012.

mapping, which can include all company transactions, not just the flow of material and production inside the company—upstream to suppliers and downstream into sales and marketing activities—as well as including a range of office and administrative activities within the company itself. There is some disagreement within the consulting community about whether transaction mapping is entirely different from VSM or can be integrated into it. Beau Keyte and Drew Lochner, for example, argue that there are effective ways to present transactional data within a VSM framework.[22]

A number of MEP centers provide case studies of successes using this methodology. One example is presented for the Delaware MEP center,[23] where VSM was applied to the operations of Pulmonary Associates, a medical office. The owner found that VSM had a substantial impact on the efficiency of his operations, in this case well beyond the factory floor.

- "26 percent decrease in the number of steps in the posting and billing process.

[22]See Beau Keyte and Drew Lochner, *The Complete Lean Enterprise: Value Stream Mapping for Office and Administrative Processes*, New York: Productivity Press, 2004, pp. 1-3.
[23]Albert A. Rizzo, M.D.1, and Heinz J. Dommert, "Value Stream Mapping: A Case Study of One Practice's Experience," *Del Med J*, May 2009, Vol. 81 No. 5, p. 185-187.

- 62 percent decrease in elapsed time (total time) from initial patient contact to collection for services.
- 315 percent increase in first time quality of insurance submissions.
- Dramatic reduction in rework time for the billing and posting staff."

The Theory of Constraints

The Theory of Constraints[24] (TOC) is based on a systems approach that simplifies improvement in complex organizations by focusing on a few physical and logical "leverage" points that constrain the system's operation as a whole. TOC provides a tool set to address selected "levers" to achieve an order of magnitude improvement in system performance.

The crucial insight of TOC is that only a few elements (constraints) in a business control results for the entire organization. By identifying these constraints, and focusing the entire organization on addressing them, TOC provides opportunities for substantial and often rapid productivity gains. TOC also has a defined methodology and is used by some MEP consultants as an alternative or complement to Six Sigma and/or TPS. The Iowa Center for Industrial Research and Service and South Carolina MEP both appear to have had success using TOC methodologies.[25]

Just in Time (JIT)

Just in Time refers primarily to the provision of parts and supplies "just in time" for their integration into the manufacturing process. A well-known characteristic of the TPS, this approach is also used by some MEP consultants and centers.

Demand flow management is another lean manufacturing methodology, or toolset, that is highly complementary with JIT: In effect, it seeks to replace traditional schedule-driven production with demand-driven production in which inventory is minimized and actual demand drives production schedules.

While some of the more systemic elements of JIT may not always be applicable, the overall focus on reducing lead and production times can pay substantial dividends. For example, a JIT project led by TechHelp in Idaho allowed Idaho Wrecker Sales Inc. to reduce the lead time on one of its key products from 30 days to 4 days; sales increased by 42 percent as a direct result, according to the company.[26] This suggests that the tension between MEP's traditional lean manufacturing focus and the new innovation-growth approach

---

[24]Dave Nave, "How to Compare Six Sigma, Lean, and the Theory of Constraints," *Quality Progress*, March 2002, p. 73.

[25]MEP Success Stories, "Iowa Spring Manufacturing Inc.," <*http://ws680.nist.gov/mepmeis/SearchSS.aspx?ID=3024*>. Accessed July 22, 2012.

[26]Idaho TechHelp, Idaho Wrecker Sales case study, <*http://www.techhelp.org/index.cfm?fuseaction=about.success&id=64&region=3*>. Accessed July 17, 2012.

maybe be overstressed. Here, at least, Idaho Wrecker found that a successful lean manufacturing project that had no overt innovation or growth component nonetheless drove new sales and rapid growth.

Similarly, Bob Browning, director of supply chain at Savage Arms, describes the project as follows, "This multi-year project is in conjunction with Savage's overall Lean initiative and had the objective to develop a process throughout the supply chain of demand pull signals from the consumer, through distribution, manufacturing, and the supply base. Simply put, when a consumer buys a gun off the rack, we want a signal that goes all the way back to the steel mill to tell them to pour more steel."[27]

LEAN MANUFACTURING: CONCLUSIONS

The wide range of possible lean tools and modalities, the large number of centers and the much larger number of providers working out of the centers mean that no simple list of programs and formats can be used to describe the work covered by the MEP program as a whole. Each center has a slightly (or widely) different view of objectives and of the modalities appropriate for addressing them.

And while as each center maintains a list of success stories upon which we draw to help describe various aspects of MEP's operations, it would be a mistake to believe that simple descriptions can be applied wholesale across the program. Below are a few conclusions drawn from our review of lean manufacturing programs and practices at a number of centers:

- While some centers have a menu of services that they sell to potential clients, other centers act primarily as brokers, identifying needs and connecting companies to third-party service providers (e.g., Oklahoma MEP[28]).
- While lean manufacturing is often presented via a menu of specific individual tools of training programs, in practice, projects often involve the application of multiple tools (see cases described above, many of which required multiple tools).
- Even where centers appear to be using the same toolset (e.g., 5S training), in practice, there are likely to be significant differences in the way the program is implemented (e.g., the addition of a sixth S for Safety in South Carolina's version).
- It is therefore very difficult to make claims that specific tools generate specific outcomes, except in individual cases. There are no simple alignments between cause and effect that can be the basis for

---

[27]Karen Myhaver, "Savage Sports: Masters of the Supply Chain," *Next Generation Manufacturing Newsletter*, Massachusetts MEP, n.d., *<http://www.massmac.org/newsline/0911/article03.htm>*. Accessed July 17, 2012.

[28]See Oklahoma Alliance for Manufacturing Excellence, Operating Plan Year 19 (FY 2013), p. 3.

generalized conclusions about the effectiveness of particular tools or approaches.

Overall, MEP's support for lean manufacturing shows evidence of success. Evidence of this overall success is found in academic reviews, as well as in the analysis, case studies and interviews of center directors and company staff conducted for this study. While there have been many success stories from MEP's 20 years as a lean manufacturing support system, numerous evaluations, center directors have indicated that lean in itself is not a sufficient platform for the growth needed to expand the manufacturing sector, and to turn companies that are sustaining themselves—perhaps through reducing their costs—into local engines of growth that are in a position to expand and hire.

And while some MEP center directors have expressed concerns about the way in which MEP is implementing new approaches and in particular about funding, they nonetheless largely support the objective of expanding offerings beyond traditional lean manufacturing.[29]

GOING BEYOND LEAN

There is ample evidence that MEP has played a significant role in introducing and spreading lean manufacturing techniques across the U.S. manufacturing base. The modalities and examples described above illustrate these impacts. It is also clear that MEP's strategy of developing core tools for deployment across the system by different centers was reasonably successful: Different applications of similar tools can be found throughout.

Other services extended by MEP centers have also been important, including those related to ISO quality systems, energy audits, industrial hygiene, web presence, industrial design, industrial marketing, and many others. Although lean manufacturing has been a cornerstone for many centers, most centers strive to be responsive to the needs of their customer base in order to generate sufficient numbers of projects with an increasing number of clients.

Given that lean manufacturing remains at the core of a successful manufacturing support program, MEP should encourage centers to maintain the current capacity in this area and integrate lean manufacturing into new initiatives, including those related to innovation. At the same time, there is evidence—including from the international programs discussed in Chapter 7—that lean is no longer enough. While it remains an important part of the U.S. manufacturing landscape, the growing globalization of manufacturing means that relying on strategies that emphasize service to existing customers and existing products at lower cost are unlikely to be successful in the long run, and are therefore not likely to fuel rapid growth.

For that, approaches focused on innovation and growth will be needed.

---

[29]Private conversations, responses to NAS information request (see Appendix B).

Chapter 4

Development of MEP Center Metrics

The Manufacturing Extension Partnership (MEP) has for many years sought to use detailed metrics as a means both of measuring the performance of individual centers and as a means of establishing the value of the program overall.

This chapter reviews the systems and approaches used by MEP in developing and implementing metrics, covering both the system used in recent years (along with criticisms made of that system), as well as the new system of metrics currently being introduced.

Figure 4-1 shows the evolution of MEP metrics. Formal metrics were developed in the mid-1990s, and were until 2000 focused on tracking individual projects. In 2000, the project-based approach was replaced by a focus on companies, taking the view that the company was the more relevant unit of analysis for impact assessment. The new approach was formalized by FY 2002. After that, adjustments were made as needed, for example, through an increase in the minimally acceptable score for individual centers in FY 2007. Starting in 2009, MEP undertook an effort to address some of the criticisms that were raised with existing metrics and their use for evaluation, notably that they rely on data from the client survey, which some critics held to be unreliable. These criticisms and the new metrics that followed are discussed in detail in subsequent sections.

Figure 4-1 charts the system of metrics against broad measures of program impact, discussed below.

MEP PERFORMANCE METRICS

MEP's performance metrics system addresses performance at three levels, shown in Figure 4-2.

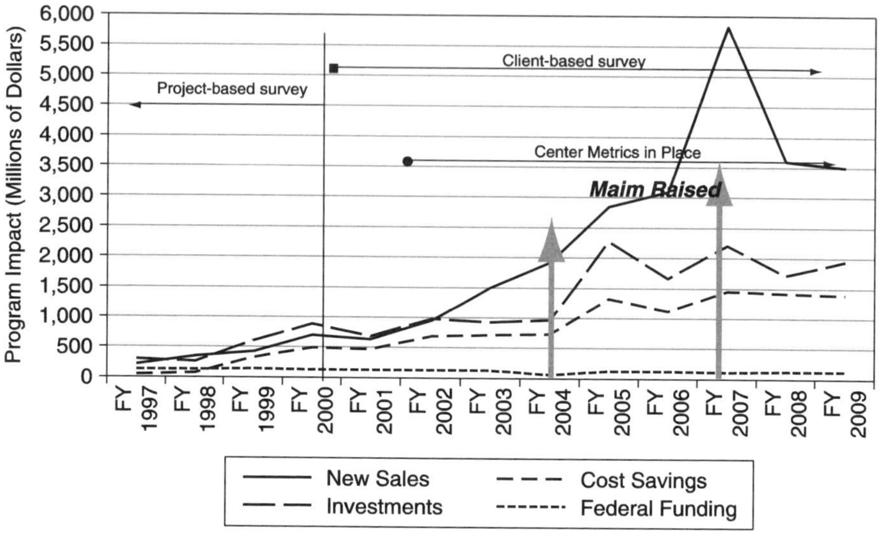

FIGURE 4-1 MEP program level metrics, 1999-2010.
SOURCE: NIST MEP Program.
NOTE: MAIM refers to Minimally Acceptable Impact Measures, against which NIST MEP has evaluated MEP center performance since the early 2000s.

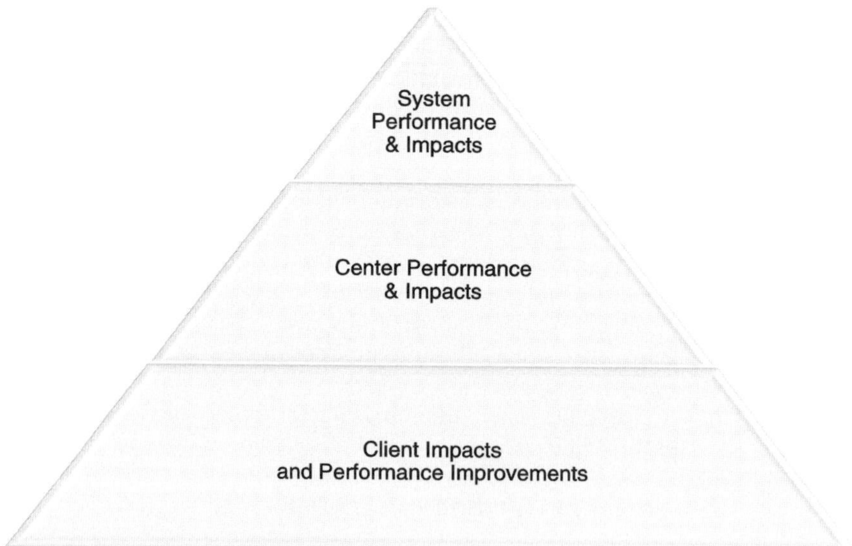

FIGURE 4-2 Program review and evaluation—three levels.
SOURCE: NIST MEP Program.

The data underpinning these evaluations come from two primary sources: The first source is a survey of MEP clients implemented by a contracted survey organization, which generated dollar estimates of MEP project impacts. The second is detailed data about center operations that has been entered into MEP's system database by staff at individual centers. This generated detailed information about center costs and activities (in effect, the input counterpart of the survey's output metrics).

System-level Metrics

At the highest level, MEP aggregates data from all centers and clients to generate cost, services, and outcomes data for the MEP system as a whole. At this level, there are three key sets of metrics:

- ***Reach or penetration***, which indicates the number of clients served and projects completed and undertaken.
- ***Bottom line impact***, which is calculated using data from the MEP survey of center clients, including both cost savings and revenue growth.
- ***Job impacts,*** which indicates direct and indirect employment creation.

Reach

MEP overall reaches approximately 7,000-8,000 clients annually, in the course of approximately 12,000 projects. MEP data indicate that both metrics spiked substantially in FY 2007, reaching 9,000 and 14,500 respectively, and have since declined, in line with the recent dramatic shifts in the United States and world economy. Most recent data is for FY 2010, where MEP served just under 7,000 clients through about 10,000 projects (see Figure 4-3).

It is clear that successful centers do work hard to reach out into the community. Magnet Ohio, for example, ran 57 events attended by 1,601 people from 791 companies in FY 2011, which contributed to acquiring 99 client projects.[1]

Bottom-line Impact

MEP has traditionally measured the program's bottom-line impact by aggregating data on client cost savings and new sales, drawn from the client survey. This is the source of MEP's claims about overall program effectiveness,

[1]James Griffith, "The MAGNET Story: From Lean Manufacturing to Partnerships for Innovation," National Academies Workshop on "Diversity and Achievements: The Role of the Manufacturing Extension Partnership in the Midwest," March 26, 2012.

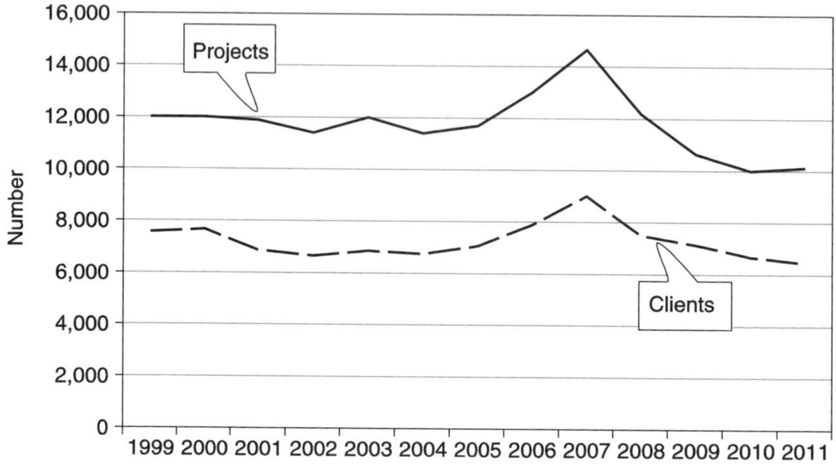

FIGURE 4-3 MEP projects and clients, 1999-2010.
SOURCE: NIST MEP Program.

for example, that its efforts led directly to $9 billion in new and retained sales for its customers in FY 2008.[2]

Jobs

MEP also utilizes aggregate metrics focused on jobs. The most recent MEP performance report leads off its summary data section by indicating that according to the MEP client survey, MEP clients in aggregate claim to have generated or saved more than 53,000 jobs in FY 2010.[3]

CLIENT-LEVEL METRICS AND PERFORMANCE ASSESSMENT

Client-level metrics are (except for success stories) based on a survey of MEP participants undertaken by an external marketing and survey research firm. Client surveys have been sponsored by the National Institute of Standards and Technology (NIST) MEP since 1996. The following section provides an overview of the FY 2009 survey and reports, which was undertaken by Turner Marketing Inc.

The survey was deployed to about 8,900 MEP participants, and generated an overall response rate of 85.7 percent—a remarkably high percentage. This high rate partly reflects the pressure on MEP centers to ensure that their clients complete the survey, as until recently a very substantial component on the center's overall assessment was pinned to the survey and included scoring based on survey response rates.

---

[2]NIST MEP, "Delivering Measurable Results," January 2010, p. 2.
[3]NIST 2010, p. 2.

The survey is deployed two quarters after completion of the MEP engagement, and does not track outcomes longitudinally. Firms with multiple MEP engagements were more likely to respond.[4]

About 40 percent of projects reported that they worked with other service providers, but a majority used only MEP.

Firms identified three top challenges:

- Ongoing continuous improvement (72 percent).
- Identifying growth opportunities (59 percent).
- Product innovation and development (45 percent).

MEP outcomes analysis tends to emphasize aggregate sales impacts (these numbers are found in many MEP documents). The survey sought to identify "increased" and "retained" sales that resulted directly from MEP engagement. The survey identified a total of \$3.3 billion in increased sales and \$5.5 billion in retained sales, as shown in Figure 4-4.

The report for FY 2009 indicated that about 29 percent of projects reported increased sales and 38 percent reported retained sales. These published data are, however, insufficient to provide a more robust analysis (leaving aside the question of control groups). Key questions might include:

- How many firms reported both increased and retained sales? I.e., what was the total universe of firms reporting some positive impact?

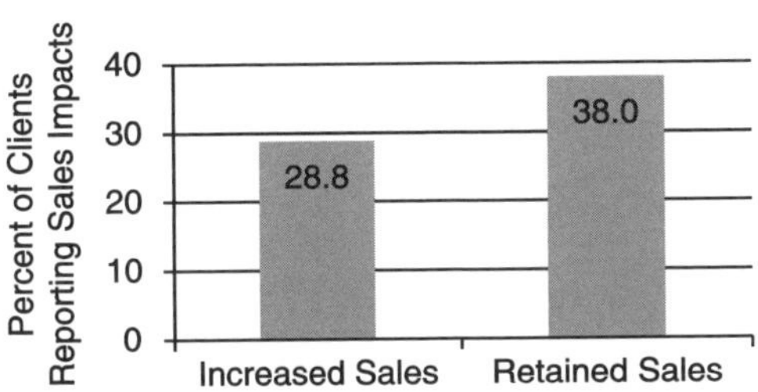

Impact	Dollar Amount
Increased Sales	\$3.6 Billion
Retained Sales	\$5.5 Billion

FIGURE 4-4 New and retained sales from MEP projects.
SOURCE: FY 2009 MEP client survey. Data provided by MEP.
NOTE: Sales impacts reported by MEP clients (*n*=7,648 clients).

---

[4]The survey methodology is described in NIST 2010, p. 3.

- What was the distribution of impacts by size of impact? These data are likely to be highly skewed.
- What was the distribution of both overall impacts and the distribution of impacts by geography?
- Are there breakouts by demographics? By women- and minority-owned firms, for example.

The report's review of employment impacts indicates that 24 percent of projects resulted in new jobs (total 14,000) and 40 percent resulted in retained jobs (total 39,000). Further breakouts are apparently available in other unpublished reports. The published report does not make the connection to the cost of these jobs—at MEP's current budget, preliminary calculations suggest that these 53,000 jobs cost a total of about $100 million in federal funds or just under $2,000 each. MEP might consider comparing the cost per job here against that from other federal or state programs with job-related impacts.

Cost savings are a very important part of the MEP mission, and reflect the priority focus of many participant firms on lean manufacturing and continuous improvement. As with other metrics, the report provides no further breakouts. Cost savings include both direct cost savings reported by 59 percent of respondents.

The report also covered investment resulting from MEP activities. Respondents reported investments in four categories, although in none did more than half of respondents report investments, which overall totaled approximately $1.75 billion (Appendix B, Table 8):

- Plant and equipment (35 percent).
- IT/software (23 percent).
- Workforce development (50 percent).
- Other (17 percent).

While encouraging, a closer examination of these numbers can provide a more nuanced understanding of how MEP has helped firms make these investments and whether the investments were successful.

Finally, the survey sought to identify "promoters" and "detractors" of the MEP program, a metric used to measure customer satisfaction. Respondents are asked whether they are likely to recommend MEP center services to other local companies. The survey found that about two-thirds of respondents are "promoters" and only 10 percent are "detractors." However, the published material does not link these results to other outcomes or program demographics.

As Luria has noted, centers are accustomed to the survey; however, the survey has a number of flaws, notably that a large number of clients cannot monetize the effects of program assistance, outliers play a significant role, and the existence of attribution is of concern.[5] There are also concerns about incentives within the scoring system for centers to ensure that the most positive possible answers are provided by surveyed clients.

MEP has sought to respond to some of these issues, for example, by applying value-added adjustments to sales results from the MEP survey in its bottom-line client impact MAIM metric.[6]

CENTER LEVEL METRICS: THE MAIM APPROACH[7]

MEP has since the early 2000s evaluated center performance against what it has called Minimally Acceptable Impact Measure (MAIM), which MEP aimed to review and adjust over time as center performance improved. It has apparently been adjusted once, in 2007. MAIM is a bar against which to measure center performance which is either minimally acceptable or not.

Each center is reviewed quarterly using four-quarter rolling averages, then again annually and biennially, using a weighted scoring system:

1. **Cost per impacted client (30 points)**. The federal investment in a center divided by the total number of clients responding to the survey.
2. **Bottom-line client impact ratio (30 points)**. This is the sum of cost savings reported by all center clients plus 15 percent of total sales impact (new sales and retained sales) reported by center clients, divided by the federal investment in the center.[8]
3. **Investment leverage ratio (20 points).** This is the sum of new investment reported by all center clients divided by the federal investment in a center.
4. **Percent quantified impacts (10 points)**. This is the number of center clients responding with some positive quantified impact to the client survey, divided by the number of center clients selected to be surveyed.
5. **Clients served per million federal dollars (10 points)**.

Items 1 and 5 above are measures of relative input cost, items 2 and 3 are outcome metrics. Item 4 can be viewed as either or both. All items except 5 rely on data from the client survey.[9]

In order to generate these benchmark metrics, MEP collects data from two sources: the centers themselves, via electronic submission of numerous data about their activities; and the client survey which is focused on client outcomes.

---

[5]See Luria, 2011.
[6]Voytek et al., 2004.
[7]Center performance metrics are currently under review at MEP.
[8]Based on an MEP analysis of data from the 1998 Annual Survey of Manufacturers published by the U.S. Census Bureau.
[9]See Appendix A for a list of data fields provided by MEP.

1. **Center data**. There are three primary input metrics:

    - **Federal funding**. Amount of federal funding provided through MEP
    - **Other funding**. Amount of other funding from state and industry sources as well as user fees
    - **Staffing patterns.** Centers report on the numbers of staff provided through the program. Data cover total full-time equivalent staff, as well as breakouts for technical specialists, sales start, and management and support staff, the number of hours worked, use of third-party service providers, etc.

2. **Center cost and utilization metrics.** MEP calculates a number of cost metrics for each center. These include federal cost per project hour, federal funds as percentage of total expenditures, federal funding per full-time equivalent position, and total expenditure per client.
3. **Output metrics** include:

    - **Center report on total number of clients.** This measure is all clients that the center claims to have touched in some way and is much larger than the number with whom centers have contracts. No further detailed information is available from MEP about these clients.
    - **Center report on impacted clients**. These are clients with whom the center has had substantial interaction based on MEP funding, and on which it has expended actual resources. These clients are all included in the client survey.
    - **Bottom-line impact drawn from the client survey,** defined as the sum of the company's direct cost savings plus 15 percent of its additional sales. This is an effort to measure the direct impact of the MEP engagement on the firm's bottom line. Fifteen percent of sales are used as part of the metric because the average margin for small manufacturing companies is, according to MEP, 15 percent. Cost savings are seen as falling directly on to the bottom line.
    - **Bottom-line impact ratio**. This measure divides bottom-line impacts by federal expenditures.
    - **Investment leverage ratio**. This aggregates additional investments reported through the client survey for the center, divided by the center's federal funding.
    - **Job impacts reported via the client survey**. The MEP survey asks respondents to indicate how many jobs were created or retained as a result of the MEP project.

Using the metrics as defined by MEP, it is possible to track median center performance over time.

Figure 4-5 shows median performance for 2001, 2004, 2008, and 2010.

Minimally Acceptable Impact Measure (MAIM)

Each of the five components of MAIM has its own minimally acceptable level of performance and weighted score (see Table 4-1). The five scores are weighted to reflect the relative importance of each.

Individual scores are aggregated across the measures to generate a MAIM score for each center.

> *A rolling four-quarter average of MAIM scores is used for measuring Center performance. If a Center meets or exceeds the minimally acceptable level of performance for a particular measure, they receive all the points assigned to that measure. If they do not meet or exceed the minimum performance level for a particular measure, they do not receive any of the points assigned to that measure.*
>
> *Aggregate scores can range from a minimum of 0 (failing to meet the minimum threshold for any of the five*

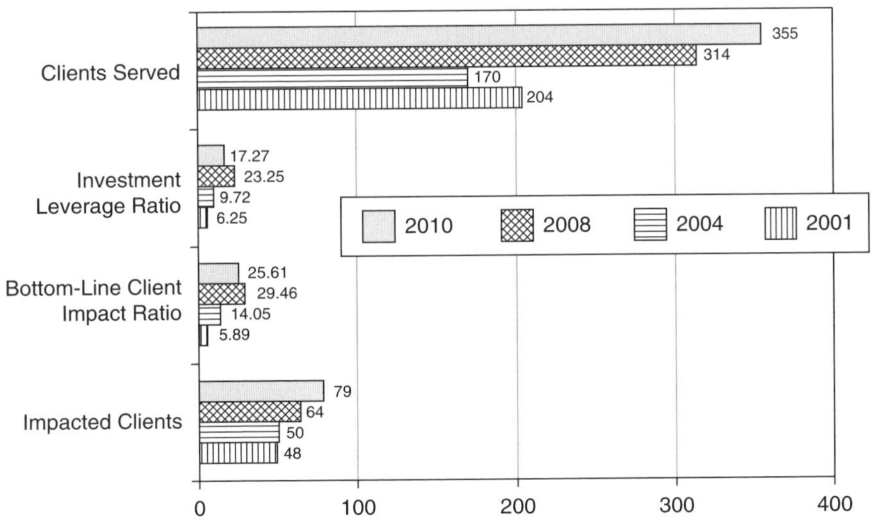

FIGURE 4-5 Median Center performance over time.
SOURCE: MEP.

TABLE 4-1 MAIM Minimum Levels and Weights

Performance Metric	Cost Per Impacted Client	Bottom-line Client Impact Ratio	Investment Leverage Ratio	Survey Response Rate	Percent Quantified Impact
Minimally Acceptable Level	$31,000.00	3.00	3.00	70.00%	32.00%
Weighted Score	30	30	20	10	10

SOURCE: Thomas, op. cit.

> *indicators) to 100 (meeting or exceeding the minimum threshold for all five indicators). A Center with an aggregate score of 70 or more is considered to satisfy the MAIM requirements.*[10]

Meeting the MAIM score benchmarks has been a real and not a formal requirement. For example, in 2008 and 2010 the Nebraska MEP struggled to meet those benchmarks, and was provided with a series of recommendations by the MEP Advisory Board. In its 2010 annual report, the Nebraska center explicitly addressed each recommendation and showed how it had or was attempting to meet them.[11] Once the Nebraska MEP got into trouble, it reached out not only to MEP headquarters staff but to other centers (such as Ohio and Arkansas) for help.[12]

Perhaps most interestingly, it appears that the single most important factor in helping MEP of Nebraska increase scores to meet MAIM benchmarks was improved reporting via the MEP survey, rather than an actual improvement in service.[13]

MAIM's Weaknesses

MEP itself has recognized that the existing MAIM approach has a number of significant weaknesses:

- It is heavily reliant on data from the client survey.
- Centers and their staff are rewarded based on high levels of positive responses, which may lead some to coach clients on their responses.
- The focus on minimally acceptable standards has numerous problems:

[10]NIST, "Report on NIST/MEP's Center's Performance Reporting," January 2007.
[11]MEP of Nebraska, "2010 Center Progress Report," May 26, 2011, section 1-12.
[12]MEP of Nebraska, op. cit., pp. 1-12.
[13]Ibid.

- It provides no guidance once centers meet the minimum.
- It is binary (pass-fail) when center performance is not.
- It relies on data that is strongly tilted toward lean manufacturing as the core suite of services.
- It provides no detailed information for centers about their relative strengths and weaknesses.

- It does not address the critical balancing act maintained by centers as they struggle to manage the imperatives of generating maximum client impact, while expanding market penetration, within the broad context of retaining financial stability (currently at a time of declining state contributions).
- It is not well designed to help assess center activities related to the MEP's new innovation-focused strategy.[14]

In light of MAIM's weaknesses, which MEP acknowledges, MEP is now shifting toward a new set of metrics which it calls its CORE standards.

CORE METRICS: A BALANCED SCORECARD APPROACH[15]

The new CORE approach is different from MAIM in a number of important respects. CORE has four elements:

- New center diagnostics.
- Revised impact metrics.
- Opportunities and challenges.
- Panel review.

These are combined to provide a much broader perspective, and MEP anticipates a more balanced view of center activities.

Figure 4-6 illustrates how the different components of the CORE metrics are designed to fit together. The assessment process will utilize data from the client survey and centers along with qualitative evaluations from new sources.

New Impact Metrics[16]

Impacts are calculated by aggregating all impacts reported by center clients, using a four-quarter rolling average. Metrics are designed to control for

---

[14]See Chapter 6 of this report for a review of this new strategy.

[15]Unless otherwise noted, this section is based on a detailed briefing on CORE provided by MEP on July 23, 2012. All graphics unless otherwise noted are from this briefing.

[16]Material in this section draws on MEP's Scoring Methodology document, privately provided by NIST MEP.

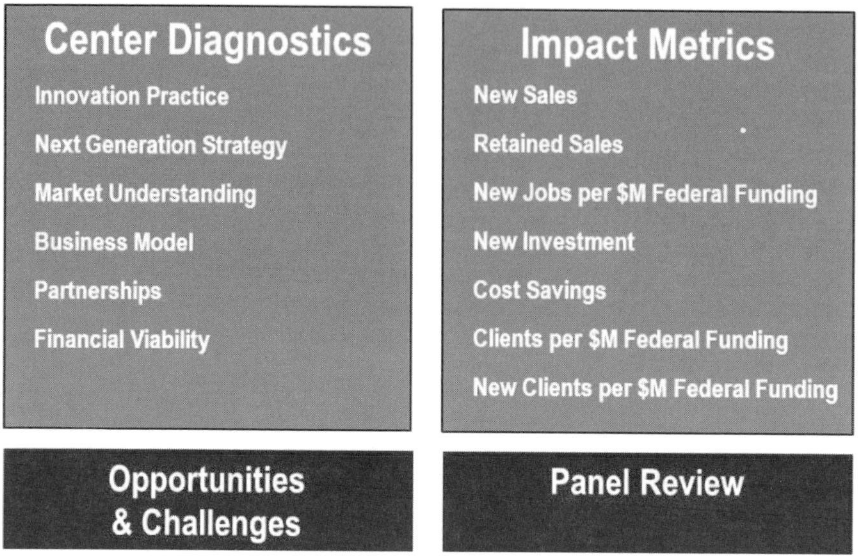

FIGURE 4-6 CORE Metrics: a balanced scorecard approach.
SOURCE: MEP.

the size of the center, using the amount of federal funding to do so.

For each metric, data collected by the center is normalized by using the amount of federal funding received, and is then compared to a threshold level established for each metric by NIST MEP. Centers scoring at more than 100 percent of the threshold are graded excellent and receive the maximum score. Centers can gain additional points by scoring at a multiple of the established threshold, up to a maximum of four multiples of the existing threshold, which would generate four additional points on the score for that metric. NIST's methodology document does not explain how threshold ratios for individual metrics are derived.

Scores for each metric are aggregated to arrive at the overall impact score for each center. The aggregate impact score accounts for 50 percent of the overall score for each center; the remainder comes from the center diagnostics (discussed below).

The individual metrics described here capture direct effects; the indirect social welfare costs and benefits are not captured.[17]

1. **New sales.**

The new sales metric focuses squarely on growth-oriented company outcomes. It is calculated by aggregating the total amount of new sales that

[17]The benefits to social welfare could include, for example, increases in profits. However, data capturing this information is not available, as small private firms are reluctant to provide data on earnings.

clients report related to MEP-funded projects, divided by total federal (MEP) funding.

2. Retained sales.
Aggregated data from clients is divided by federal funding to generate a ratio. The retained sales ratio is then compared to a threshold established by NIST MEP. Centers scoring at more than 100 percent of the threshold are graded excellent.

3. New jobs.
The primary metric is new jobs per $1 million of federal funding. Total new jobs reported are divided by federal funding expressed in millions of dollars.

4. New investment.
New investment is calculated by aggregating all related new investments reported by clients for a center.

5. Cost savings.
MEP centers are still expected to maintain a substantial focus on lean manufacturing, where impacts are best captured through cost reduction. Here the aggregated amount of cost savings is divided by federal funding.

6. Manufacturing clients.
For this metric, the total number of manufacturing clients is divided by federal funding expressed in millions of dollars. This metric aims to capture at least in part market penetration. It should be noted that the amount of federal funding is not closely tied to the number of manufacturing companies in an MEP center's population.

7. New manufacturing clients.
For this metric, the total number of new (i.e., nonrepeat) manufacturing clients for a center is divided by federal funding expressed in millions of dollars.

A number of observations can be made about the new impact metrics.

First, they are heavily weighted toward top-line growth rather than the existing lean consulting base of the centers. Four of the seven metrics are explicitly focused on "new" sales, jobs, and clients.

Second, the impact metrics continue to be based entirely on impact data drawn from the client survey. There are, as we have seen elsewhere, some significant issues with the survey, and these issues may be exacerbated by the likelihood that innovation-oriented projects may take much longer than cost-cutting projects to generate positive impacts.

Third, it is encouraging to note that NIST MEP is moving away from simple pass-fail metrics even in the impact assessment; the notion of adding points for multiples achievement recognizes enhanced performance without strictly grading on a curve.

Center Diagnostics

The center diagnostics component of the balanced scorecard offers a dramatic contrast with MAIM. All of the metrics involved are assessed using data and information other than the client survey; and in many cases these involve qualitative assessments of center progress and activities, rather than quantitative metrics.

- Many of the center diagnostic components are based on qualitative assessment, not quantitative data. In some cases, it would appear that the criteria could be highly subjective; for example, item NGS d) requires that "Centers work with a sense of urgency that transitions reactive clients to proactive."
- Those that do have a quantitative component—e.g., the level of state funding—are based on data beyond the traditional sources used for evaluation.
- Assessment is comparative. While MEP is not directly grading centers on a curve, it is grading them against benchmarks that are presumably derived from center performance.

The new metrics are best understood as capturing six broad areas of Center activity—Innovation Practice, Next Generation Strategy, Market Understanding, Business Model, Partnerships, and Financial Viability—each of which has several subcomponents. Assessment is not simply focused on a limited number of outcomes; it also covers design and implementation of the center's strategy as well as its implementation of Next Generation Strategy services.

Combined CORE Assessment—Impacts and Center Diagnostics

The interplay of these variables is captured in the summary scorecard developed by MEP that includes both impact metrics—which remain largely unchanged from the MAIM system that CORE replaces—and the center diagnostics, which are new. Both the impact metrics and the center diagnostics are tracked quarterly by MEP for each center, and are measured against the performance of the median center, as show in Figure 4-7.

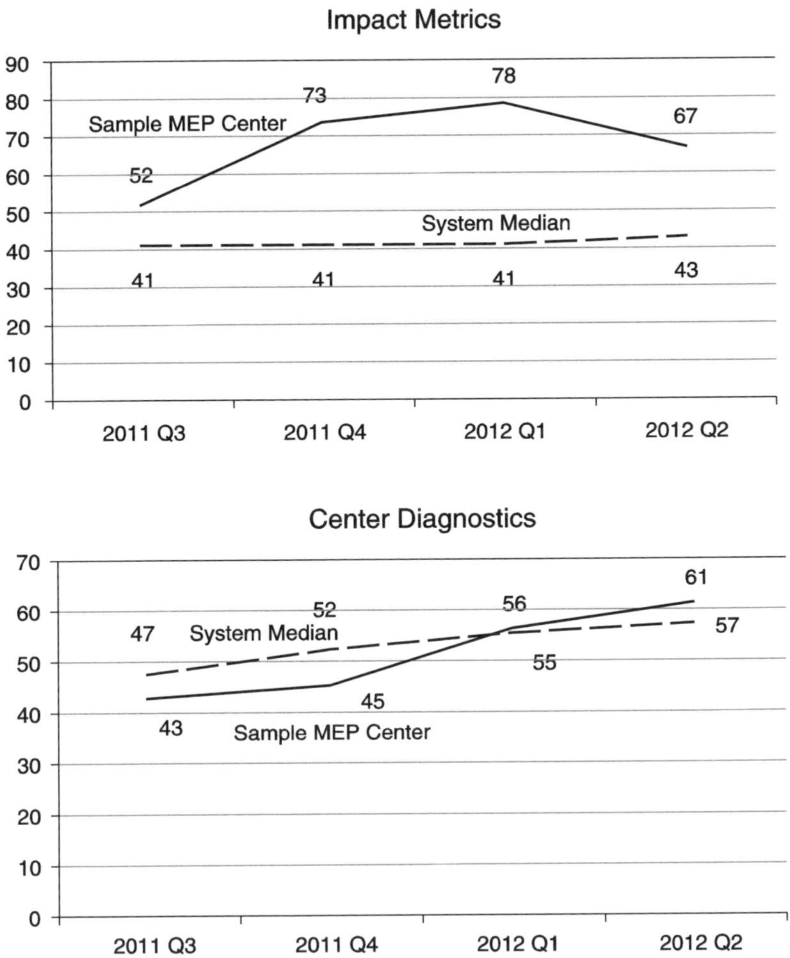

FIGURE 4-7 Center performance over time.
SOURCE: NIST MEP, private communication.

CORE Metrics—Conclusions

These changes do reflect an effort to address previously identified weaknesses in the MAIM approach. However, the CORE metrics bring some questions and concerns of their own.

- NIST MEP is using CORE metrics to drive the transition from the traditional lean-oriented consulting practice to an innovation and growth practice at every center; for example, the first metric listed requires a "transition from a focus on process improvement to the development and implementation of an innovation practice." The stress is heavily on organizational transformation, rather than the extension of new services.

- NIST MEP is clearly seeking to drive the new strategy through the rapid adoption of the specific innovation- and growth-oriented tools developed by NIST MEP. A number of the metrics make this abundantly clear. As noted elsewhere in this report, many of these tools are still at the experimental or at least unproven stage, making adoption potentially risky for organizations with scarce resources such as MEP centers.
- NIST MEP is seeking to drive centers to adopt all components of the new strategy—a one-size-fits-all approach which ignores the likelihood that centers will need to match different offerings to their differing client populations.

Despite some concerns about the new assessment approach, it should be noted that feedback from center interviews has generally been positive. As one center director observed, "If we are doing things right on the ground, we are confident that this will be reflected in the metrics."

However, it should be noted that if at some future point NIST MEP is able to allocate additional funding to individual centers, or to reallocate existing funding between centers, the qualitative aspects of the balanced scorecard approach will require further bolstering: The transparency and independence of the metrics will undoubtedly attract close scrutiny.

Chapter 5

MEP Center Performance Measures and Evaluations of Program Outcomes

As detailed in Chapter 4, NIST-MEP collects a range of metrics on the performance and initial impacts of centers within the MEP system. These metrics provide information that can be used for the monitoring and reporting of MEP center performance, including by activities, staffing, funding, clients served, and intensity and type of service. Additionally, there have been numerous efforts to evaluate the MEP over its more than two-decade history. Many of these studies are sponsored by the program itself, or by state economic development agencies or federal oversight bodies. The evaluations themselves have usually been carried out by various third-party private consultants, university researchers, policy foundations, or government examiners. This chapter begins by considering the broader challenges involved in measuring MEP center performance and evaluating program outcomes and by discussing how the committee addressed these challenges. This is followed by a section that draws on NIST's MEP metrics to gauge center performance and discusses the value and limitations of these observations. The final section of the chapter reviews external evaluations of the MEP and summarizes their main findings and implications.

CHALLENGES OF MEASURING MEP PERFORMANCE AND IMPACTS

It is never straightforward to measure the performance and impact of individual program interventions in a dynamic system with multiple internal and external factors influencing firm performance and economic change. The MEP is not an exception here. There are a number of challenges in measuring and interpreting program and impact data and in undertaking evaluations. Given these challenges, listed below, the committee has been careful in interpreting the outcomes of the MEP program.

Multiple Stakeholders

The MEP involves a mix of federal, state, organizational, and private actors, who have varying goals and interests. Thus, while at the federal level, NIST seeks national impacts from the MEP on the competitiveness and capabilities of SMEs across the country, including improvements in productivity, innovativeness, and exporting, states are naturally most interested in effects within their own jurisdiction. States are especially interested in ensuring that their firms are served and in job retention and creation. Individual firms are typically interested in gains they can capture, including added sales and improved profitability, while industry organizations seek improvements in performance across sectors and supply chains.

Reliance on Self-Reported Data

NIST's own metrics of center impacts on firms seek to incorporate many of these goals, including efficiency improvements, added sales, and jobs retained or created. While NIST does sponsor studies with a range of methods, the major ongoing way through which NIST collects this data is via the MEP survey tool (discussed in Chapter 4). The data collected are undoubtedly valuable, but they must also be regarded with caution. A number of center directors have observed that the very high level of survey responses that MEP headquarters considers to be acceptable has the effect of ensuring that local MEP centers expend considerable efforts and resources on ensuring that clients comply and complete the survey. Current average response is 85 percent, and the amount of effort expended by centers to ensure compliance is reported to range from 5 percent to 35 percent of total staff time.[1]

Incentives and Performance Indicators

In addition, several center directors have noted that the reliance on survey data as a primary metric of center performance introduces related incentives for centers (whose performance evaluations largely rest on survey data) and individual staff (whose compensation in at least some cases depends on both the level of client compliance with the survey and the degree of positive outcomes recorded). Staff thus have incentives to encourage clients to respond and to present the maximum possible outcome. This is therefore somewhat different than relying on self-reported data from a nonincentivized population.

Survey Complexity

If clients must undertake fairly complex calculations in order to complete the survey—which they do—and if there is minimal guidance within

---

[1]Information provided by center directors in response to NRC request, June 2012.

the survey itself on the correct methodology for completing these calculations—which appears to be the case—then the data collected could be flawed: that even where firms are seeking to provide a completely honest and unprompted response, identical outcomes at different firms may result in different responses to the survey.

Identifying the Relevant Controls

The NIST survey data is collected from firms that participate in the program, and not from those who do not. Comparison group studies have been undertaken, sponsored by NIST and by other organizations, with the aim of discerning any additional effects associated with MEP program participation compared with nonparticipation. As with other government support programs for business, care must be taken in identifying relevant control groups to ensure that comparisons made are appropriate. The fact that a company is willing to engage with an MEP center and to seek improvements in its processes may in itself be a distinguishing characteristic, which separates such firms from essentially identical firms that are *not* prepared to take this step. A few controlled studies have sought to address this concern by matching comparison groups and adjusting for performance prior to program entry. However, this is a complex procedure, and many of the available studies of the MEP do not attempt to use formal control groups. This is a known limitation. However, some of the available studies do seek control through the application of counterfactual methods and logic models, while in other cases the intent of the study has been to probe other aspects of the MEP using qualitative approaches and simulation techniques.

These methodological issues provide the backdrop against which the various assessments made to date can be addressed, as described in Box 5-1.

CENTER-LEVEL REVIEW OF PERFORMANCE AND OUTCOMES

In this section, we examine NIST's own data on MEP performance and outcomes. MEP is currently in the midst of a major transition from traditional metrics which drew on the client survey for 85 percent of outcomes to a new approach that relies on a balanced scorecard approach in an effort to better capture, in particular, the services included in the MEP new innovation-focused strategy.[2] As these new metrics were not available to the committee when it undertook its review, we rely on the previous metrics to review center activities.

---

[2]See Chapter 6 for a discussion of MEP's innovation-focused Next Generation Strategy (NGS).

Box 5-1
Data Limitations and the Use of Multiple Data Sources

NIST has made a substantial effort to acquire data with regard to both the operation and the performance of the program. Most federal research and development programs have not collected similar levels of data. Engaging an outside contractor to collect data in a systematic manner through survey deployment is commendable. Nonetheless the committee believes that the data collection methodology could be substantially improved and suggestions to this effect are contained in the Findings and Recommendations.

Despite these reservations, some of the data collected are of high quality. The data on program inputs, collected from the centers, appears to be accurate, and is collected into a database through a well-designed standard process. With regard to outcomes data collected by NIST, it provides valuable information, although we identify challenges that can and should be addressed. There are significant difficulties in encouraging client companies to report on their outcomes at some time after the event. As noted, the use of an independent third-party survey company to collect data is a positive practice; however, the committee has reservations about the impact of the push by NIST MEP for high levels of survey compliance, and also about the role of center staff in encouraging that compliance. These reservations limit, to some extent, uses to which these data can be put, and these limitations are identified. At the same time, there are methodologically sound ways in which these data limitations can be taken into account.[a]

In the context of this study, multiple steps were taken to address challenges in data collection and evaluation:

Balancing through multiple sources of information: To balance potential limitations in the NIST data, the committee drew on a wide range of information sources when conducting the study. Data from MEP recipients is used only as one source of information and it is not the sole source for any recommendation. Importantly, the committee made sustained efforts to develop its own information base, and to draw on other sources of information. This included:

a. Carrying out multiple site visits and interviews with MEP center directors and staff
b. Commissioning a comprehensive review of previous MEP evaluations in the professional literature
c. Developing an information request to all center directors that generated responses from more than 40 centers
d. Convening five NRC workshops—two in Washington and three held regionally
e. Conducting multiple interviews with NIST MEP staff

f. Drawing on input data from the MEP awards database
g. Considering examples from programs outside the United States
h. Conducting interviews with academics and other experts
i. Analyzing internal NIST documents and presentations, as well as and previous external reports, including those from the Government Accountability Office (GAO), and the National Research Council (NRC).

The committee's findings and recommendations therefore reflect the sum total of the knowledge gleaned by the committee, which itself reflects significant program-related expertise.

[a]See, for example, Todd D. Jinck, "Mixing Qualitative and Quantitative Methods: Triangulation in Action," *Administrative Science Quarterly*, Vol. 24 No. 4, Dec. 1979.

As described in Chapter 4, MEP maintains a number of standard metrics related to center performance. These focus primarily on three core aspects of center activity:

1. **Penetration or scope.** Centers are required to serve the local manufacturing community, and penetration metrics are used to measure how well they reach out to their potential client base. This is usually normalized for various possible independent factors, such as the size of the small and medium business population (we use the European term SME [small- and medium-sized enterprises] to describe potential MEP clients, although in some cases MEP works with larger companies).
2. **Effectiveness.** The relative effectiveness of centers, again normalized for various possible independent factors, is critically important to the program as a whole. MEP gathers extensive data on the outcomes of projects, used here to construct appropriate outcome-efficacy metrics.
3. **Efficiency.** The substantial differences in the cost per client and per project among the centers suggest variation in the effectiveness of centers, with implications for improving transmission of best practices between centers.

Numerous intervening variables can be used to help explain variation in the performance of centers against these metrics. However, it does not appear that MEP has focused on this question. Such factors metrics might include demographics of the client population, including size of company, skills levels within client companies and in the local population, degree of existing integration into global supply chains, role of local universities and colleges, nature of existing industry clusters, current average value added per employee, and transportation links and ease of center access to clients.

Given the limited nature of this review, we have focused on one factor that is at least potentially under the control of MEP—the size of the center and its potential client population. Some centers are simply much larger on every dimension than others, and it seems reasonable to suggest at a minimum that the largest centers—which account for a proportionately large percentage of MEP funding—should be of particular interest in addressing MEP impacts.

CENTERS IN CONTEXT: REVIEW OF SIZE DIFFERENCES

Centers differ on many dimensions, including the composition of the local manufacturing sector by industry, the existence or otherwise of major research universities, company demographics, and startup rates. One of the more influential variables is the size of the center. Size can be measured in several ways: by the size of the service area, by the number of clients, by institution revenues, and in revenues or funding per SME in the service area. In this section we review data related to the size of the MEP center, the gross number of clients, and revenues and funding per SMEs in the area.

Population Size of Service Area

MEP centers vary enormously in the size of the SME population that they serve. Based on MEP estimates, the range is from more than 31,000 companies in Southern California to approximately 500 in Alaska. The median number of manufacturing SMEs per service area was 3,353.[3]

Gross Number of Clients

Unsurprisingly, the number of potential clients contacted and indeed the number of clients served is correlated positively with the size of the service area: Overall, centers in the top quintile for size of service area were almost all also in the top quintile for clients contacted and clients served, as shown in Table 5-1.

Centers in Kansas and South Carolina appear to be attracting more clients than would be the median. Puerto Rico claims to touch a much higher percentage of potential clients than the median, but converts few into actual clients.

Gross Size of Funding

Overall, median federal funding in FY 2010-2011 was $3.1 million. Average funding was $4.9 million. In terms of funding, there are four large MEP centers (see Table 5-2). Even the smallest of these—Texas—is almost twice as big as the next largest center.

Thus in terms of size, the four biggest centers are to some degree outliers.

---

[3]MEP center database, FY 2010.

TABLE 5-1 SME Populations and Clients by Center (Top Quintile by Number of SMEs Contacted)

State	Center Name	SMEs Contacted	Impacted Clients
CA	California Manufacturing Technology Consulting	3,563	571
OH	Ohio Manufacturing Extension Partnership	3,273	434
NY	New York Manufacturing Extension Partnership	1,817	404
FL	Florida Manufacturing Extension Partnership	1,633	212
TX	Texas Manufacturing Assistance Center	1,422	282
MI	Michigan Manufacturing Technology Center	1,043	225
MA	Massachusetts Manufacturing Extension Partnership	1,005	155
KS	Mid-America Manufacturing Technology Center	992	85
WI	Wisconsin Manufacturing Extension Partnership	928	170
NH	New Hampshire Manufacturing Extension Partnership	867	41
SC	South Carolina Manufacturing Extension Partnership	854	158
GA	Georgia Manufacturing Extension Partnership	852	152

The rankings for the gross numbers above change substantially when adjusted for size of service area (see Table 5-3). Of the four highest-funded centers identified, none appears in the top ten centers for funding per SME, suggesting there are some economies of scale in servicing firms.

Conversely, many of the remainder are clustered at a much smaller scale, as shown in Figure 5-1.

There is a limited but distinct negative correlation between the amount of funding per SME and the total number of SMEs in the service area. Of the 14 centers funded at $1,500 per SME or above, all have fewer than 5,000 SMEs in their area, as shown in Figure 5-2.

Figure 5-2 indicates that median spending per SME is under $1,000, and the median number of SMEs in the service area is just under 4,000. While there are outliers on both dimensions, with a maximum of just under $4,000 per

TABLE 5-2 Largest Centers by Funding Volume, FY 2010

California Manufacturing Technology Consulting (CMTC)	CA	$28.8M
Ohio MEP	OH	$22.4M
New York MEP	NY	$21.2M
Texas Manufacturing Assistance Center (TMAC)	TX	$16.8M

SOURCE: Data prepared for GAO by MEP, private communication.

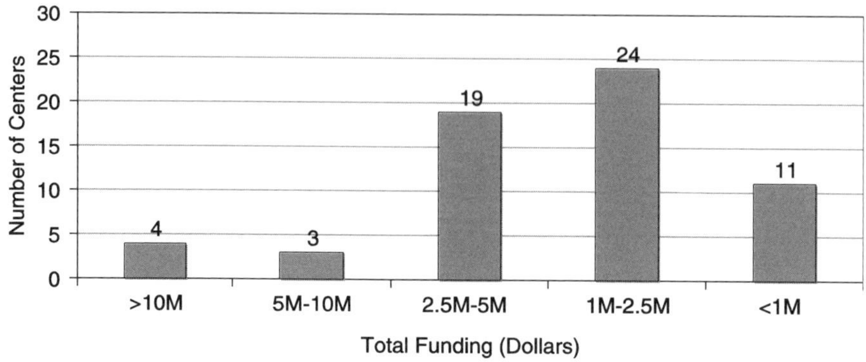

FIGURE 5-1 Number of centers, by funding level, FY 2010.
SOURCE: MEP data, FY 2010.

SME, and a maximum of about 31,000 SMEs in the service area, no large SMEs spend much above 150 percent of the median. Conversely, 12 centers with smaller than the median number of firms in their area spend more than 150 percent of the median.

It may also be worth noting that eight centers spend less than $500 per firm, but all serve more than the median SME population. This suggests that there may be economies of scale. However, while smaller centers utilize more funding per SME, the four biggest centers still account for about 30 percent of total MEP expenditures.

TABLE 5-3 Highest Funding per SME, FY 2010

Recipient Name	State	Total Income (Dollars)	#SMEs in Service Area	Dollars per Service SME
Alaska MEP	AK	1,960,534	501	3,913
North Dakota Manufacturing Extension Partnership	ND	2,370,782	752	3,153
MEP Utah	UT	9,143,913	3,312	2,761
Northwest Pennsylvania Industrial Resource Center (NWPIRC)	PA	3,590,377	1,633	2,199
IMC-PA	PA	2,095,214	996	2,104
Manufacturing Works	WY	1,193,002	568	2,100
MRC	PA	3,206,326	1,537	2,086
New Mexico MEP	NM	3,110,313	1,535	2,026
Mid-America Manufacturing Technical Center (MAMTC)	KS	5,476,200	3,143	1,742

SOURCE: MEP data, FY 2010.

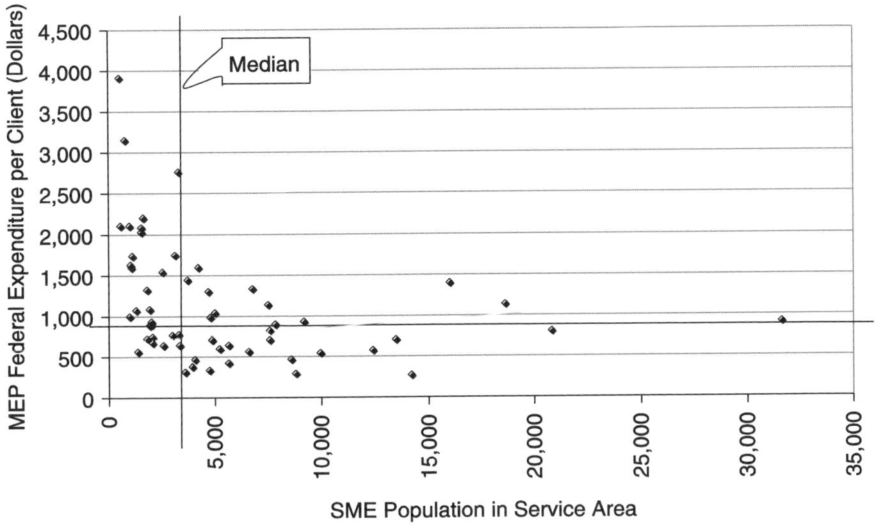

FIGURE 5-2 Centers by federal MEP funding per client served and size of SME population in service area, FY 2010.
SOURCE: MEP dat,a FY 2010.

EFFECTIVENESS

MEP makes a considerable effort to identify and capture outcomes data. Here the key metric is the bottom-line impact, calculated as the sum of cost savings plus 15 percent of additional or retained sales resulting from an MEP engagement. This metric generates widely different outcomes for the centers: Impacts per client range from $3.6 million in Mississippi, to just over $100,000 in northwest Pennsylvania.

Overall, 3 centers claimed bottom-line impacts of over $1 million per client, while 14 claimed less than $250,000 (see Figure 5-3). The large variation in outcomes suggests that determination of why this might be—and how low-impact centers might improve their results—would be a top priority of MEP management.

The three centers with more than $1milion in claimed impacts per client were:

Mississippi Technology Alliance	$3,628,316
GENEDGE ALLIANCE	$1,777,412
Arkansas Manufacturing Solutions	$1,752,289

The next highest claimed just over $900,000.

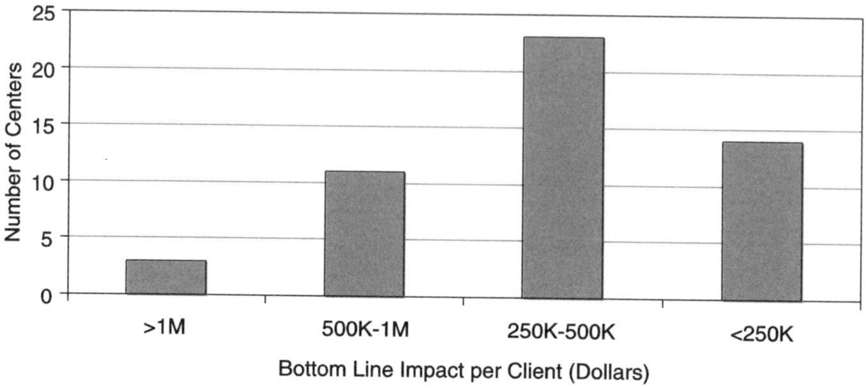

FIGURE 5-3 Centers' bottom line impacts per client, FY 2010 Dollars.
SOURCE: MEP data.

Given the need to impact the overall SME population, not just clients, we also consider the bottom-line impact per SME in the center service area. While there is overlap between rankings, the correspondence is far from conclusive. There are a number of centers in the top quintile for impacts per client who score much lower when considering their impact on the overall SME population; and conversely, some that score low on client impacts score high in the context of the overall SME population, as shown in Table 5-4.

This review shows that while Alaska scores only moderately in terms of bottom-line impact per client, normalizing by the number of SMEs indicates that Alaska had a relatively high impact on the overall SME population. Similarly, while MANEX in Northern California generated more than $200,000 per project in bottom-line impacts, the impact of these projects on the overall SME population was low because MANEX has a low client take-up rate.

However, there are large disparities in impacts per client served and in impacts per SME between the highest-scoring and lowest-scoring centers. Mean impacts per area SME for the top quintile are a multiple of about 4 compared with impacts from the lowest quintile. This is a substantial difference in outcomes.

Jobs Impact

MEP also measures program outcomes through jobs—which includes both new hires and jobs retained. MEP does not appear to address jobs lost as a result of the program.

MEP data again show large differences between the best and worst performing centers. Ohio claims more than twice the per-client job impact of any other center, while there are 14 centers claiming fewer than 3 jobs per client—less than one-thirtieth the claims made by Ohio, as illustrated in Table 5-5.

TABLE 5-4 Bottom-line Impacts—Top and Bottom Quintiles, FY 2010

State	Center Name	Dollars per Impacted Client	Impact per SME in Area (Dollars)
MS	Mississippi Technology Alliance	3,628,316	83,687
AR	Arkansas Manufacturing Solutions	1,752,289	48,367
PA	Northeastern Pennsylvania Industrial Resource Center	406,158	42,016
VA	GENEDGE ALLIANCE	1,777,412	35,122
MT	Montana Manufacturing Extension Center	903,744	30,588
AK	Alaska Manufacturing Extension Partnership	258,127	22,155
UT	Utah Manufacturing Extension Partnership	787,255	20,680
PA	Catalyst Connection	449,238	19,026
ME	Maine Manufacturing Extension Partnership	501,969	17,309
PA	IMC-PA	227,337	15,977
PA	MANTEC	350,835	15,460
IA	Iowa Center for Industrial Research and Service	357,066	14,598
CA	California Manufacturing Technology Consulting	236,000	4,265
DE	Delaware Valley Industrial Resource Center	153,771	4,207
MI	Michigan Manufacturing Technology Center	249,156	4,161
WI	Northwest Wisconsin Manufacturing Outreach Center	172,018	3,582
OR	Oregon Manufacturing Extension Partnership	295,514	3,186
MD	University of Maryland Manufacturing Assistance Program	224,834	3,184
NE	Nebraska Manufacturing Extension Partnership	142,483	3,145
RI	Rhode Island Manufacturing Extension Services	128,873	2,935
FL	Florida Manufacturing Extension Partnership	181,849	2,716
CA	Corporation for Manufacturing Excellence (Manex)	316,356	2,272
CO	Colorado Association for Manufacturing and Technology	199,414	2,131

SOURCE: MEP data, FY 2010, NRC staff calculations.

Those centers with the largest jobs impact per SME are also to a considerable degree the centers with the largest jobs impact per client. One possible question concerns the population served: Larger companies will

TABLE 5-5 Jobs Impact—Top Quintile

State	Center Name	Sum of Job Impacts	Jobs per Client	Jobs per SME
OH	Ohio Manufacturing Extension Partnership	4,163	101.5	7.33
NY	New York Manufacturing Extension Partnership	4,871	41.6	4.31
FL	Florida Manufacturing Extension Partnership	2,022	47.0	4.04
TN	Tennessee Manufacturing Extension Partnership	2,205	31.5	2.21
IL	Illinois Manufacturing Extension Center - Downstate	3,160	51.0	1.76
MA	Massachusetts Manufacturing Extension Partnership	1,642	58.6	1.57
AR	Arkansas Manufacturing Solutions	4,624	55.7	1.54
TX	Texas Manufacturing Assistance Center	4,751	34.2	1.45
PA	Catalyst Connection	1,071	48.7	1.42
OK	Oklahoma Manufacturing Alliance	823	17.5	1.30
MS	Mississippi Technology Alliance	2,851	48.3	1.11
CT	Connecticut State Technology Extension Program	1,390	31.6	1.07

SOURCE: MEP, NRC staff calculations.

obviously tend to generate larger gross job impacts from successful projects. Where jobs per impacted client are 30 or more, the center is probably not primarily focused on smaller firms.

EFFICIENCY

MEP gathers a considerable amount of data related to efficiency, which we define as federal resources per unit of defined output. Federal funding is in turn used to buy inputs that include, for example, full-time equivalent (FTE) positions at the center and hours of engagement per project.

Some of these inputs might be potential independent variables, explaining the internal efficiency with which centers address their resources to tasks at hand.

Such detailed efficiency analysis is, however, beyond the scope of this analysis. Here we focus instead on broader measures of input—notably federal funding.

Table 5-6 shows two input measures—federal funding per client impacted and federal funding per SME in service area. In this case, rankings are reversed: Centers scoring lowest (i.e., with least cost per unit) are performing best.

Box 5-2
The Challenges of Evaluating Impact

MEP services potentially have a range of impacts on firms, including improving knowledge and capabilities, enhanced strategic and business planning, training and skills development, improved efficiency, yield, quality or waste reduction, increased productivity, and improved product and service development. These impacts may in turn lead to measurable economic outcomes, including increased sales, although they may also result in greater business stability and resilience. It is appropriate to use a range of evaluation methods to probe the range of impacts and outcomes associated with the MEP, including qualitative and quantitative methods. Indeed, this has been the case.

As Appendix B on MEP evaluation indicates, there have been a large number of evaluations of the MEP using diverse methods and metrics. This includes studies where controls are applied to ascertain the "additionality" of MEP services compared to the situation that would have occurred if MEP services had not been received (noting that the firm might otherwise have changed in some way). In some cases, controls have been applied through logic-based counterfactual methods (studies by Cosmos Corporation in the 1990s and SRI International and Georgia Tech in 2009). In other instances, it has been possible to develop formal comparison groups of MEP clients and nonclients, including controlling for selection bias in the kinds of clients served by the MEP. A first round of comparison studies was undertaken in the 1990s. These studies include Jarmin (1999) use of national data from the Longitudinal Research Database of the Annual Survey of Manufacturers to construct a comparison group for MEP clients, and comparison group studies by Shapira and Youtie (1998) and Nexus Associates (1999) at the state level.

More recently, longitudinal comparison group studies of MEP clients and nonclients have been undertaken at the national level by Cheney et al. (2009) and at the state level by SRI and Georgia Tech (2009) and Youtie et al. (2008, 2010). These control group studies generally find that MEP clients performed better than comparable firms who did not receive MEP services, as measured by changes in value added, productivity, or resilience during recession, although the Cheney et al. (2009) study found that additional value added per worker impacts from MEP services were sensitive to econometric modeling methods. These control group studies are useful in that they explicitly include comparisons with nonassisted firms and can adjust for selection bias.

At the national level, there has been a reliance on available U.S. Census Bureau data. While this data is collected through robust and stringent procedures, there are time delays in reporting, relatively few variables are available, and access procedures (to ensure confidentiality) are complex. Short of undertaking a costly independent survey of MEP clients and nonassisted firms, these are limitations of current national control group methods.

TABLE 5-6 Federal Funding by Center, FY 2010 (10 Most Efficient Centers)

State	Center Name	Federal Funding (Dollars)	Total Expenditures (Dollars)	Fed Funding per client (Dollars)	Fed Funding per SME (Dollars)
PA	Northwest Pennsylvania Industrial Resource Center	702,744	1,984,123	13,140	1,215
PA	IMC-PA	363,412	1,017,078	14,530	1,021
PA	MANTEC	674,820	1,941,301	17,489	771
PA	Northeastern Pennsylvania Industrial Resource Center	504,000	2,179,905	18,632	1,927
ID	Idaho TechHelp	508,800	1,014,826	18,793	521
KY	Manufacturing Extension Partnership of Louisiana	701,868	1,540,952	22,661	460
PA	Catalyst Connection	1,092,004	3,325,741	23,926	1,013
HI	Hawaii HTDC - MEP	464,416	698,992	24,964	721
WY	Manufacturing-Works	372,048	1,031,756	25,165	1,816
PA	Manufacturers Resource Center	588,000	2,573,850	25,234	1,675

SOURCE: MEP.

The data indicate that there are massive differences between the most and least frugal centers. The Northwest Pennsylvania Industrial Resource Center uses less than $14,000 in federal funding per impacted client, while the Maine MEP utilizes more than $87,000. Further research is needed to explain these large differences.

There is some overlap between the two metrics among the most resource-intensive centers, though not so among the most frugal centers, only one of which falls into the bottom spending groups on both metrics. It is noticeable that all 6 Pennsylvania centers are among the 10 lowest spenders per client, though not among the lowest spenders per SME.

MEP data show that these differences tend to persist over time. Figure 5-4 shows that the amount of federal dollars per job created at the Alabama Technology network remains more than four times the cost required per job in Texas or Wyoming, even though the cost in Alabama declined by a quarter in FY 2010.

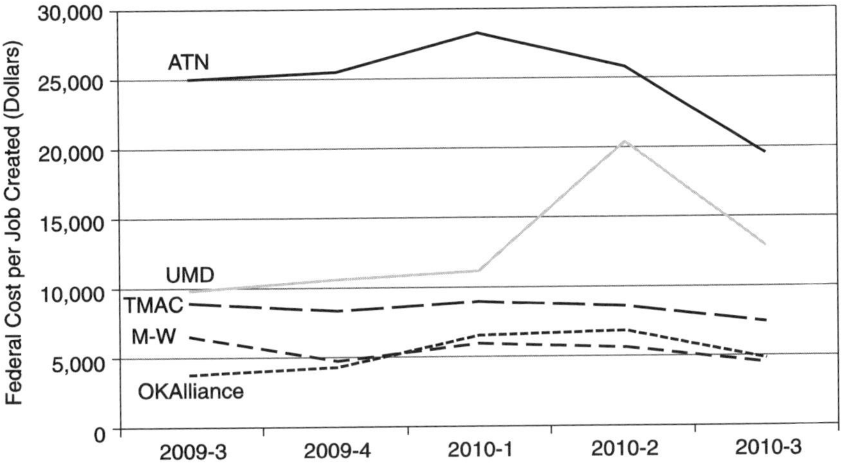

FIGURE 5-4 Cost in federal dollars per job created, selected MEP Centers, FY 2010 (4-quarter moving average).
SOURCE: MEP.

Looking at inputs alone, even on a per-client basis, provides only a limited view of efficiency. What may matter more is the relationship between inputs and eventual outputs—which MEP measures primarily in terms of bottom-line impacts for the client.

Figure 5-5 charts the relationship between federal funding per client and bottom-line impacts per client. It illustrates the wide range of efficiency demonstrated by the centers. For our purposes here, we have excluded results from the centers in Arkansas, Virginia, and Mississippi where the 2010-2011 outcomes were so much higher than for other centers that they would distort the overall picture.

The chart above illuminates the opportunity to focus on centers where the highest returns have been generated at the lowest cost. The chart shows that 12 centers fall into the high cost-low return quadrant, while 8 are in the low cost-high return quadrant.

SCOPE AND OUTREACH

Centers have been under some pressure to improve penetration, and perhaps in part as a result have been reporting a substantial increase in "touches"—contacts with potential clients. However, the number of actual client engagements has been essentially flat since at least 1999. It should be noted that while the population of firms of a given size is well understood and tracked via the Census Bureau (and the Small Business Administration), not all firms within

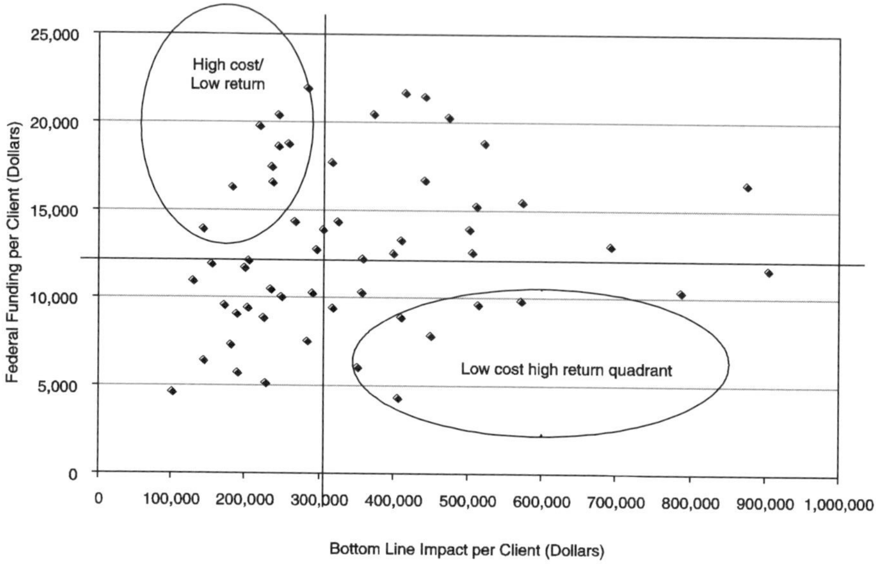

FIGURE 5-5 Center impacts per client and federal funding per client.
SOURCE: MEP data.

a given size population may be addressable by the local MEP center—they may, for example, be uninterested in MEP services, or already have access to alternative services.

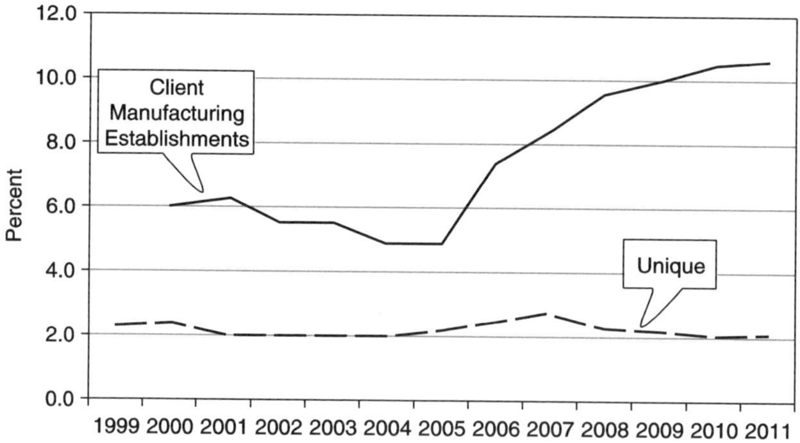

FIGURE 5-6 MEP market penetration trends.
SOURCE: MEP.

There are many possible explanations for the doubling of touches since 2005, although the lack of additional engagements raises the questions as to whether seeking additional touches with new potential clients is a useful way to expend scarce resources.

It is certainly true that in the context of the entire manufacturing sector, MEP reaches only a small percentage in a given year. However, this debate does to some extent appear to be based on questionable assumptions:

1. **Annual figures**. Given that MEP projects are challenging and sometimes expensive, at least in terms of time and resources for companies, and that MEP is focused on SMEs with limited capacity, it makes little sense to focus on a single year. Companies that have been served by MEP may be considerably less likely to return the next year. So penetration could be measured over a period of years, not just one.
2. **Limited client capacity**. MEP itself suggests that 80 percent of all SMEs are not sufficiently engaged in strategic thinking to be ready for MEP services.[4] Chuck Spangler of South Carolina MEP observed that most of his clients are in the range of 100 employees.[5] Other center directors have made similar observations. MEP centers are understandably cautious about expending efforts to engage potential clients who are not ready for their services: There is longer-term value in mentoring and readying potential clients, but such efforts may well impinge on the immediate delivery of measurable and revenue-generating projects.
3. MEP centers indicate that 40 percent to 70 percent of clients are returning customers.[6] This suggests that about on average at least one-third are new customers in any given year. Overall then, the data suggest that over a 5-year period a given MEP center provides in-depth services to 4 percent to 5 percent of SMEs in its service area; this is about a quarter of all the SMEs who are—according to MEP—sufficiently engaged to provide opportunities for center services. While this share could certainly be increased, and some centers manage to reach out into the community much more effectively than others, their data do not appear to provide strong evidence that market penetration rates are low.

These points are reflected in data from Vermont, for example, where between January 1996 and June 2011 the Vermont Manufacturing Extension Center served more than 865 of the state's approximately 2,000 manufacturing companies.[7]

---

[4]Yakimov and Woolsey, 2010, op. cit. p. 19.
[5]NRC/Georgia Institute of Technology Workshop on "Diversity and Achievements: The Role of the Manufacturing Extension Partnership in the South," April 20, 2012.
[6]Responses to NAS information request to MEP centers, May 2012.
[7]Vermont MEP 2011 Impact Report, p. 3.

A Balancing Act

MEP centers must perform a complex balancing act, weighing the need for revenue (and especially recurrent revenue, which tends to come from larger and more established firms) against the mission need to support smaller businesses and the need to enhance innovation (which encourages a focus on larger firms with more resources) against the need to support small manufacturing businesses in what they actually do (which covers many small nongrowing businesses). MEP should recognize that an assessment of possible trade-offs should not automatically result in decisions favoring support for larger and more innovative firms. The need to meet the federal match through fees and the reliance on in-kind contributions may also be pushing the program away from its presumptive target group—i.e., away from smaller companies and rural firms that should be a part of MEP's public vocation. MEP centers that work with the target audience of SMEs also tend to have smaller jobs impacts. Centers in markets with a greater percentage of smaller companies, or centers that choose to focus on the SMEs, tend to be at a disadvantage in the calculation of results.

Several metrics help to indicate the extent to which centers reach out into the manufacturing community in their region.[8] MEP measures both "touches," i.e., contacts with manufacturers, as well as engagements. Both can be measured against the total local population of manufacturers, and both provide a useful data point against which to evaluate community engagement: Neither wide outreach with lower levels of engagement nor high levels of engagement with a relatively small percentage of manufacturers can be seen as best practice in connecting to the community.

However, while client engagements is a well-verified number backed by actual contracts between centers and clients, touches are unverified by MEP and are open to different measurements by different centers. It is therefore a figure to be used with considerable caution.

That being said, it may be useful to provide a blended score for these two metrics, scoring each by quintile and adding the quintiles to provide an overall relative ranking for each center. This approach limits the impact of outliers while providing sufficient detail to indicate the relative success or otherwise of each center. Complete tables with both raw and normalized numbers are found in Appendix C. Here we provide only the 10 highest and 5 lowest scoring centers (and ties). See Table 5-7.

Table 5-7 shows that a number of centers score well, both on touches and clients, while others score poorly on both. Leaving aside the special cases of Alaska and Hawaii,[9] median touches for the top scoring group was just under a fifth of the target population, 19.4 percent, while that for the bottom group was

[8]As noted in Chapter 4, the committee is aware of the limitations of MEP metrics.
[9]Where the small-manufacturing sector is closely interconnected and hence is easy to touch.

4.3 percent. Similarly, the median impacted client percentage for the top group was 6.8 percent, and that for the bottom group was 1.1 percent.

These data show that there are large differences in the degree with which different centers reach out into the community. Accordingly there is likely room for the further adoption of best practices in this area. In addition:

- On this preliminary basis, it appears that the best predictor of eventual client acquisition (percent impacted) is not a center's success in converting initial touches into clients; it is the number of initial touches themselves. The more effectively the center touches its local community, the more likely it is to acquire clients and engagements. However, as the data on touches are unverified by MEP, further review is warranted before this point can be confirmed.
- Conversely, all of the centers that scored low on touches also scored low on client acquisition, although several had respectable conversion rates.
  - This is to some degree surprising, as many centers have observed in interviews that a preponderance of their client engagements are with clients with whom they have previously engaged.[10]
  - Size of region appears to play a role—all of the centers with the lowest conversion rates had at least 3,500 manufacturing companies in their region; only 2 of the top-scoring group did so, and none had more than 5,000 companies in their region, while 6 of their 8 bottom centers did so.
- Of the top 10 positions, 5 are occupied by Pennsylvania centers, including the top 2.

The data suggests that where regions have large SME populations, centers may struggle to touch a substantial share of these companies, and hence to generate a high rate of client acquisitions relative to the population.

Among other things, this suggests that MEP might consider whether more centers with smaller manufacturing populations might be a better option in some areas.

Finally, and perhaps most important, it suggests that MEP should perhaps pay additional attention to touches—to preliminary outreach, in other words. It appears that some centers are more active and use different modalities for outreach, so there should be opportunities to spread best practice in this area.[11]

[10]NRC information request to center directors, July 2012.
[11]Ibid.

TABLE 5-7 Measure of Outreach Success, FY 2010

Center Name	#SME in service area	Touches/yr	Impacted Clients/year	Touch Percent of SMEs	Impacted clients Percent SMEs	Impacted clients per touch	Percent Touch quintile	Percent Impacted quintile	Sum of quintiles scores
Highest scores									
PA Catalyst	3,282	635	139	19.3	4.2	21.9	5	5	10
PA NE IRC	1,131	219	117	19.4	10.3	53.4	5	5	10
NM MEP	1,535	306	73	19.9	4.8	23.9	5	5	10
PA IMC	996	292	70	29.3	7	24	5	5	10
AK MEP	501	397	43	79.2	8.6	10.8	5	5	10
WY Mf-Works	568	124	41	21.8	7.2	33.1	5	5	10
SC MEP	4,260	854	158	20	3.7	18.5	5	4	9
IA CIRC	3,718	710	152	19.1	4.1	21.4	4	5	9
PA NW IRC	1,633	223	151	13.7	9.2	67.7	4	5	9
MRC PA	1,537	256	102	16.7	6.6	39.8	4	5	9
DE MEP	634	82	47	12.9	7.4	57.3	4	5	9
HA MEP	970	467	28	48.1	2.9	6	5	4	9
Lowest scores									
CA Manex	12,395	506	89	4.1	0.7	17.6	1	1	2
WA Imp	7,582	435	86	5.7	1.1	19.8	1	1	2
IN Purdue MEP	8,790	355	68	4	0.8	19.2	1	1	2
OR MEP	5,658	285	61	5	1.1	21.4	1	1	2
AZ MEP	5,005	211	59	4.2	1.2	28	1	1	2
CO MEP	5,240	253	56	4.8	1.1	22.1	1	1	2
MD UMD MA	3,601	124	51	3.4	1.4	41.1	1	1	2
KY MAC	4,064	174	32	4.3	0.8	18.4	1	1	2

SOURCE: MEP data, FY 2010; NRC staff calculations.
NOTE: MEP definitions. Touch penetration means potential SME manufacturing clients with which the center has been in contact during the previous 12 months. No further details are provided. Impacted clients means clients with which the center had an engagement. Quintiles refer to the quintile in which the individual center metric falls when centers are arrayed from highest to lowest scoring, with the highest quintile scoring 5 and the lowest scoring 1.

EVALUATIONS OF THE MEP

Aside from regular collection of client and center data, and the assessment of individual centers, MEP and other sponsors have invested in a range of evaluation studies. We review them in brief below (Appendix B provides a detailed review of this literature). A diverse range of MEP evaluation studies have been conducted over the past two decades, in the following categories.

- **Logic-based studies**. MEP has used logic-based case studies carried out by independent evaluators to examine clients with exceptionally high impacts from MEP services (Cosmos Corporation 1997, 1998, 1999). More recently, similar studies have been undertaken to improve understanding how new services are being implemented (SRI and Georgia Tech 2009a). The logic models describe MEP services, showing inputs, work processes, intermediate outcomes, and impacts. Counterfactual probes are used to discern what might have happened without program participations.
- **Comparison group studies.** These studies of clients and nonclients have been used partly in an effort to control for selection bias (Jarmin 1997, 1997; Nexus Associates 1996; Oldsman 1996; Nexus Associates 1999; Youtie et al. 2008, 2010) and to account for observable and unobservable factors (Cheney et al. 2009).
- **State evaluations.** Several states have conducted evaluations of their particular state centers or system of centers. New York evaluated their centers in the 1990s (Oldsman et al. 1996, Nexus Associates 1996). Pennsylvania conducted two highly regarded assessments of the centers in the Industrial Resource Center system (Nexus Associates 1999, Deloitte 2004). Ohio included MEP program analysis as part of their assessment of the Third Frontier and other technology-based economic development programs (SRI and Georgia Tech 2009).
- **Center evaluations.** Most centers do not have additional evaluation programs beyond what is required by the national or state sponsor. However, a few centers have developed and maintained special capabilities to support their evaluation efforts. The Michigan Manufacturing Technology Center created a longitudinal dataset of survey-based metrics for full-scale comparison, the Performance

Benchmarking Service (Luria 1997, Luria and Wiarda 1996). The Georgia MEP has administered a survey of nonclients and clients which has been used for evaluation, including controlled studies, as well as conducting studies to address particular needs of the center from time to time (Shapira and Youtie 1998; Youtie et al. 2008, 2010).

- **Trade association studies.** Trade associations have been responsible for several evaluations of aspects of the program. The National Association of Manufacturers included a question about service use as part of its survey of members' technology adoption practices (Swamidass 1994). The Modernization Forum (a former association of MEP centers) sponsored several studies in the early years of the MEP. In the 2000s, the Advanced Small Manufacturers Coalition (AMSC) (the association of MEP centers), along with other nonprofit organizations, has turned the emphasis of MEP evaluations towards strategic redirection of the program (NAPA 2003, 2004; AMSC 2009; Stone & Associates and CREC 2010; MEP Advisory Board, Yakimov, and Woolsey 2010) and international benchmarking efforts (Ezell and Atkinson 2011).
- **Service delivery reviews.** Some independent evaluations have reviewed particular types of services or service delivery pilots including interfirm networking (Welch et al. 1997) and SBDC-MEP partnership programs (Yin et al. 1998).

EVALUATIONS RESULTS AND OUTCOMES

These evaluations have resulted in a substantial body of studies related to the MEP program. The table in Appendix B identifies 39 evaluative studies in the 1990s and another 26 in the 2000s. Thirty percent of these studies are published in academic journals or books, including in two journal special issues in the 1990s (*Research Policy* 1996 issue and *Journal of Technology Transfer* 1998 issue). More than 10 percent are official federal government publications, with the remainder comprised of state government publications, conference proceedings, "gray" literature reports and white papers, dissertations, and informal memoranda.

The MEP client survey is used by approximately one-third of the studies. Six studies used case study methodology (although a few others had case examples within primarily quantitative papers), while five linked client data to administrative databases at the state or national level. Fourteen studies utilized comparison groups. Seven used cost-benefit and/or fiscal impact methodologies to represent public and private returns from the program.

Evaluation studies in the 1990s used diverse and sometimes novel methods to understand program processes and effects. This mix of studies was heavily influenced by the MEP's evaluation working group 1993-1999, which produced a formal evaluation framework.[12] Subsequent studies attempted to measure penetration, intermediate effects, and short and longer-term outcomes on clients and the broader economy.[13]

The studies conducted in the 2000s reflected a different climate. Whereas the 1990s was a period of program expansion and experimentation, the 2000s saw substantial fluctuations in the program's budget, a systematizing of services, and consolidation of the number of centers, as some centers were combined into statewide or regional programs. The MEP evaluation system itself became more standardized, reflected in the MEP evaluation plan and metrics published by Voytek et al. in 2004. Evaluations in the 2000s utilized more international comparisons as well as expert panelists and document review.

The following sections draw upon the studies to highlight key findings and conclusions.

Competition and Complementarity with Private Service Providers

There have been questions about whether MEP competes with or complements private service providers. This was the subject of a major study sponsored by the Modernization Forum through surveys of MEP clients, a comparison group of manufacturers, and private consultants (Modernization Forum and Nexus Associates 1997; Oldsman 1997). The report indicated that MEP complements private services. Seventy-three percent of manufacturer responses indicated that MEP complemented consultants' work, while only 7 percent of MEP clients reported that the MEP offered the same services as private consultants. MEP clients also appeared more likely to experience benefits, in that the probability of a typical MEP customer improving its performance was 5.4 times higher than a typical private service client.

Four governmental assessments—OMB (2002), National Commission on Fiscal Responsibility and Reform (2010), Schact (CRS) (2011) and GAO (2011)—addressed this issue. OMB's Program Assessment Rating Tool (PART) concluded that "It is not evident that similar services could not be provided by private entities," but the National Commission on Fiscal Responsibility and Reform concluded that MEP provides services that exist in the private sector. GAO's 2011 review of the cost-share match requirements reported that rural areas were harder to serve as centers sought to find matching funds.

---

[12]Atlanta Workshops on the Evaluation of Industrial Modernization, Aberdeen Woods, GA, 1993, 1994, 1996, 1997.

[13]Swamidass (1994) surveyed National Association of Manufacturers (NAM) members and found that only 1 percent indicated that government was an important source of assistance in technology investment decisions. However, many MEP centers are known through their university or center name rather than as a source of government assistance, making it difficult to measure use of the program beyond counts of manufacturers served.

To address OMB concerns, the National Academy of Public Administration (NAPA) (2003) used a panel and document review and interview process, and concluded that small- and medium-sized manufacturers still faced barriers to productivity improvement, and these firms were underserved by the private market.

Collaboration and Partnerships

MEP has engaged in several formal efforts to collaborate with other service-providing organizations. Shapira and Youtie (1997) found that MEP sponsorship led to greater service coordination than individual center efforts alone or state government demands would have provided, which in turn generally improved the services to MEP clients, albeit at a significant expenditure of resources for validating and coordinating with these providers.

Partnership with state governments has been an important element for some centers. MEP's own analysis (MEP 1998) concluded that two-thirds of the states would not provide state funds if federal funding was discontinued.

Although most MEP technical assistance services are delivered on a one-on-one basis to a single manufacturing client at a time, MEP did invest in a networking service delivery pilot from 1996 to 1998. Evaluation results indicated that the median net benefit of network participation to the firm was $10,000, while some members experienced significantly higher benefits, raising the mean to $224,000 (Welch et al. 1997).

Differentiation within the Program

The mix of services and delivery methods was the subject of various evaluations in the 1990s. The GAO's study identified a misalignment between the legislation establishing the centers—which emphasized technology transfer from federal laboratories—and the needs of small manufacturers (primarily assistance with proven technologies) (GAO 1991). This conclusion was reinforced by the 1993 National Academies study (National Academy of Sciences 1993).

Youtie and Shapira (1997) observed that outcomes are associated with service mix: Marketing and product development services were 60 percent more likely to lead to sales outcomes, energy products more likely to lead to cost savings, plant layout and environmental projects more likely to generate capital expenditure avoidance, and quality projects were not strongly associated with any type of outcome. Oldsman and Heye (1998) concluded that services which enabled a manufacturer to raise piece prices generated more profits than services that reduced the scrap rate. Luria (1997) maintained that the program's service mix attracted cost-minimization/lean companies that were not on the path to increasing productivity. Cosmos Corporation led case studies of high-impact projects with 25 manufacturing clients (1997), then 6 more manufacturing clients (1998), and finally 7 different highly transformed manufacturers (1999).

The results indicated the importance of integration of services, and that leadership by top management was required to make the necessary discontinuous changes across multiple systems.

Center-to-Center Comparisons

Center-to-center comparisons were the subject of a few evaluation studies. Chapman (1998) showed that different centers were at the frontier of different service areas, with no one center consistently in the lead. Wilkins (1998) compared 14 centers, similarly finding that no one center excelled on all measures.

Center-to-center variability continues, indicating the existence of stronger and weaker performing centers. An analysis of Minimally Acceptable Impact Measure scores in 2001, 2003, and 2005 concluded that there were no consistent top-performing centers from period to period, although a few centers landed near the top in many of the periods under analysis (Youtie 2005). NAPA (2004) found strong performance differences between centers. The report identified a substantial positive association between high-performing centers and number of clients served, years in operation, number of FTE employees for the center and per million dollars of federal investment, and ratio of state dollars.

Market Penetration

Market penetration has been addressed in several studies. Stone & Associates and CREC (2010) found penetration to be a concern in that the MEP only serves 10 percent of manufacturers, 2 percent with in-depth assistance. One question is whether this reflects cherry picking of clients, but Deloitte (2004) reported that a comparison of the credit rating of IRC PA clients and a matched group did not generate statistically significant differences. However, GAO (2011) found that 80 percent of MEP centers were very or somewhat likely to give preference to revenue-generation projects with larger clients.

A number of reports have suggested that the MEP needs to expand its reach and to attract a wider share of the manufacturing sector. Stone notes that "currently the MEP national network only provides in-depth assistance to 9 percent of the available market of companies with 20-499 employees that are willing to seek out and invest in outside support."[14] Indeed, the need for a new business model and strategy for MEP is in part based on the perception that the current range of offerings is too narrow, and that a broader range—especially including growth-related services—would expand MEP's reach by bringing in new clients.

---

[14]Stone et al., 2010, op. cit. p. 7. It should be noted that this report provides no reference or basis for this claim. It is hard to determine where this data could have come from, as there are no known surveys or reports that reference such a population.

Intermediate Outcomes

Given the difficulties of measuring the final impact from support programs like MEP, several studies focused on intermediate outcomes. Several client survey-based studies provided nonquantitative evidence that a higher percentage of companies engage in implementation following MEP assistance. Two-thirds of Georgia MEP customers took action on center recommendations (Youtie and Shapira 1997). Nearly 30 percent of Massachusetts center customers would not have carried out changes without MEP assistance (Ellis 1998). Many client surveys also suggested positive views of performance improvement, with the GAO (1995) finding that 73 percent of manufacturing survey respondents across the nation had better business performance and Ellis (1998) indicating that 71 percent of Massachusetts manufacturers who participated improved their competitiveness as a result of center assistance.

Technology Adoption

Shapira and Rephann (1996) observed that clients of a manufacturing technology assistance program in West Virginia were more likely to adopt individual technologies and were also more amenable to technological investments than nonparticipants. However, these clients did not have significantly higher aggregate adoption across a range of technologies. The Luria and Wiarda (1996) Performance Benchmarking database analysis indicated that MEP customers adopted most technologies (with the exception of information technologies) more quickly than non-MEP customers. However, evidence from case studies of centers in Northern Pennsylvania, Michigan, and Minnesota conducted by Kelly (1997) led to the conclusion that the use of one-on-one services militates against widespread diffusion of knowledge and skills important for advanced technology usage.

Overall, these studies indicate that MEP was associated with accelerated technology adoption by clients in these studies, but that the aggregate effects were modest at best.

Productivity

The relatively small size of the MEP means that it is difficult to measure any aggregate effects at the level of national economic growth or productivity. However, a number of studies have measured productivity impacts at the level of participating firms, including with controls for nonparticipating companies. Measuring productivity—usually by value added—is challenging for several reasons. There is a high degree of skew, with the majority of impacts coming from a small number of clients (Oldsman 1996). Indeed, Luria notes that the *median* productivity impact is zero.[15] Most manufacturers had difficulty

[15]Telephone interview, March 15, 2012.

calculating productivity impacts, and the timing of measurement led some customers to overestimate benefits, especially sales impacts, and underestimate costs, except for a small number of high-impact projects (Youtie and Shapira 1997).

While customer surveys tended to present positive outcomes, quantitative business outcomes tended to be more moderate when comparison groups were used to control for factors other than program assistance. Some comparison group studies surveyed all manufacturers in a particular region (e.g., Youtie et al. 1998) or in a national sample (e.g., Luria 1997 and Luria and Wiarda 1996). Others linked MEP customer information to administrative datasets at the Census Bureau or Department of Labor and selected enterprises from these datasets to match client profiles (Jarmin 1997, 1999; Oldsman 1996; Nexus Associates 1996).

Most of these studies focused on productivity as measured by value added per employee, although other outcomes metrics were used as well. Jarmin (1997, 1999), Shapira and Youtie (1998), and Nexus Associates (1999) found clients to have higher growth in value added per employee than nonclients. Jarmin's analysis of eight MEP centers from 1987 to 1992 found productivity increases in clients over nonclients ranging from 3.4 percent to 16 percent. Nexus Associate's analysis of Pennsylvania centers reported higher labor productivity of 3.6 percent to 5 percent in clients as compared with nonclients. The average Georgia client had between $366,000 and $440,000 more in value added than nonclients (Shapira and Youtie 1998). Note that these analyses tended to focus on a few centers or a network of centers (Georgia in the case of Shapira and Youtie, and Pennsylvania in the case of Nexus Associates).

Comparison Group Analysis

Other comparison group-based evaluations found fewer differences between served and unserved manufacturers. Analysis of the Performance Benchmarking dataset showed that MEP clients do better than nonclients in sales growth, job growth, and adoption of some process improvements, but clients are not significantly better than nonclients in growth in wages, profits, and productivity (Luria 1997).

MEP and Jobs

The effect of the MEP on job creation and retention is not clear. Although NIST survey data reports positive employment impacts (in part because of the inclusion of jobs retained), this association is not necessarily found when comparison groups are used. For example, Oldsman (1996) found that participating manufacturers in New York added 5.7 percent fewer workers than similar, nonparticipating companies. Because MEP seeks to enhance productivity, efficiency measures may eliminate some factory worker positions. Given the lack of relevant and robust data, the committee has not addressed the use of jobs as a system metric, or the connection between employment and productivity.

Evaluating Growth-oriented Services

SRI and Georgia Tech (2008) found that customers of the E!WW pilot program were more likely to be in industries with job losses, to be private family-owned firms, have a concentrated structure for product development, and have a history with technological implementation.[16] These observations suggests that different kinds of firms may have very different patterns of demand for growth-oriented services.

State-focused Evaluations

Survey-based analyses of Georgia clients and nonclients in 2008 and 2010 maintained prior findings of higher increases in value added per employee for clients. Clients in Georgia were also more apt to engage in product and process innovation than similar noncustomers (Roper et al. 2010), and while rural and urban manufacturers were similar in their adoption of hard technology, rural manufacturers were less likely to use soft technologies and supply chain integration than their urban counterparts (Youtie and Shapira 2005).

SRI and Georgia Tech (2008) found that Ohio MEP customers were more likely to have retained jobs than noncustomers, controlling for industry, size, and other factors.

In 2009, MEP commissioned an effort to update the original Jarmin study. The resulting SRI and Georgia Tech study (Cheney et al. 2009) used two different models to assess productivity impacts of clients and nonclients from 1997 to 2002. Results from these two models diverged sharply, indicating positive and negative effects, respectively. However, both models found significant positive productivity outputs associated with MEP assistance for smaller establishments, certain types of MEP services, and a short delay (one or two years) after receipt of service.

---

[16] E!WW was designed by an MEP contractor to "help MEP clients grow their revenues and profits through identification of ideas that have higher probabilities of success and through faster movement of selected ideas into implementation." See Lynne Manrique, David Roessner, Jan Youtie, and Philip Shapira, "Logic Model and Case Study Design Eureka! Winning Ways Client Engagement Case Studies," prepared for the National Institute of Standards and Technology Manufacturing Extension Partnership, January 2008. See also Lynne Manrique, Kamau Bobb, David Roessner, Jan Youtie, and Philip Shapira," Eureka! Winning Ways: Analysis of Early Client Experiences," report for National Institute of Standards and Technology, Manufacturing Extension Partnership, SRI International and Georgia Institute of Technology, August 2008.

Aggregate Outcomes: Cost-Benefit and ROI

Results from most cost-benefit studies have been positive, although the scale of impacts is sensitive to the methods and assumption used. Shapira and Youtie (1995), Nexus Associates (1996), and Michigan Manufacturing Technology Center (1996) demonstrated that net public and private benefits of MEP assistance outpaced costs by a ratio between 1:1 and 3:1. A Pennsylvania study (Nexus 1999) reported much more positive net returns to the state investment of 22:1. Thompson (1998) found that taxpayer payback in Wisconsin varied from slightly below break-even to positive.

Several of these studies addressed some of the methodological issues involved in cost-benefit analysis. These include accounting for the full range of private and public costs and benefits, addressing returns and investments over time, and considering zero-sum redistribution of benefits and value-added capture through downward adjustment of sales impacts for export sales and value added (e.g., Shapira and Youtie 1995).

RECOMMENDATIONS FROM EVALUATION STUDIES

Twelve evaluation studies included recommendations, and these studies form the basis of the following analysis.[17] As described in Table 5-8, the main recommendations of these studies were:

- The need to transition toward services that involve new product and market offerings.
- Integration of advanced technology into products.
- Changes in federal funding, with two studies recommending a substantial increase in federal funding (Stone & Associates and CREC 2010; Helper and Wial 2010), while NAPA (2004) called for greater flexibility in the funding formula. In contrast, the National Commission on Fiscal Responsibility and Reform argued for sun-setting federal funding (NCFRR 2010).
- Increased attention to exporting and global supply chains (NAPA 2004; Ezell and Atkinson 2011).
- More green service offerings (MEP Advisory Board (2010)
- Workforce retention and development (MEP Advisory Board 2010; Deloitte 2004).

Table 5-8, summarizes the key recommendations from these evaluation studies.

---

[17]The National Commission on Fiscal Responsibility and Reform (2010) recommended the elimination of federal funding for the MEP. This was as part of a broad review of government business support programs with the aim of seeking fiscal savings, rather than a specific evaluation of the MEP.

TABLE 5-8 Summary of Recommendations in Evaluation Studies

Author/Year	Recommendations
Luria (1997)	Nurture distinctive manufacturers with proprietary or design-intensive products and encourage other manufacturers to follow this strategy.
Modernization Forum and Nexus Associates (1997), Oldsman (1997)	Enhance work with consultants through using resources to identify them, application materials, project proposals; selecting consultants; developing standard contracts for working with them.
Oldsman and Heye (1998)	Help companies become more distinctive as well as more efficient.
NAPA (2004)	Focus more on technology diffusion, product development, supply chain integration, and integrating the national network with other state and university providers and private-sector firms. Provide greater flexibility in funding.
Davila (2004)	New evaluation metrics and approaches should be developed to measure adaptive learning, worker benefits, public benefits and costs, spillover effects, and how to allocate scarce MEP resources.
Deloitte (2004)	Lower barriers to access; offer more business strategy, product innovation, worker retention services; continue process innovation; support and grow advocacy and research for SMEs.
Youtie (2005)	Examine consistently high-performing centers' service mix and survey practices.
SRI and Georgia Tech (2008)	Focus marketing on target customer characteristics; think strategically about participant mix; select and train growth coaches; use team-based approaches and fewer idea limitations in service design; evaluate after one year.
Helper and Wial (2010)	Enhance funding for MEP to provide more product and market development services and greater coordination between MEP and other federal programs for manufacturers.
Stone & Associates and CREC (2010)	Expand program scale from current levels of 7,000-8,500 firms served to 30,000 with substantially greater funding.
MEP Advisory Board, Yakimov, Woolsey (2010)	Streamline innovation and growth services, target green services, emphasize exporting, develop talent.
Ezell and Atkinson (2011)	Offer technology-acceleration programs and practices, exporting, energy efficiency, quality, standards, and design.

CONCLUDING OBSERVATIONS

In general terms, the following points may be drawn from the extensive set of evaluations discussed above, bearing in mind caveats about the limitations of data and methods.

Researchers included a considerable range of outcomes as possible metrics:

- **Productivity**. Value added, though difficult to calculate, has been used as a primary program metric. Here control group analysis has found that center clients have been associated with modest increases on average productivity, at levels ranging up to about 5 percent. The most recent study (Cheney 2009) used two models which generated contradictory outcomes (positive and negative productivity changes, respectively).
- **Market penetration**. Recent studies find that MEP has reached out to 10 percent of its potential market of manufacturing firms (with 2 percent client engagements). They argue that this rate of penetration is low, and is also focused on larger firms.
- **Return on investment** (ROI). Overall ROI analysis leads to generally positive but wildly diverging outcomes, with some studies showing 1:1 or 2:1 returns on federal investments; others showing returns of 20:1 or more.
- **Intermediate outcomes**. Earlier studies were more likely to focus on intermediate outcomes such as client action following center recommendations. They tended to find that clients did pursue related actions.

Like some other government support programs focused on small firms, results are highly skewed, with a few projects providing most of the value added.

Researchers identified some significant issues with the MEP survey:

- Methodologies required by MEP for firms to calculate value added are quite complex and may lead to undervaluation of outcomes.
- Incentives from MEP for survey completion and for positive results reported through the survey likely provide strong incentives for center staff to encourage the most positive view possible of firm results.

Limited research has been conducted on the differentiation within the program. What there is indicates that centers can be differentiated across a range of variables, including geography, service mix, client mix, organization structure, and size.

Next generation services have been subject to only limited evaluation and assessment:

- Growth-oriented services were subject only to initial evaluation. It appears that no external assessment exists for the fully implemented program.
- Similarly, no assessment exists of other aspects of the new strategy, green manufacturing, and supply chain integration and export-led growth.

There does not appear to have been a significant review of program management since NAPA (2004).

Chapter 6

New Approach: Next Generation Strategy

Starting in 2007, and based on conclusions from numerous evaluations and assessments (described below), the Manufacturing Extension Partnership (MEP) has worked to develop a new strategic approach. This approach is based on an expanded vision of its role in the manufacturing sector, in which MEP seeks to help firms grow by becoming more competitive and innovative.[1] MEP calls this the Next Generation Strategy, and it is built on findings from a business model study sponsored by MEP and conducted by a longtime MEP consultant, Mike Stone.

Published in 2010, the Stone study stated that:

> *"MEP needs to provide a broader range of services, especially services that foster growth, innovation and sustainability.[2] These include in particular those that focus more on innovation, beyond the process improvements that were a traditional MEP focus."[3]*

Based on this view, MEP has recast its mission statement:

> ***Vision:*** *MEP is a catalyst for strengthening American manufacturing—accelerating its ongoing transformation into a more efficient and powerful engine of innovation driving economic growth and job creation.*
> ***Mission:*** *To act as a strategic advisor to promote business growth and connect manufacturers to public and private*

[1]NIST, The Future of the Manufacturing Extension Partnership, October 2008, p. 5.
[2]Stone and Associates, "Re-examining the MEP Business Model," October 2010, p. 7.
[3]Stone and Associates, op. cit. p. 4.

resources essential for increased competitiveness and profitability.[4]

A Change in Direction

This new strategy represents a massive and ambitious change of direction for MEP. It orients MEP to the difficult task of encouraging small- and medium-sized manufacturers to adopt and implement growth-oriented strategies, seeking to encourage continuous innovation instead of just continuous improvement. In doing so, it relegates the core of existing MEP business—lean manufacturing—to a supporting role at best.

Such a strategic shift will undoubtedly require time to implement, and will likely require the adoption of midcourse corrections. For now, though, MEP is focused on implementing its vision through the provision by MEP centers of new services, and the adoption of new metrics.

In its new strategy document, MEP translates the new vision into five service categories:[5]

- Technology acceleration.
- Supplier development.
- Sustainability.
- Workforce development.
- Continuous improvement.

This chapter describes each element of this strategy in turn. Given that the new strategy is focused on improving the competitiveness and innovative capacity of SMEs, this chapter also addresses how MEP is helping SMEs enter the global marketplace by exporting more effectively and discusses the role of MEP centers in the development of innovation clusters and networks. While we describe each of these elements separately for analytical clarity, all of these service categories can and do overlap to a greater or lesser degree in practice. Each of the new initiatives is discussed in the following sections.

TECHNOLOGY ACCELERATION

MEP's Technology Acceleration strategy has four principal components:[6]

- Connecting manufacturer technology needs with technology sources.

---

[4]MEP, "The Future of the Hollings Manufacturing Extension Partnership," December 2008, pp. 7-8.
[5]MEP, 2008, op. cit.
[6]MEP "Technology Acceleration," *<http://www.nist.gov/mep/manufacturers/tech-acceleration.cfm>*. Accessed July 20, 2012.

- Technology scouting.
- Supplier scouting.
- Product development and commercialization assistance.

We describe each of these components in turn below:

Connecting Manufacturers to Technology Providers

MEP's signature initiative in the area of technology acceleration has been the National Innovation Marketplace (NIM).[7] This is an online tool that MEP anticipates will connect manufacturers to technology and business opportunities by facilitating connections between technology requestors and potential suppliers. NIM is expected to encourage technology translation and adoption, and provide estimates for the potential for business growth. It is also expected to facilitate the building of technology-based supplier networks.

Reaching Critical Mass

While NIM focuses attention on the need for networks to connect SMEs better to technology and to business opportunities, there are real challenges in building tools that can do so effectively. Most importantly, all online markets require that they contain a critical mass of buyers and sellers. Once beyond the beta or pilot stage, online markets have only a very limited time in which to attract this critical mass. Without that, visitors will come to the site, try a search, find few or zero relevant results, and never return. This has been true of online information markets since the earliest online communities.[8]

Effective Service Delivery

Moreover, online marketplaces must explicitly provide users with information and/or services that they cannot readily access elsewhere. To date, this does not appear to be the case with NIM. Currently, there are about 2,500 entries, with widely varied characteristics.[9] For example, a search for a "magnetometer" vendor on the NIM website generated zero results, while on Google it generated 2.4 million hits and on eBay, 77 offers. While the concept of NIM appears to be sound, what matters is whether it works in practice. While this is only one example, this sample search suggests that NIM needs to mature further before it can serve as a unique and useful resource for manufacturers.

The point here is that building successful online marketplaces is difficult, and usually requires considerable investment in outreach to potential

---

[7]See <*http://www.usainnovation.org*>.

[8]For several examples, see Tharon W. Howard, *Design To Thrive: Creating Social Networks and Communities that Last*, Elsevier, 2010.

[9]Accessed July 19, 2012.

participants and partners. It would be useful to determine whether the NIM website is attracting significant repeat business from vendors or customers.

Technology Scouting

In 2009-2010, MEP partnered with RTI International to pilot a technology scouting program. Unlike traditional "push-based" technology transfer in which the initiative comes from the technology provider, technology scouting is pull based, finding solutions for a small manufacturing client's unmet technology need.[10]

NIST-MEP has been researching potential tools and services that can provide technology scouting support for SMEs by pulling technologies from government laboratories, universities, and private-sector sources. MEP is also testing the efficacy of a "light" version of technology scouting that utilizes a "self-service" technology needs write-up approach with subsequent distribution to technology sources. "This effort is intended to provide another Technology Scouting opportunity for clients, and possible service offering for MEP Centers, requiring fewer resources from both MEP Centers and client manufacturing companies."[11]

Partnering with RTI[12]

MEP's engagement with RTI has been implemented in four phases starting in mid-2008. It focuses on two core efforts, with one additional recent initiative:

- Technology Scouting (TS): Helping clients find technologies they need.
- Technology Driven Market Intelligence (TDMI): Finding markets for client technologies and technology-related products.
- RTI is now introducing training in Lean Product Development (LPD).

RTI began work on technology scouting for MEP in 2008 with pilot projects leading to a full series of Technology Scouting Training Workshops in 2009.

The objective of the Technology Scouting Training Program is to train MEP center staff members to be able to acquire and undertake successful technology scouting mission assignments for MEP client companies. It is not yet

---

[10]MEP, "Technology Scouting," <*http://www.nist.gov/mep/upload/Technology-Scouting-One-Pager_v3.pdf*>. Accessed July 10, 2012.

[11]NIST, "Next Generation Strategy," April 2011. Access at <*http://www.nist.gov/mep/technology-acceleration.cfm*>.

[12]This box is based on interviews with Tom Culver, RTI manager to MEP center clients, and Kirsten Reith, RTI manager NIST to MEP, October 24, 2012, and on an RTI presentation, RTI Technology Acceleration Services Development: Supporting business growth and innovation initiatives for NIST MEP, September 2012.

clear what benchmarks MEP and RTI have in place to gauge the success of this training program: While some intermediate metrics have been identified—e.g., number of center staff trained—we do not yet have a clear view of the outcome metrics.

Technology Driven Market Intelligence is a parallel effort focused on training center staff to provide enhanced marketing services to clients, focused on the sale of technology or technology-driven products.

RTI staff note that TS/TDMI assignments require an entirely different level and type of engagement with client companies. Existing lean manufacturing assignments are largely based on sales to project managers and line supervisors; TS/TDMI assignments require connections to C-level executives (e.g., chief executive officers [CEO], chief financial officers [CFO]) prepared to discuss strategic issues. This in turn requires a very different set of skills for center staff.

While RTI's primary role is to train MEP center staff, RTI also provides more direct support for newly trained center staff, helping them to make sales and to implement projects. RTI staff indicates that to some extent at least, RTI's role is evolving significantly beyond training to active support for centers. This may indicate that even where successful, training MEP staff is insufficient to support successful transition to the new services.

As of September 2012, RTI had delivered a total of 14 workshops on TS/TDMI, had trained more than 300 MEP center staff, and had more than 25 pilot projects running at different centers (21 pilot TS projects and 5 pilot TDMI projects).

Outcomes

Two sets of outcomes will define the success or otherwise of the TS/TDMI component of the new strategy:

1. Can centers successfully train or acquire capacity to successfully engage client companies (and potential clients) at the C level?
2. Can centers successfully find and attract an economically sustainable market for these services?

The first of these concerns the successful implementation of the program; the second relates to the successful development of a sustainable market for these services.

RTI notes that the program is still rolling out and that it takes time to make what is a substantial transition in center staff capabilities and orientation. As of September 2012, RTI claimed the following outcomes:

- 6 TS MEP client-funded pilot projects running.
- 13 TDMI MEP client-funded pilot projects running.

- Center success stories:
    - The Manufacturing Resource Center, a Lehigh, Pennsylvania, based MEP center sold more than $300,000 in projects to clients in 2010-2011.
    - The Texas Manufacturing Assistance Center sold more than $85,000 in projects to clients in 2010-2012.
    - Five other centers report selling a total of nine projects to clients in 2011-2012.

RTI is also engaged in working on other aspects of Technology Acceleration with MEP. These include partnerships with Edison Nation, an online clearinghouse of innovative ideas,[13] and with the U.S. Patent and Trademark Office on an IP Awareness Assessment Tool.[14]

Supplier Scouting

In line with the new strategic emphasis on connecting SMEs to supply chains and networks, MEP is developing approaches to leverage MEP center-based knowledge of local manufacturer capabilities and even prequalify supplier capabilities.[15] Pilots were under way in 2009 with the Defense Logistics Agency Defense Supply Center Columbus, and with BAE Systems, Inc.

It is not entirely clear whether the objective of this strategic emphasis is to enhance the visibility of SMEs and other lower-tier suppliers to Original Equipment Manufacturers (OEMs) and tier 1 suppliers, or to engage OEMs and tier 1 suppliers to seek out SMEs with specific capabilities.[16] Perhaps this will be clearer once the pilot projects are concluded and more documentation is available. The importance of supplier development in the new MEP strategy is discussed further in the section on supplier development below.

Lean Product Development and Commercialization Assistance

MEP is developing three sets of tools for supporting product development and commercialization related to the Technology Acceleration initiative:

- Lean Product Development uses existing lean manufacturing tools to focus on product development to support a waste-conscious, disciplined approach.

---

[13]See <*http://www.edisonnation.com*>.

[14]See <*http://www.uspto.gov/inventors/assessment/index.html*>.

[15]MEP, "Technology Acceleration," op. cit.

[16]A tier 1 (or prime supplier) submits products and services and their invoices to the final customer, e.g., a major corporation or government agency. Lower-tier suppliers provide inputs to the prime supplier.

- MEP has partnered with the U.S. Patent and Trademark Office to offer basic training in intellectual property for MEP field staff.
- MEP is partnering with national organizations and other government agencies to help small companies identify financing options for product development and commercialization processes.

Finally, aside from these initiatives, technology acceleration may also be considered to include MEP's innovation engineering initiative, which is discussed in the context of workplace training below.

SUPPLIER DEVELOPMENT

MEP argues that it is well positioned to understand the rapid changes in supply chains now under way, and to help small firms position themselves as suppliers by providing assessments of industry dynamics and market intelligence. In recent years, SMEs have faced relentless price pressure, both domestic and especially from abroad. But this pressure has also brought opportunities. After dealing for years with tier 1 suppliers that monopolized all contacts with the eventual purchaser (e.g., a major auto manufacturer such as Honda or Ford), some SMEs are now finding that larger firms are increasingly seeking to partner with companies that can add value, regardless of their location in the supply chain.[17]

Irene Petrick suggests, moreover, that as higher-tier suppliers OEM-led vertical supply chains are challenged or even replaced by networks of collaborating suppliers, there will be more opportunities for SMEs to engage at levels from which they would previously have been excluded.[18] Not only is negotiating leverage more widely distributed across the network, firms are increasingly valued for the specialized capabilities that they bring. Petrick also stresses that increasingly, value is driven by the ability to innovate, which references other aspects of the MEP new strategy.[19]

Honda of America, for example, is seeking ways to reach out beyond tier 1 suppliers, especially in the Midwest, where its suppliers are heavily concentrated: Almost half of Honda's U.S. suppliers (308) and about 60 percent of its procurement from suppliers ($11.8 billion in 2011) are in Ohio, Kentucky, Michigan, and Indiana.[20]

OEM concentration geographically is matched by the way that some centers focus on particular industries, perhaps especially within supplier development. For example, while Ohio MEP assigns specific industries to

---

[17]Irene Petrick, "Supply Chain Globalization," in *Alliance for American Manufacturing, Manufacturing a Better Future for America*, 2009.
[18]Petrick, op. cit. p. 291.
[19]Petrick, op. cit. p. 295.
[20]Fred Braun, Honda of America Manufacturing, The MEP in the MidWest, NAS Ohio workshop, "Diversity and Achievements: The Role of Manufacturing Extension Partnerships in the Midwest," March 26, 2012.

individual centers, in Oklahoma supply chain activities are focused on three industries in particular across the state: wind energy, aerospace, and petrochemicals.[21]

MEP now has a supply chain initiative, led by MEP centers in Illinois, Virginia, South Carolina, and Texas. This initiative has drawn on extensive research undertaken by NIST MEP into the needs of customers at the top of the supply chain.[22] It identifies areas of potential weakness as well as improvement among tier 1 and even tier 2 suppliers, for connecting tier 1 with lower-order suppliers. This research has included a substantial mapping and planning exercise, involving questionnaire replies from 49 manufacturing companies at various levels of different supply chains.

The centers leading the study eventually proposed a series of possible activities that could address the needs identified in the course of the study. This approach is summarized in Table 6-1.

In May 2012, the working group of MEP centers presented a framework through which to implement the supply chain initiative to the MEP Advisory Board.

SUSTAINABILITY

MEP's signature initiative in this area is E3: Economy, Energy, and the Environment initiative. This is a collaborative effort of five federal agencies, including MEP.[23] The concept is that each agency will bring specific tools and expertise to support innovation, energy saving, and job growth in manufacturing to support improvement at the firm level.[24] Typically, E3 begins with a one-day energy audit and a two- to three-day lean and green assessment, after which specific steps for improvement are recommended to the company and tracked.

Pilot projects were completed in Columbus, Ohio, and San Antonio, Texas, and about a dozen projects have now been completed. In Alabama, the E3 project is supported by a team comprising the Alabama Technology Network, Alabama Power, and the mayors of Montgomery, Huntsville, and Tuscaloosa. The projects have utilized more than $2 million in funding from multiple sources[25] to focus on 40 automotive industry suppliers in the area that have committed to take part in the project. As of 2012, the project has identified $263,200 of potential energy savings and savings of more than 26,700 metric tons of greenhouse gases. The project has also lined up financing for

[21]Oklahoma MEP, op. cit. Exhibit I-17—Key Products.

[22]NIST MEP Supply Chain Development Initiative: *Improving the competitiveness of U.S. supply chains*, May 9, 2012.

[23]The other agencies are Department of Commerce, Small Business Administration, Department of Labor, Department of Energy, and Environmental Protection Agency.

[24]NIST, "Economy, Energy and Environment: Going Green in the Black," July 21, 2010.

[25]E3 "E3 Alabama," <*http://www.e3.gov/accomplish/alabama.html*>. Accessed July 26, 2012."

TABLE 6-1 Supply Chains: SME Needs vs. Proposed MEP Offerings

	Supply Chain Needs						
Supply Chain Opportunities	Supply Chain Strategy Workshop for Prime	Supply Chain Strategy Workshop for Supply Chain Partners	Total Cost of Ownership Support Workshop & Services	Demand Planning/Volatility Reduction	Failure Mode Effects Analysis (FMEA) Supply Chain Risk Mgmt	Collaborative New Product Development	Technology Scouting/Innovation Engineering
Identify and Address Choke Points	X	X					
Increase ERP/MRP Effectiveness				X			
Mitigate Global Risks			X		X		
Document Supply Chain Strategy	X	X					
Reduce/Manage Volatility				X	X		
Align Supply Chain Metrics with Long-term Business Focus	X	X					
Optimization of Supply Chain		X				X	X
Recognize Emergent Needs for Future Supplier Capabilities							
Expand Supplier Matching Capability							X
Better TCO Decision			X				
Enhance Value Chain Collaboration Among Suppliers						X	
Improve OEM/Supplier Product Development Collaboration						X	

SOURCE: NIST MEP, op. cit.

improvements from Alabama Saves, a $60 million program to provide low interest loans to support energy initiatives in manufacturing.[26]

As E3 suggests, lean manufacturing may provide a foundation for improving the environmental impact of the company or factory. As the High Liner case described in Box 6-1 indicates, many energy-saving improvements

[26]E3 "E3 Project Summaries," *<http://www.e3.gov/ground/summaries.html>*. Accessed July 25, 2012.

result from incremental changes in operations. These may be facilitated by the ways in which lean manufacturing empowers front-line workers to take the initiative in suggesting improvements. At the same time, lean manufacturing provides tools through which senior management can focus on the details of the production process, e.g., through the Six Sigma toolkits.

What is increasingly apparent is that there are overlaps between different elements of MEP's new strategy, and with other aspects of individual center strategies. For example, Vadxx Inc., a Cleveland-based firm whose plants convert plastic and rubber scrap into synthetic crude oil, has implemented projects that address technology acceleration and green manufacturing, but also falls within the Ohio strategic focus on polymers. The company manufactures and sells crude oil made from polymer-based waste, such as scrap tires, scrap plastic, and medical waste.

As with any startup, finding product development and marketing partners is a huge challenge. Jim Garrett, the Vadxx CEO, explicitly credits Magnet MEP Ohio with enhancing the company's credibility, helping it to partner successfully with Rockwell International.[27]

There is direct overlap as well. Lean manufacturing emphasizes the reduction and elimination of waste, which is in itself a step toward greener manufacturing. Efforts to reduce waste are in themselves green initiatives, whether or not they are labeled as such.

A recent study of companies that have implemented lean manufacturing practices suggests that there is substantial synergy between lean and green.[28] Bergmiller and McCright found that there was a strong positive correlation between the introduction of green management practices and lean management outcomes. On reflection, this does not seem surprising. As the High Liner case, described in Box 6-1, suggests, many energy-related improvements are found through incremental improvements in production processes. The introduction of lean manufacturing also often has the effect of empowering workers on the factory floor to identify and recommend such improvements. At the same time, it encourages management to become much more vigilant in seeking out these opportunities.

Some of the findings from Bergmiller and McCright's paper further illustrate the synergies between green management and lean management:

- "Substitution of less toxic, more recyclable, or more easily processed materials is significantly correlated to customer satisfaction, profitability, and improved cost performance.
- Extending the life of products correlates with improved customer satisfaction and profitability.
- Returnable packaging is highly related to improved cost performance.

---

[27]Jim Garrett, Vadxx, NAS Ohio workshop, "Diversity and Achievements: The Role of Manufacturing Extension Partnerships in the Midwest," March 26, 2012.

[28]Gary G. Bergmiller and Paul R. McCright, "Are Lean and Green Programs Synergistic?" *Proceedings of the 2009 Industrial Engineering Research Conference*.

- Waste segregation is significantly correlated to customer satisfaction and profitability and highly related to improved company cost performance."[29]

The point here is that in some ways green manufacturing initiatives are a natural extension of—and complement to—existing lean manufacturing tools and programs.

Green manufacturing is also increasingly tied in to "green" courses offered by colleges and universities—overlapping with workforce development. For example, Purdue University now offers three levels of "Green Ed," including a certification as an SME Green Specialist, as part of its green enterprise development program.[30]

WORKFORCE DEVELOPMENT

NIST MEP's identification of workforce development as one of the five strategic dimensions signals an effort to reinforce and extend efforts already under way in some MEP centers. Under this strategy, MEP proposes to help firms take a more deliberate approach to developing their workforce as they adopt new technologies and new business models. This focus on workforce development also addresses a key challenge raised in recent studies, which show that manufacturing companies are behind those in other sectors on a range of workforce development criteria.[31]

MEP's strategy for workforce development has focused on efforts to formalize and extend training programs for use within a company, as well as new efforts to partner with community colleges and other educational organizations—including professional organizations—to develop new certifications relevant to manufacturing and especially advanced manufacturing.

MEP's attention to workforce development has traditionally included management training in lean manufacturing. These programs, as implemented by MEP centers, however, did not result in additional certifications. A number of universities and management training programs, however, offer a range of certifications in lean manufacturing to middle management and production workers.

In addressing this component of new strategy, MEP focuses on four core areas:[32]

- Layoff aversion.
- Skills standards and certification.

[29]Ibid.
[30]Purdue University, Green Enterprise Development, *<http://www.greenmanufacturing.purdue.edu/>*.
[31]Sloan Center on Aging and Work, Talent Management Survey 2009.
[32]MEP, "Creating an Innovation Practice Next Generation Strategies: Workforce," briefing presented at MEP Advisory Board Meeting, May 6, 2012.

Box 6-1
Environment and Energy Case Study: High Liner Foods, a Pilot Project with Maine, New Hampshire, and Massachusetts MEPs.

One of North America's largest seafood processors with additional lines of business, High Liner Foods participated in a test pilot developed by the New Hampshire, Massachusetts, and Maine MEPs in collaboration with the U.S. Environmental Protection Agency (EPA), designed to field test EPA's energy and environment toolkit and integrate it into the lean manufacturing techniques employed by the MEPs. After preliminary review, the team decided to focus on the company's mozzarella cheese stick line because it utilizes more equipment than any other product line, and therefore seemed to offer the greatest potential for energy savings.

An initial walkthrough of the production facility quickly identified areas for initial savings. For example, in a 100-foot area, lighting far exceeded recommended levels, and the area was delamped for immediate energy savings.

Other issues that emerged from the value stream mapping exercise were tackled in a series of "kaizen" or rapid process improvement events. Four of the largest energy-saving proposals were:

* Changing the initial handling of the raw product, as cold storing mozzarella would be replaced by flash freezing to eliminate the energy required for 10 days of refrigeration.
* Minimizing the use of compressed air, used to blast each carton of mozzarella on the packaging line to ensure that it was full. This could be replaced by a less expensive alternative involving photo-sensor detection and a mechanical rejection system. The project also repaired leaks in the compressed air lines causing thousands of dollars of lost energy.

* Creating variable frequency drives for the hydraulic pumps, compressed air systems, and portable heater units. Previously, these systems had no way to incrementally increase their output. Additional demand required putting on another 50 hp compressed air unit or 70 hp hydraulic pump. Variable drive systems could instead dial up additional compressed air, heat, or hydraulic volume as needed.

* Recycling heat from the deep fryer. The project team devised a heat recycling system that utilized deep fryer energy to preheat the 30,000 gallons of water used daily in the production process to 100°F, leaving only an additional 40°F or 50°F of heating required, thereby creating substantial energy savings. A new water recovery system also reduced water consumption.

Overall, the pilot project identified more than $200,000 in energy savings opportunities for High Liner Foods, including electricity ($108,800), compressed air ($48,000), water consumption ($21,500), hydraulic power ($13,000), and heat recovery ($26,600).

SOURCE: Excerpted from New Hampshire MEP, <http://*www.nhmep.org/press/pr2010_jan11.pdf*>. Accessed July 17, 2012.

- Innovation Engineering Leadership Institute.
- SMARTalent.

These four core areas are further described below.

Layoff Aversion

Layoff aversion refers to efforts to identify and assist manufacturers that are at risk before they are forced to suspend or dismiss their workers. According to MEP, layoff aversion programs have been operating in a number of states over the past decade—including California, Pennsylvania, South Carolina, Oklahoma, Missouri, Michigan, and New York. CMTC, a California MEP center, reports that its programs have had a range of positive results:

- 126 manufacturers engaged.
- 1,805 jobs retained.
- 349 jobs created.
- $59 million in sales increases.
- $49 million in sales retained.
- $8 million in cost savings.
- $17 million in investments created.

The CMTC layoff aversion effort is part of a partnership with the Los Angeles County Economic Development Corporation and the City of Los Angeles.[33] Other centers have also had impressive results. In Mississippi, for example, the Mississippi Technology Alliance feared that Cooper Tires, a global tire manufacturer, might close its factory in Tupelo, which would have had a substantial negative impact on the Northeast Mississippi economy.

MTA implemented a lean production improvement program with management at the plant, focused on continuous improvement, which it believes led directly to Cooper's decision to expand the Tupelo plant by 7,000 square feet

[33]See Los Angeles County Economic Development Corporation, City of LA/LAWIB Monthly Layoff Aversion and Business Retention Initiative Progress Report, <*http://www.wiblacity.org/images/stories/PDF/laedc_report_may_2011.pdf*>. Accessed July 19, 2012.

and add $16 million in investment in two phases. Overall, the project resulted in the addition of 150 jobs and further investment of $22 million.[34]

MEP appears now to be working to engage other centers in developing similar programs.[35]

Skills Standards

The development of skills standards is in part an effort to address skills gaps in the workplace, identified (among others) by the National Association of Manufacturers (NAM).[36] A key aspect of MEP workforce development strategy is the development of widely recognized standards for certifying the skills and professional capacities of workers. Accordingly, MEP has adopted the widely used skills standards developed by the NAM.

The NAM standards provide for a range certification from what NAM calls "personal effectiveness competencies" to "industry-specific certification." The Manufacturing Skills Certification System endorsed by NAM validates specific skills certified by its partners ACT (a nonprofit organization focused on career planning and workforce development),[37] the American Welding Society, the Manufacturing Skill Standards Council, the National Institute of Metalworking Skills, and the Society of Manufacturing Engineers. This system of certification provides workers with "a system of stackable credentials applicable to all sectors in the manufacturing industry" that are nationally portable and recognized across industry.[38]

In June 2011, President Obama announced a major expansion of Skills for America's Future, an industry-led initiative that aims to improve industry partnerships with community colleges for skills development.[39]

The expansion includes an expansion of the Manufacturing Skills Certification System, designed to give students access to manufacturing credentials valued by employers. The expansion should allow students and

[34]MEP Success Stories, "Cooper Tires," <*http://ws680.nist.gov/mepmeis/SearchSS.aspx?ID=3143*>. Accessed July 23, 2012.

[35]It should also be noted that CMTC is among the largest MEP centers, and the data do not indicate how long a time period is covered. So in the context of a 10-year period in a huge region, these results might not been seen as quite so impressive.

[36]NAM "2011 Skills Gap Report," <*http://www.nam.org/content/Institute/Research/Skills-Gap-in-Manufacturing/2011-Skills-Gap-Report/2011-Skills-Gap-Report.aspx*>. Accessed July 19, 2012.

[37]Formerly known as American College Testing, but simply ACT since 1996.

[38]NAM, "NAM-Endorsed Manufacturing Sills Certification System—Executive Brief," n.d. <*http://www.themanufacturinginstitute.org/~/media/E5A6A4DBCEBD4F579636DAFAA9952F18/Executive_Brief.pdf*>. Accessed July 19, 2012.

[39]White House, Office of the Press Secretary, President Obama and Skills for America's Future Partners Announce Initiatives Critical to Improving Manufacturing Workforce, June 8, 2011, <*http://www.whitehouse.gov/the-press-office/2011/06/08/president-obama-and-skills-americas-future-partners-announce-initiatives*>. Accessed July 29, 2012.

workers to access programs for manufacturing credentials in community colleges in 30 states as a for-credit program of study.[40]

MEP is expected to play an important role in this initiative:

> *"The 60 Centers of the national MEP system will serve as the "boots on the ground" with local manufacturers to educate them about the value the NAM-endorsed skills certification system to their business so that they utilize the skills certification system in their recruitment and hiring efforts. In addition, the MEP will provide input to The Manufacturing Institute about aggregate skill needs of manufacturers by industry and geography so that certification systems can remain dynamic and evolving."*[41]

Innovation Engineering

The effort under the new strategy to enhance innovation among manufacturing SMEs represents, perhaps, the most radical departure from previous MEP practice. The name of the Innovation Engineering initiative is itself an important clue as to the driving vision: Just as lean manufacturing can be seen as the application of methodological tools to well-known problems—a kind of systemic engineering of the production process—innovation engineering anticipates the development of a toolset focused on stimulating more innovation. Indeed, it utilizes a similar engineering metaphor. However, unlike lean manufacturing which dates back to the 1990s in the United States and the 1960s in Japan, there is as yet limited evidence to show that this "innovation through engineering" approach is successful.

To date, the primary approach to innovation engineering at MEP revolves around a program presented by an organization called the Eureka Ranch, renamed the Innovation Engineering Leadership Institute (IELI). The program was initially called Eureka! Winning Ways (E!WW), and has been renamed by IELI as the Innovation Engineering Management System. Here, we use primarily E!WW as this nomenclature is more familiar to centers and other MEP stakeholders.[42] MEP is strongly encouraging MEP centers to adopt this program as the core of their innovation engineering curriculum—utilization of

---

[40]According to the MEP, "This program was designed by the Manufacturing Institute, an affiliated non-profit of the National Association of Manufacturers, in partnership with leading manufacturing firms, the Gates Foundation, and the Lumina Foundation, and key players in education and training including ACT, the Society of Manufacturing Engineers, the American Welding Society, the National Institute of Metalworking Skills, and the Manufacturing Skills Standards Council."
[41]White House, June 8, 2012, op .cit.
[42]IELI, <*http://innovationengineering.info/*>. Accessed July 28, 2012.

E!WW is one of the new measures of center performance under the new CORE metrics system.[43]

The E!WW website states that it provides "a proven system of tools developed through scientific inquiry—grounded in hard data and proven reliable in real world application" and that is now available to smaller companies where once it was reserved for large companies like Nike, HP, Proctor and Gamble, and Disney.[44]

The E!WW program is usually presented in conjunction with MEP centers by staff from IELI. The initial approach is through a one- to three-day seminar. This stresses tools for brainstorming and thinking new product ideas, as well as providing a considerable amount of encouragement for innovation itself. The program appears to provide much less guidance in implementing new ideas, although it is probably fair to say that this would be the place where the MEP center itself steps in to provide additional consulting services.[45] Based on the published schedule for fall 2012, one to two workshops are held each month by IELI, usually in conjunction with MEP.[46]

As we have seen above, the methodologies for lean manufacturing are fairly well known. Numerous studies have addressed their effectiveness, and they are now a staple of every MBA course. This is not yet the case with the innovation program encouraged by NIST.

- The staff at the MEP centers are trained in lean manufacturing, not innovation. Discussions with center directors have indicated that many see the need to replace existing staff with new staff that have a different skill set.
- New curricula will need to be developed. There are some questions as to whether E!WW is as universally applicable as MEP appears to believe, and in particular whether it is sufficient to cover the entire product development cycle for innovative products, and whether there may be alternative programs that are in some respects superior, particularly in light of the diversity of the MEP client base.
- Innovation and product development can take much longer than the implementation of lean manufacturing. Thus metrics designed to capture the short-term outcomes from lean oriented projects will almost certainly not be sufficient to capture outcomes from innovation-related projects that often take years to reach fruition. The new core metrics are a good start in this direction.

It should be noted that not all MEP centers are using the E!WW toolset. For example, MEP of Nebraska is using the Gallup Corporation's

---

[43]MEP Private communication, July 23, 2012.
[44]IELI website, <*http://innovationengineering.info/*>. Accessed July 28, 2012.
[45]The standard agenda covered by IELI is provided in Appendix C.
[46]IELI, <*http://innovationengineering.info/location.php*>. Accessed July 28, 2012.

Entrepreneurial Acceleration System, which it is implementing through a partnership with the University of Nebraska Omaha and the Omaha Chamber of Commerce.[47]

SMARTalent

SMARTalent is a new set of tools (developed by MEP) which aims to integrate different aspects of workforce development into a coherent strategy for clients. It pulls together four elements of workforce development:

- Planning and strategic alignment.
- Recruitment.
- Development and management.
- Retention and succession planning.

In 2012, several MEP centers (Illinois, New York, Texas, Pennsylvania, Washington, and Ohio) are helping to pilot the functionality and value of the SMARTalent tool. At the same time, all centers are being encouraged to investigate how workforce development and talent management can support the growth of manufacturing businesses, and the integration of workforce components into strategic business planning.[48]

CONTINUOUS IMPROVEMENT

The new strategy calls on MEP to promote operational excellence within manufacturers by providing services that minimize expenses and reduce wasted activities. This strategic thrust reflects a continued focus on lean manufacturing.[49] This component of MEP's new strategy is discussed in detail in Chapter 3.

EXPORTING AND GLOBALIZATION

In addition to the five service categories identified by MEP and described above, this chapter also describes two additional activities—support for exporting and support for clustering—that also advance the MEP's new strategic thrust.

At present, export-led growth is not identified as one of the five strategic thrusts of the Next Generation Strategy, although there is ample evidence that supply chains in manufacturing have globalized. This means that

---

[47]MEP of Nebraska, op. cit. 1-12. More details can be found at <*http://www.gallup.com/se/152741/gallup-entrepreneur-acceleration-system.aspx*>. Accessed July 20, 2012.

[48]MEP Advisory Board presentation, May 2012.

[49]Lean manufacturing is discussed in detail in Chapter 3.

efforts to improve supply chain integration for U.S. SMEs will involve more exports and even potentially direct investment abroad.[50] MEP nonetheless appears to be paying considerable attention to the need for increased exporting from manufacturing SMEs. A recent report from the U.S. International Trade Commission (USITC) highlighted some comparative characteristics of exporting and nonexporting SMEs, as described in Table 6-2.

The table shows that exporting is strongly associated with positive economic outcomes in terms of higher revenue, growth, and more value added per employee. It is therefore not surprising that enterprising SMEs in manufacturing are increasingly seeking out additional markets, although the overall percentage of SMEs that export remains below 14 percent of all manufacturing SMEs, as shown in Figure 6-1.

The USITC survey also addressed perceived barriers to exporting among SMEs and found that these firms had substantial concerns and that they were different from those of larger enterprises, as shown in Figure 6-2.

MEP, with its new innovation-focused strategy, seeks to help SMEs link up with a wider ecosystem of firms and research organizations and to promote the development and manufacture of innovative products and services, many of which could have potential markets overseas. The International Trade Administration (ITA) can offer specialized assistance at the final stages of this

TABLE 6-2 Comparing Exporting and Nonexporting SMEs in Manufacturing Sectors

Indicator	Exporters	Non-exporters	Key Finding
Average Revenue per Firm (Millions of Dollars, 2009)	3.9	1.5	SME manufacturers that export earned more revenue than non-exporting SME manufacturers
Manufacturer Revenue Growth (Percent Change, 2005-2009)	36.8	-6.8	Exporting SME manufacturers' revenue grew faster than that of non-exporting SMEs
Average Revenue per Employee (Thousands of Dollars, 2009)	281	163	SME manufacturers that export are associated with higher labor productivity than non-exporting SME manufacturers

SOURCE: U.S. International Trade Commission, *Small and Medium Sized Enterprises: Characteristics and Performance*, November 2010, <*http://www.usitc.gov/publications/332/pub4189.pdf*>.

[50]Bernard and Jensen have long noted the superior performance characteristics of exporting plants and firms relative to nonexporters. Andrew B. Bernard and J. Bradford Jensen, "Why Some Firms Export," *The Review of Economics and Statistics*, 2004, MIT Press, vol. 86(2), pp. 561-569.

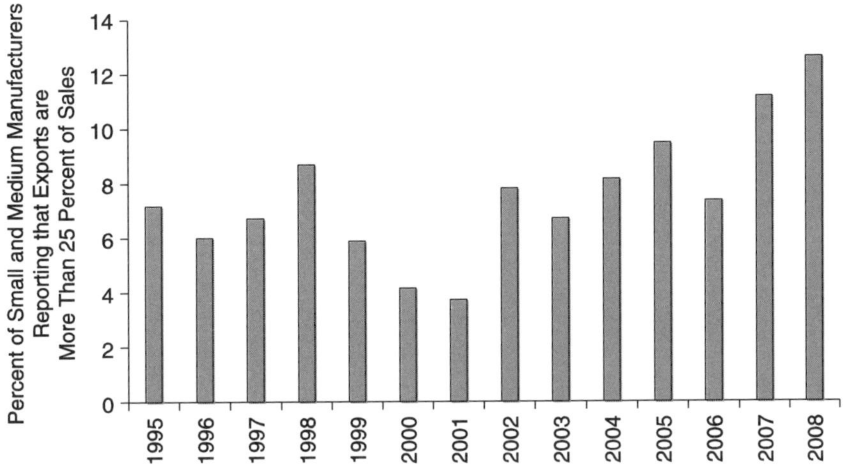

FIGURE 6-1 SMEs in manufacturing with >25 percent revenues from exports.
SOURCE: NAM Annual Survey of Manufacturers, 2009.

development. In particular, ITA's ExportTech program runs a three-day workshop in which companies with interest in exporting their products are provided with information and expertise that will help them develop effective international growth plans and get into international markets.

According to MEP, "ExporTech is deployed nationally as a collaboration between the Manufacturing Extension Partnership, U.S. Export Assistance Centers, and other partners including District Export Councils, State Trade Offices, Ex-Im Bank and SBA. These partners help to recruit participants, line up speakers, and in many cases serve as speakers themselves in the ExporTech sessions."[51]

Typically, it runs as three full-day sessions in which groups of eight companies (usually C-level executives) are provided with the tools and background needed to develop a customized international growth plan, which is the eventual result of the exercise (see Figure 6-3).

As operated in North Dakota, the first ExporTech session provides broad background on international transactions, including a review of successful export strategies. Content and expert speakers for the second session are customized to the specific needs and composition of the group of companies. In the final session, executives present each company's international growth plan for feedback from a panel of experts.

Between sessions, executives gather information and develop plans with the aid of experts from the sessions who include government export

[51]MEP, The ExporTech Program, <*http://www.nist.gov/mep/exportech.cfm*>. Accessed July 19, 2012.

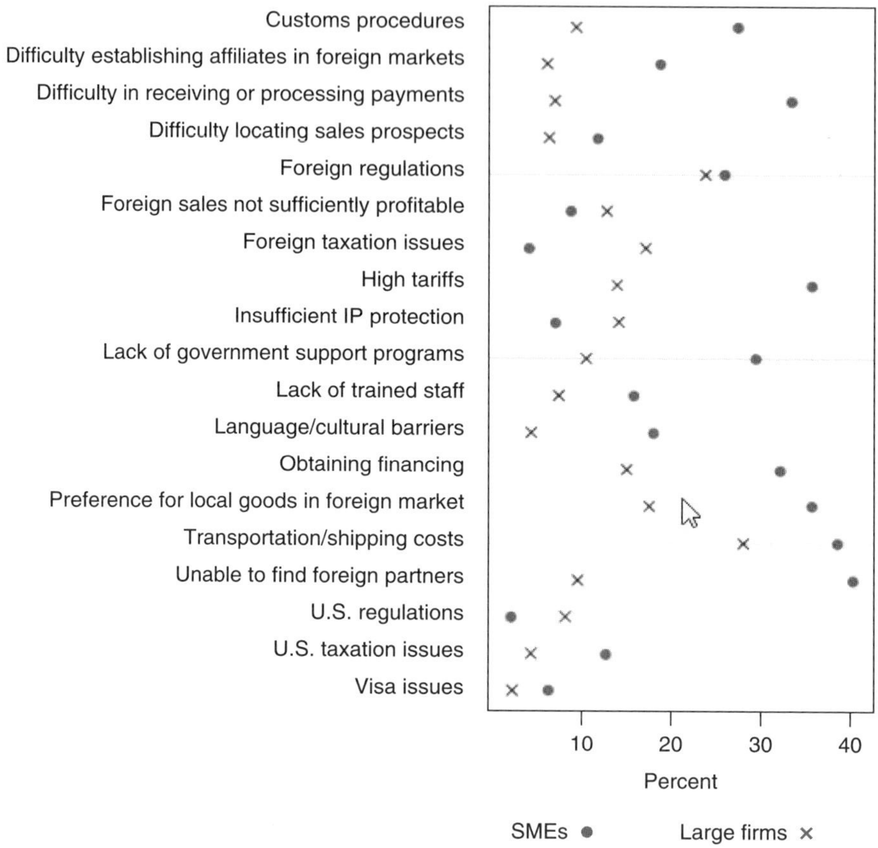

FIGURE 6-2 Concerns about exporting among SMEs and multinational corporations.
SOURCE: U.S. International Trade Commission, op. cit., Table ES2.

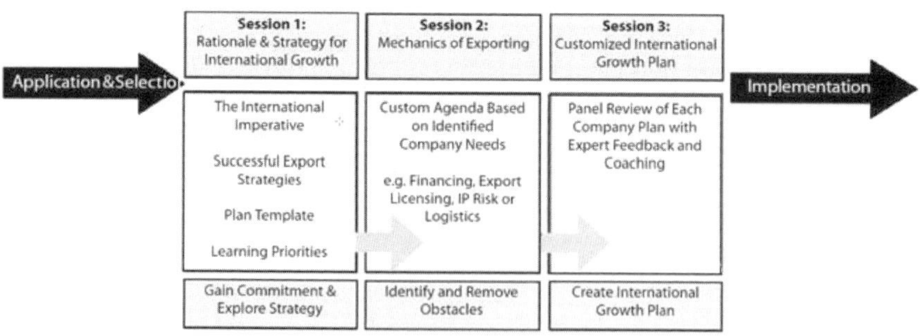

FIGURE 6-3 ExporTech three session outline.
SOURCE: North Dakota MEP center, ExporTech 2011.

assistance providers and private-sector resources. North Dakota MEP estimates that a minimum of 40 hours of work outside the sessions will be needed to develop an effective international growth plan. [52]

Evidence from other centers suggests that ExporTech has gained traction and is now a regular service offering (see Box 6-2 for a profile of ExporTech). For example, Oklahoma reports that since the program was first implemented in 2008, 6 programs have been delivered serving 28 companies.[53]

CLUSTERS AND NETWORKING

Given that, in many cases, MEP support for cluster development will require access to a wider range of resources and expertise than do traditional lean-focused projects, a number of MEP centers are developing new approaches. Some MEP centers have developed strategies that focus heavily on networking and the development of local clusters, while other remain primarily focused on identifying and serving individual clients. NorTech, "a regional nonprofit technology-based economic development organization serving 21 counties in Northeast Ohio," represents the cluster end of this continuum.[54] Its focus is on building regional clusters for selected manufacturing sectors, specifically flexible electronics and advanced energy products.

This reflects the dense set of regional networks that impact support for manufacturing in the region:

- JobsOhio Network partners.
- Edison Technology Centers.
- Ohio MEP center.
- TechSolve and MAGNET.
- Ohio's Manufacturing and Technology Small Business Development Centers (MTSBDC).
- Community colleges.
- Local economic development organizations.[55]

In Ohio, a dense network of support programs encourage partnerships and networking. The state-led MEP program operates in conjunction with Edison Technology Centers.[56] The market for support services has been

---

[52]North Dakota MEP, op. cit. Once again, the precise mechanics are likely to differ between centers. The North Dakota program outlined above is an example; we cannot conclude that it is precisely typical.

[53]Oklahoma MEP, op. cit. p. 20.

[54]See <*http://www.nortech.org/*>.

[55]Schmenk, op. cit.

[56]The Edison Technology Centers were founded in 1984 to provide a link between manufacturers and university resources. Berth Colbert, Ohio MEP, National Academy of Science MEP Workshop, November 14, 2012.

Box 6-2
ExporTech Case Study: Raloid Corporation, Reisterstown, Maryland

Company Profile

Raloid Corporation is a contract manufacturer of close tolerance components and assemblies for the defense industry. The company has 75 employees and approximately $10 million in annual sales, and competes with much larger companies. Prior to ExporTech, its customer base was 100 percent domestic. However, margins in the United States are under pressure, so international sales became more attractive.

International Potential

Raloid identified a specific international opportunity: The 51 percent Buy American requirement for the Department of Defense opens up supplier business to international companies who partner with American manufacturers to meet the 51 percent requirement.

Discovery of Opportunities through ExporTech

Raloid's Carl Livesay participated with four other companies in the initial ExporTech class. *"The program required a lot of homework from each of us, but the guidance and resources we received through it saved us tons of time and tens of thousands of dollars,"* said Livesay.

"The program gave us access to every resource we could need—legal, exporting, financial, logistical, and more. At the end of the program, we presented our plans to a panel of experts, who poked and prodded us to refine our plan and expectations," explained Livesay. "Now we have a detailed document—and the knowledge we needed—to guide our progress into these new markets."

Results

A key ExporTech benefit for Raloid was accelerating the ITAR compliance.[a] After two years of failed efforts, personal introductions made through ExporTech provided approval within days. Raloid was then able to sell internationally. Starting from zero, the company achieved $250,000 in international sales last year and expect to hit $350-500,000 this year. Raloid is now developing more opportunities in the United Kingdom, Turkey, Italy, and Australia.

"The resources from ExporTech have far surpassed my expectations, and were not available anywhere else in such an efficient and cost-effective package. And the connections we've made—with the USEAC, MEP, DEC, and state Business and Economic Development staff—give us instant credibility in the international business community. They have introduced us directly to many of the same companies that may have never returned our calls previously.
Months after the program ended, we continue to receive referrals, ideas, and introductions that are incredibly valuable." Bruce Livesay

[a]"International Traffic in Arms Regulations (ITAR) is a set of United States government regulations that control the export and import of defense-related articles and services on the United States Munitions List. ... For practical purposes, ITAR regulations dictate that information and material pertaining to defense and military related technologies (for items listed on the U.S. Munitions List) may only be shared with U.S. Persons unless authorization from the Department of State is received or a special exemption is used." Source: Wikipedia.
SOURCE: Mtech, University of Maryland.
<http://www.mtech.umd.edu/ummap/documents/Raloid_Client_Story.pdf>.

segmented by size and function, with small manufacturers who need general assistance being served by the state's Manufacturing and Technology Small Business Development Centers to provide low-cost or free "business consulting, market and product development, workshops and seminars, defense transition and commercialization, and resource referrals."[57] Larger manufacturers are sent to Magnet in Northern Ohio and TechSolve in the southern part of the state,[58] with the two centers also providing coordination and training for the MTSBDCs.

Both Magnet and TechSolve are also Edison Technology Centers with specific sector expertise: Magnet in automotive and TechSolve in aerospace and aviation.[59] These are two of the industries targeted by Ohio. Edison funds are used to provide the state match for MEP, and are focused on three areas: (1) manufacturing assistance; (2) technology advancement, and (3) cluster development.[60]

At the same time the Ohio Edison Center/MEP system assigns specific functional expertise to selected sectors. For example, Magnet is the center focused on product development, and benefits from the Product Development Center created with Third Frontier funding in 2006.[61]

[57]MAGNET, "The Incubator at MAGNET," July 2011.
[58]Christiane Schmenk, Ohio's Economic Development Strategy, NAS Ohio workshop, "Diversity and Achievements: The Role of Manufacturing Extension Partnerships.
in the Midwest," March 26, 2012.
[59]Schmenk, op. cit.
[60]Daniel E.Berry, Strengthening MAGNET as an Element of the Ohio MEP Center, NAS Ohio workshop, "Diversity and Achievements: The Role of Manufacturing Extension Partnership in the Midwest," March 26, 2012.
[61]Berry, op. cit.

The PRISM Initiative: Partnership for Regional Innovation Services to Manufacturers

On the ground, these initiatives and others will need to be integrated and tap into further resources—for example, sources of capital—if they are to have the effect of powering up innovation in local SMEs. One ambitious attempt to provide an integrated approach is provided by Magnet Ohio though its PRISM initiative.

The PRISM initiative can be seen as a means of focusing multiple resources on promising small companies, with the aim of moving the company into new markets, generating new growth and new jobs. It also has the powerful backing of one of the best innovative companies in the region, the Timken Company, whose CEO has been closely involved with Magnet for many years and strongly endorses the PRISM concept.[62]

PRISM aims to connect companies to the right resources in a faster and more systematic fashion by developing an engaged network of higher-education-based assets that link small- and medium-sized manufacturers (SMM) market knowledge and University research and development capabilities for the development of innovative new products, markets, and services. Magnet sees this as a having the following characteristics:[63]

- MAGNET-managed and facilitated.
- Curriculum and programs offering unique hands-on experience for associate, bachelor, and master's degree students who want to "Make Things."
- Active network with northeast Ohio SMMs and support resources.
- Mechanisms to apply university-sourced technologies, skills, facilities, and equipment in innovation.
- Infrastructure for skunkworks, spin-offs, internal development, and market-based research collaborations.
- Flexible capital pool tailored to projects.

The four- to five-year objectives for PRISM are outlined below.

PRISM is now well into the implementation phase. The first batch of eight companies are in the program, having paid PRISM $87,000 in fees and generated $314,000 in external funds. They are focusing on 8-10 products, which by 2014 they project to $233 million in revenue and 740 jobs. These are hugely ambitious goals. It remains to be seen whether PRISM can meet them.

However, Magnet says it is meeting carefully designed metrics, and feedback from companies is positive. For one company, Magnet found initial funding: "The first $50-100,000 seed capital was critical. We wouldn't be here without it" (client CEO); for another, early design assistance means that "The

---

[62]James W Griffith, private communication, NAS Ohio workshop, March 2013.
[63]Berry, op. cit.

CAD Models gave us credibility" (client manager). A further 25 projects are in the PRISM pipeline.

CONNECTING TO C-LEVEL CLIENTS

At the heart of the new strategy is a fundamental shift in the relationship of the center to its clients, along several dimensions:

- From a focus on lean production and improving the bottom line to an emphasis on growth and expanding the top line.
- From point solutions to continuous connection.
- From a single line of services to a strategy that provides multiple offerings covering all aspects of company performances.
- From connection to line supervisors and plant managers to continuing consultations with C-level executives (C level includes CEO, CFO, chief operating officer [COO] and other senior executives).

The latter is perhaps the single pivotal point that will determine the success or failure of new innovation-based strategies. Without effective ongoing access to C-level executives, the new focus on top-line growth and company strategy is essentially irrelevant, given that plant and line-level staff do not typically have those responsibilities. .Indeed, there is a concern that MEP's growing focus on C-level interaction, which calls on providing services to larger companies that actually have C-level executives, may have the unintended consequence of directing centers away from smaller firms with fewer resources for innovation, but as of yet, we have no evidence.

This focus on the C level also created new challenges. As our interviews with center directors have made very clear, attracting and retaining the interest of C-level executives requires an entirely different type and level of marketing effort for MEP centers. In particular, it requires that MEP center staff gain competency in providing strategic advice to a firm's leadership. This in turn requires either that existing center staff be heavily retrained, or that new senior staff with additional capabilities, especially in strategic marketing, be brought on board.

It also requires new marketing efforts. Some centers have developed extensive executive education programs to make the initial connection. For example, the Manufacturers Resource Center (MRC) in Pennsylvania offers what is in effect a mini-MBA program through which executives meet at the center regularly for six months.[64] Others focus on semiformal networking: The Delaware Valley Industrial Resource Center (DVIRC) in Philadelphia runs executive forums tailored to the size of the company, which provide executives with a sounding board of peers, and are facilitated by senior center staff.[65]

---

[64]Interview with Jack Pfunder, director, MRC.
[65]Interview with Barry Miller, director, DVIRC.

Others have tried to develop low-cost introductory evaluations, and others still have adopted approaches from other organizations. For example, one center director said that "We follow a process we refer to as 'Value-based relationships.'. It is a 10 step guide from first contact through project success and measurement."[66] A review of the information provided by center directors strongly suggests that most see this as a long-term project which requires painstaking efforts to connect and maintain the connection over time (see Appendix C).

Finally, it should be noted that both the sales cycle and the impact cycle for strategic services is much longer than it is in general for lean services. Strategic services require the development of trust by C-level executives, which takes time and is often based in positive word of mouth from peers (perhaps accelerated by the development of forums or other groupings of C-level executives who can act as champions and references for center services). And outcomes can take much longer: For example, development and implementation of a plan to enhance global marketing might take many months, and the positive impacts of implementing the plan may take years to fully emerge. These kinds of impacts are therefore difficult to capture in center metrics.

CHALLENGES IN IMPLEMENTING THE NEW STRATEGY

The new strategy is a very ambitious set of objectives, taking a well-known program with a 20-year track record and turning it in an almost entirely

Box 6-3
Attributes of leading MEP Centers

Many of the MEP centers that appear to be embracing the initiatives of the new strategy share similar attributes.

1. They have developed or acquired services and strategies to help clients think broadly about how they profitably grow their company: not just focus on operational and lean services.

2. They have leadership with some level of for-profit business acumen.

3. They build client relationships at the C level.

4. They develop relationships to more broadly promote understanding of the value of manufacturing to their state. The overall system can and should build upon these "better" centers and leaders.

---

[66]Responses to NAS Information Request. See Appendix C-1.

new direction. This new mission, however, imposes a series of significant challenges.

Building the Required Capacity

Each of the strategic components listed above will require development of new tools and capabilities for the MEP centers and in some cases for NIST MEP. It is certainly worth noting that new services required by both NIST MEP and other center funding partners place a substantial burden on what may be a small center staff. In Nebraska for example, Table 6-3 describes the requirements imposed on the center by the various funding organizations involved, with each requirement being mandated in some form. The MEP of Nebraska has 10 administrative staff and 8 customer agents.[67]

Developing New Markets

Creating demand for the suite of new innovation-focused services is a key challenge for MEP. MEP centers face the need to provide services that clients are willing to pay for, in order to ensure that they meet matching funds requirements. At the same time, MEP performs a valuable public service in assisting companies to adopt new and often unfamiliar practices in order to improve their performance and, in the aggregate, the nation's competitiveness. MEP thus faces an important challenge in encouraging the centers to be proactive given a largely risk-averse client base.

Coping with the Decline in State Funding

MEP centers require funding and/or revenue to stay in business. As state funding support has declined or vanished, centers need to offer services to larger companies and branch plants of large companies to pay their bills. This development, however, may be leading some centers away from fulfilling an important public role of serving those SMEs that private consultants do not cater to.

Funding New Strategy Implementation

The new strategy offers a substantial opportunity for centers to become much more important players in the economic life of their regions. But it also requires that they effectively build an entirely new business, providing new services to what will likely be new clients (or at least different staff at the same client companies). Typically, building a new business requires capital, and it is not at all clear in a world of shrinking state funding and limited NIST MEP

[67]MEP of Nebraska, op. cit. 1-6.

TABLE 6-3 Funding Organization Requirements—MEP of Nebraska

Funding Organizations' Requirements	Near-term Responses	Longer-term Responses
Ind. Cluster Development (DED) (NMEP)	• Cluster teams assigned. • Advanced Mfg., Biosciences, Wind Energy, R&D, Software clusters in development. • "Transportation Warehousing and Distribution Logistics" in place.	• Implement cluster developed strategies/goals.
Supplier Development (NGS/Growth) (NMEP) (DED)	• Implement Battelle "Manufacturer to Manufacturer" strategies. • Expand existing Supply Chain initiative (pilot in process).	• Integrate with Clusters and Mentoring initiatives. • Pursue more HMEP National Accounts. • Seek support to create a mfg. "core competencies database."
Technology Acceleration (NGS/Growth) (NMEP)	• Gallup Entrepreneurial Acceleration System (business mentors). • Continue Dawnbreaker Commercialization. • Reapply for continuation of FAST grant & continue "Phase 0" SBIR funding assistance.	• In conjunction with a University, State or Private College, establish a prototyping capability.
Innovation (NMEP) (DED)	• Implement Nebraska Business Innovation Act components. • Expand support/resources for SBIR/STTR services.	• Seek out new "innovation" partnerships.
Export (NGS) (NMEP)	• Continue efforts to deliver ExporTech services.	• Expand efforts to identify international business opportunities.
Sustainability (NGS/Growth) (NMEP)	• Formalize partnership w/ P2RiC Center a NBDC. • Establish an Energy Audit service and implementation program.	• Explore opportunities for implement and E^3 initiative.
Workforce (NGS/Growth) (DED)	• Partner with NE DOL to respond to identified worker training gaps. • Expand worker training	• Expand worker training services via the new Nebraska Internship Initiative.

	services provided by Community College partners.	
Continuous Improvement (NGS/Growth) (NMEP)	• Maintain strong capacity to deliver high-impact "Lean services." • Implement "Innovation Engineering and Jump Start" principles. • Continue to support the Nebraska Performance Excellence Center.	• Develop new strategies for transforming existing mfrs.

SOURCE: MEP of Nebraska.
NOTE: Organization abbreviations are as follows: Nebraska Department of Economic Development (DED); HMEP Next Generation Strategy (NGS); Nebraska Manufacturing Extension Partnership (NMEP).

resources where that capital will come from. As one center director observed, "Centers are going to have to develop additional funding sources for some of the new initiatives from banks, state EDA funding, partnerships, etc." Another indicated that success would require additional MEP and federal funding: "MEP and other public funding will be needed to launch new strategy services until they become self-sufficient from client revenue. The initial seed capital for these new services will likely be from federal and state grants, but only until self-sufficiency is reached. This may take 2-4 years." [68]

This is a major challenge in part because almost all centers rely on existing revenue streams to meet the require 1:2 MEP federal funding match. They cannot afford to allow existing revenue streams to decline without risking a spiral of decline with revenue shortfalls driving the withdrawal of matching money, triggering further cuts in services.

Current revenues are tied to current consulting contracts. Most centers generate at least a third of their funding from consulting, which until recently has been almost all focused on lean. As no center has spare resources, it is difficult to see how centers can successfully perform the difficult tightrope act of shifting resources to generate new innovation- and growth-oriented consulting and hence funding, without reducing existing consulting revenue streams.

The inescapable conclusion is that centers must find additional grants and investments. It is very telling that the centers which appear furthest along in developing innovation and growth practices—notably in Pennsylvania—benefited from a substantial investment of state funding some years ago, which effectively funded the transition.

[68] Responses to center directors information request. See Appendix C-2.

Developing C-level Relationships

Centers will have to make substantial steps in developing numerous relationships with senior staff at clients and potential client companies. This will require a new level of effort, new tools and strategies, and most likely new staff. Based on our interviews with center directors, we believe that developing these relationships is the single most critical aspect of implementing the new strategy successfully.

Again, it is important to stress just how big a jump is being asked of the centers. Organizations which have primarily been focused on day-to-day improvement, often in partnership with line supervisors, are now being asked to offer—and in fact sell—strategic advice services to C-level clients. This requires a fundamental shift in orientation, and likely a major shift in staffing. It is hard to see how the skill sets of existing lean consulting translate into innovation and growth strategy. Again, those centers furthest along have made substantial new hires with new skills, and have downgraded the resources focused on lean. It simply appears unlikely that most existing staff can effectively be retrained for the new strategy.

At the same time, it takes a major effort to develop a substantial number of C-level relationships. Both of the successful Pennsylvania centers have developed extensive outreach efforts for C-level staff, including development of an entire new training curriculum for CEOs. The successes and failures drawn from this massive effort could at a minimum be shared much more widely with other centers as they attempt to tread the same path.

Working with the Centers to Effect Organizational Change

The centers are operated by a wide range of institutions including both university and non-university organizations. .Not only are the centers careful to defend their autonomy from MEP, in some cases it appears that they are comfortable with traditional lean manufacturing focus and are neither prepared for nor committed to a wider mission. MEP is now working harder to reach out to advisory boards, and especially fiduciary boards, to expand the network of stakeholders committed to new strategy. However, interviews with center directors indicate that the transfer of expertise between peers is very limited, and that this imposes unnecessary costs and strains as centers seek to move forward. A more proactive approach from NIST MEP to enhance peer-to-peer communication through a variety of formats would be very welcome, and we believe would be welcomed by many center directors and staff.

Outreach and Expansion

While the centers and MEP have good data on companies with which they have worked closely, these account for only a small percentage of possible support recipients. Centers report to MEP that they "touch" about 10 percent of possible target firms (i.e., small manufacturers in their regions), but are apparently not required to provide further support for this claim (e.g., a list of contacts made). One key challenge for MEP is to expand its reach while working with flat or potentially declining budgets.

Our analysis suggests that there are widely different outreach strategies in play, and that there does not appear to have been a systematic effort to identify and share best practice. This will be especially important as MEP centers seek to find not just new clients but new C-level clients.

It should also be noted that traditional metrics for penetration are in some respects misconceived. Indeed they may be driving centers to adopt ineffective outreach strategies in order to meet goals imposed via metrics. For any business—and centers are best understood as subsidized consulting businesses that reach hard-to-serve but important populations—it is critically important to identify the market segments that can best be served by the local MEP organization. NIST MEP itself argues that only a fraction of small manufacturers—maybe 15 percent—are "innovation-ready." Centers freely acknowledge that their sweet spot is with firms ready for innovation and with the capacity to invest. Thus, measuring outreach as percentage of all manufacturing SMEs in a region may incentivize centers to waste resources chasing contacts with firms that may never become clients. We believe that NIST MEP could serve centers more effectively by developing a much more narrowly drawn set of penetration metrics. Given that the new CORE metrics provide a more comparative framework for assessment, this should still impose pressure on centers to expand outreach—but it would allow them to focus their efforts and resources more effectively.

NIST MEP Innovation and Growth Tools

NIST MEP has invested a very substantial amount of money (apparently at least $30 million) in developing tools which when adopted and fully implemented are designed to help centers transition to the new strategy. This ambitious effort follows to some degree on a previous successful initiative at NIST MEP to develop a suite of lean-oriented tools in the late 1990s.

However, there are some important differences between these initiatives. Lean consulting had emerged somewhat earlier with exhaustive analyses of changes in the automotive industry. Some of the core principles were well known and had been implemented at some companies (e.g., Toyota) for decades. That is not the case with the innovation and growth tools. In each case, these tools can at best be described as experimental: there is no track record of success, and early indications from the field suggest that gaining traction with them will be a long and difficult journey. Moreover, the inability of either NIST MEP or the most heavily funded contractor, the Innovation Engineering Leadership Institute, to provide any substantial analysis of program impacts is worrying. A center-piece program that has received substantial federal funding but provided no documented results indicates that, at best, the continued heavy reliance of the innovation and growth strategy on sole-source contractors could be reconsidered. Similarly limited outcomes appear to result from other initiatives even when implemented by more established providers, such as the supplier scouting tools developed by RTI. Here too, despite extended efforts, results appear underwhelming

It is possible that there are problems with the tools and the consultants. However, it is also possible that the lack of results simply reflects lack of demand: There has been no study at NIST aimed at establishing the size and scope of the market for the new services. With regard to the new mantra of business in America of "fail fast, fail cheap," NIST MEP should at a minimum be looking to develop benchmarks that will provide some objective information as to when experimental programs like those implemented in recent years are failing. All exciting initiatives carry the risk of failure: That is the case here. The issue is when and how NIST MEP will recognize it and make necessary adjustments.

Existing Metrics

MEP is considerably more advanced than agency Small Business Innovation Research (SBIR) programs were at the start of the National Academy assessment in terms of metrics.[69] Each MEP center is required to report extensively on its activities, and undergoes an annual review and assessment. Each completed project is surveyed directly by MEP's survey contractor. To date, the data collected are more extensive than the reports generated, so enhanced use of existing data would be appropriate and useful.

Adjusting the New CORE Metrics

While the CORE metrics describe in Chapter 4 are a substantial change and a real effort to capture the kinds of data needed to assess the new strategy, it is apparent that they will require close monitoring and ongoing adjustment as they are implemented. These may also require changes to the program's data collection processes.

CONCLUSIONS

There is a need and opportunity to enhance and reshape the landscape of manufacturing support in the United States to address new competitive challenges and to take advantage of opportunities in emerging technologies. In general, the United States offers limited support for civilian manufacturing in comparison to many of its major trading partners. As described in the next chapter of this report, many advanced countries have significant programs to

---

[69]National Research Council, *An Assessment of the SBIR Program*, C. Wessner, ed., Washington, DC: The National Academies Press, 2008.

support the development and commercialization of new technologies, including fostering innovation, scaling-up production, and the creation of related supply chains.

MEP is one important element of a national manufacturing strategy. MEP centers are widely dispersed geographically, covering the entire country. They are in general well connected to their local economies, and to the small manufacturing sectors within them. Many have developed a positive record of achievement in serving their clients. In addition to the federal contribution, many MEP centers also enjoy support from their state government, as well as from a variety of state and regional organizations.

MEP's plan to enhance the innovation capacity and growth of small manufacturers under the new strategic orientation is in line with recent academic and policy analyses that call for U.S. manufacturers to become more innovative and more focused on growth and international competitiveness. At the same time, as this report documents, many MEP centers face significant risks as they seek to transition from a tight focus on lean production to a much wider range of services that require new clients, new contacts, new kinds of client conversations, new services, new toolsets and capabilities and, often, new staff as well. Some centers may also not realize sufficient demand for the new services.

MEP's transition is therefore likely to be uneven. Some centers—with strong state support—have already moved forward and have approached sustainability with at least half their revenue coming from these new services. Other centers with fewer resources may face significant difficulties. They are seeking ways to adjust their reporting to meet the new CORE metrics without taking, what are in their view, extensive risks with their core businesses.

This differentiated take-up could, in the end, become an advantage for NIST MEP.[70] If new modalities for sharing best practices—and failures—can be implemented effectively, the risks for adoption laggards can be reduced, and a catch-up phase can be encouraged. Thus, the inevitable spread between early adopters and late adopters can be used by the program to help smooth the path for late-moving centers. As NIST encourages the expansion of MEP centers into new growth and innovation services, it will be especially important to ensure that best practices from leading centers are widely adopted, and that center staff have more and improved capacities to share their experience and expertise.

Looking beyond the challenges of this transition, NIST and the wider community of stakeholders in manufacturing and innovation should also consider how MEP centers might contribute to the emerging landscape for manufacturing support and hence to the development of robust innovation and

[70]MEP has not provided—and may not have—accurate data on the take-up of innovation services at the center level. In some cases, centers may have relabeled what they already do as "innovation services." For example, one center with a large information technology consulting business now calls this "growth services."

manufacturing ecosystems across the United States. For example, further consideration in national and state manufacturing support and innovation strategies is needed to address how MEP centers can fruitfully cooperate with other individual public and private organizations, including the institutes of the proposed National Network of Manufacturing Innovation, in translating research into commercial products, prototypes, and industrial production processes and in enhancing demand and capability among small- and medium-sized manufacturers for such engagements.

Chapter 7

Foreign Programs to Support Applied Research and Manufacturing

Addressing the second facet of the Statement of Task, we now describe the experience and practice of programs and organizations that support applied industrial research in the manufacturing sector in five leading countries. We review Germany's Fraunhofer-Gesellschaft, Canada's Industrial Research Assistance Program (IRAP), and Taiwan's Industrial Technology Research Institute (ITRI). In addition, we describe two recent initiatives in the United Kingdom and France to reorient, rebrand, and reinvigorate existing applied research organizations.[1]

This chapter, supplemented by detailed historical and programmatic analysis found in Appendix A, documents the significant scope and commitment of their efforts and provides an important comparative perspective for understanding U.S. policies and programs for supporting applied research and manufacturing. Some elements of these foreign programs may be relevant to the Manufacturing Extension Partnership (MEP), and some to other U.S. programs. Still others may not be applicable to the United States given that national innovation systems differ in significant ways. Rather than make direct comparisons, the objective of this review of foreign programs is to learn from international practice, which in itself is shaping the competitive environment of U.S. firms. A number of these institutions appear to deploy exemplary practices that could be considered in establishing or reinforcing U.S. partnerships for applied research and manufacturing. This chapter concludes by listing what the committee believes are their exemplary practices.

All of the foreign organizations surveyed here have the same basic mission, which is to translate knowledge from the research base into commercial

[1]These leading national programs are not an exclusive list. There are other national examples, such as the Korea Institute for Advancement of Technology (KIAT) and Brazil's FINEP. The selection of programs in this review is based on the committee's interest and knowledge about these programs.

products and industrial processes utilized by the private sector. All of them grapple with the challenges of globalization and seek to ensure that knowledge generated by national research institutions is applied domestically, giving rise to domestic value added and jobs.

THE RESEARCH ORGANIZATIONS

The organizations reviewed in this chapter include some of the finest applied research institutions in the world. Germany's Fraunhofer-Gesellschaft and Taiwan's ITRI have won widespread acclaim for their achievements and their methods have long been the subject of extensive study. The Catapult and Carnot groupings, although bearing designations that have been established only recently, include storied research institutes of strong reputation in the science and technology community, such as INRIA of France and the UK's Advanced Manufacturing Research Centre. Canada' IRAP has been recognized as "one of the best managed and effective government programs to facilitate R&D and commercialization by small and medium enterprises (SMEs)."[2]

Germany's Fraunhofer-Gesellschaft

The Fraunhofer-Gesellschaft (Fraunhofer Society) is a not-for-profit association based in Germany comprised of a headquarters organization, 60 research institutes, and 20 other domestic research centers, and international research centers in the United States, Chile, Austria, Portugal, and Italy. Its mission is to perform contract research for public organizations and companies that transforms basic research from university and nonuniversity laboratories into commercial products and industrial processes. Each Fraunhofer institute specializes in a particular technological competency and is paired with a German university that has relevant scientific knowledge. The institutes perform research and development (R&D) on their premises utilizing their own professional staff as well as undergraduate students and postdocs. Pursuant to contractual arrangements with companies, the institutes develop product prototypes and processes and enable companies to test equipment and processes on pilot manufacturing lines and simulation platforms. A byproduct of the institutes' research for industry is a continual flow of Fraunhofer alumni with theoretical knowledge and practical skills into German industry. The institutes occasionally spin off small start-up companies.

The Fraunhofer derives roughly one-third of its funding from so-called core funds provided by the German federal and *Länder* (state) governments, one-third from research contracts with government agencies and other public organizations, and one-third from contract research for private companies. Of

---

[2]The 2011-2012 Departmental Performance Report for the National Research Council of Canada. Access at <*http://www.nrc-cnrc.gc.ca/eng/reports/2011_2012/dpr.html*>.

the latter, some of the revenue derived from the companies represents incentive payments by state, federal, and European Union government entities.

Taiwan's Industrial Technology Research Institute

Taiwan's Industrial Technology Research Institute is a not-for-profit government research organization conducting applied industrial research for Taiwanese industry. It draws upon Taiwan's university research base as well as a highly sophisticated network of international relationships to secure basic research that it transforms into commercially relevant products and manufacturing processes. ITRI transfers technology to industry through the conduct of joint research with companies; licensing of technology; sale and auction of intellectual property; movement of ITRI personnel into jobs in the private sector; and spin-offs of start-up companies. Some of ITRI's spin-offs have been spectacularly successful, including the United Microelectronics Company (UMC) and Taiwan Semiconductor Manufacturing Corporation (TSMC), today two of the most competitive semiconductor manufacturers in the world.

France's Carnot and Britain's Catapult Initiatives

The Carnot (France) and Catapult (United Kingdom) initiatives both are recent attempts by European countries to emulate what they see as the best features of Germany's Fraunhofer-Gesellschaft to improve the flow of technology from the national research base to private industry. Both of these initiatives assign a designation ("Carnot" in France and "Catapult" in Britain) to a group of existing institutes of applied research with the hope that the designation will develop into a recognized symbol of excellence, as has been the case with the Fraunhofer name. The Carnot and Catapult institutes are also being provided with core funding from their respective governments, with the expectation that most of their revenue will be derived from contract research. Although the institutes in these programs are being accorded a considerable degree of autonomy, they are expected to align their activities with coherent national innovation strategies.

Canada's Industrial Research Assistance Program

Canada's 50-year-old Industrial Research Assistance Program (IRAP) is a government program administered by the National Research Council of Canada (NRC), which provides technical and business advisory services to small businesses and distributes financial assistance (1) to small companies to develop technologies, (2) to third-party innovation organizations supporting SMEs, and (3) to companies to facilitate the hiring of new graduates for R&D projects. Unlike the other programs surveyed, IRAP does not perform research for companies itself but provides advice and dispenses government funds in support of private innovation efforts.

IRAP connects companies to Canada's innovation research base through the Industrial Technology Advisers (ITAs) and Innovation Network Advisors (INAs), who focus on creating and improving regional innovation system relationships and working with innovation organizations to provide technological assistance to SMEs. The ITAs and INAs collaborate to determine where gaps exist in regional innovation infrastructures and where to locate research resources to address unmet research requirements.

CROSS-COUNTRY COMPARISONS

Hermann Hauser, a UK-based entrepreneur, produced a seminal survey of global institutes of applied research in 2010 that provided much of the intellectual foundation for Britain's Catapult initiative.[3] He emphasized that the role of each institute was "context dependent" based on a given country's industrial structure, innovation culture, and other research organizations, and that the variation in national contexts might inhibit transfer of a model from one country to another. As the Hauser Report made clear, the institutes tend to share certain common features, including public financial support, a training function, and an intermediate role in the innovation ecosystem between the research base and private industry.[4] But beyond such similarities, the institutes are significantly differentiated in their objectives and basic approach.

ITRI's greatest achievements have been the creation of entirely new high-tech industries that did not previously exist in Taiwan. It obtained the necessary technologies from foreign multinationals, assembled teams of researchers to master the requisite product and process technologies and spun off start-up companies from its own staff, in some cases forming entire industry supply chains. As new industries became established and competitive, ITRI curtailed its support and moved on to develop another generation of new industries. In this way, ITRI fostered Taiwan's semiconductor, LCD, computer, and photovoltaic industries and is now developing the country's electric vehicle, flexible electronics, and biotechnology industries. ITRI places large bets on what it sees as industries of the future, committing substantial resources to large-scale, long-term R&D projects. While this involves substantial risk, the payoffs for success have been enormous, helping to transform Taiwan from a developing country to a major technology-intensive economic power.

---

[3]Hauser, Hermann, 2010, *The Current and Future Role of Technology and Innovation Centres in the UK*, for Lord Mandelson, Secretary of State, Department of Business Innovation and Skills, March 2010. Hermann Hauser is a Vienna-born entrepreneur closely associated with Scotland's "Silicon Glen," and was a cofounder of Acorn Computers Ltd. He holds a PhD in physics from Cambridge University's Cavendish Laboratory.

[4]Hauser Report, 2010, op cit. pp. 8-10.

The Fraunhofer provides support to industries that already exist and that, in many cases, were well established in Germany before the Fraunhofer was founded in 1949—medium-tech industries like machinery, autos, chemicals, pharmaceuticals, and engineering. While the Fraunhofer engages in some large R&D projects, most of its work involves comparatively small projects of short duration (six months to two years) designed to produce incremental improvements in existing products and processes that will have an immediate market impact. Thousands of these projects are launched every year across an extremely broad range of industry sectors. Their cumulative effect over time has been to enable German firms to produce high-quality products very efficiently and to compete successfully with low-cost Asian manufacturers. In particular, the Fraunhofer's incremental R&D projects have supported Germany's *Mittelstand*—medium-sized companies that dominate niche industrial subsectors and are frequently characterized as the backbone of the German economy.[5] In contrast to ITRI, which phases out its support of new industries as they mature, the Fraunhofer has maintained high levels of R&D support for successful individual companies and industry sectors over decades. While it has not given rise to entirely new industries, Fraunhofer has enabled Germany to retain a leading position in its traditional industries over the very long run.

Britain's Catapult initiative seeks to capitalize on that country's superb capabilities in basic science. The concept underlying Catapult is that the initiative will pick a few thematic areas where British science is strong and where an industrial capability exists in Britain to commercialize the fruits of that science. The effort will be limited to sectors where major and rapidly growing global markets are expected to exist in the future. The role of the Catapult centers will be to serve as an intermediary between Britain's science base and British companies. Like ITRI and Fraunhofer, the Catapults will perform contract research for companies to turn scientific ideas into products and industrial processes.

The primary objective of France's Carnot initiative is to foster stronger ties and partnerships between that country's extensive public research organizations, on the one hand, and "other socio-economic actors," on the other hand, mainly private companies. The Carnot designation, as well as additional government money, is bestowed based on a given institute's demonstrated willingness and ability to engage in contract research work for industry. Underlying the Carnot initiative is the recognition that France's strengths in basic research do not necessarily translate into enhanced innovation capacity in French industry, a reflection of the historically weak links between public research organizations and the private sector.[6]

---

[5]See *Wall Street Journal*, "The Engine of Growth," June 26, 2011.
[6]Blanca Vakova, "Reconceptualizing Innovation Policy: The Case of France," *Technovation* 26, p. 453, 2006.

Although all of the organizations surveyed here cite support for small and medium enterprises) as an important part of their mission, none are as singular in their focus on SMEs as IRAP. IRAP offers technical and commercial advisory services to SMEs, helps them recruit young people, and dispenses government funds to them directly. It is currently administering a CN $80 million pilot program to support SME adaptation and absorption of digital information and communications technologies.

Physical Sites

The German, Taiwanese, British, and French organizations all feature substantial physical sites with the infrastructure necessary to conduct research and to create a manufacturing environment with onsite production lines and simulation platforms. Company personnel can work on the premises to conduct research and to prove equipment and industrial processes. The institutes are staffed by professionals with deep scientific and engineering competencies. The Fraunhofer and ITRI develop product prototypes for the industrial clients, and ITRI proudly displays scores of recent prototypes in its visitors' center in Hsinchu. These characteristics make the organizations particularly important for small and medium enterprises that cannot afford extensive R&D expenditure or equipment purchases.

Governance and Supervision

All of the organizations surveyed are subject to government supervision. Although in a strictly legal sense the Fraunhofer is not answerable to the German government, given its dependency on government core funding and contract research, it is not surprising that its policies and practices tend to align with the priorities of the German federal government and, to a lesser extent, the EU authorities. ITRI's strategic direction is determined by Taiwan's Ministry of Economic Affairs (MOEA), but the process involves extensive consultations between and among MOEA, ITRI, and a system of highly competent advisory bodies. The Technology Strategy Board, a nondepartmental public body that is largely staffed by individuals with business experience, supervises Britain's Catapult centers. The individual Catapults are run by autonomous, business-led management boards and headed by individuals with business or science backgrounds. The Carnot institutes are selected, monitored, and funded by France's National Agency for Research. Canada's IRAP is a government program managed by the National Research Council, a government organization.

Although governance structures of the research organizations surveyed vary considerably, most of them accord a substantial degree of autonomy to individual institutes within larger networks. The Fraunhofer has a complex internal organization directed by a headquarters that sets the organization's strategic priorities, allocates resources and oversees performance evaluations.

Although sometimes criticized for being too bureaucratic, the Fraunhofer lets its institutes establish their own research strategies, build relationships with industry, and spend their money with little interference from headquarters. France's Carnot initiative features an association to which each Carnot institute belongs. Carnot provides support services, branding and networking, but no common governance or direction. Britain's Catapult centers are to be governed by autonomous, business-oriented management boards under the loose supervision of the Technology Strategy Board (see Table 7-1).

Geographic Footprint

The German, French, British, and Canadian applied research organizations are comprised of a number of institutes that are distributed around the respective countries' geographic space. A number of individual institutes in these countries operate facilities in multiple locations.[7] By contrast, most of ITRI's operations are concentrated at one alpha site in Hsinchu, with only one other satellite location, ITRI South, in Tainan. ITRI managers are critical of the geographic dispersion of research locations, which they see as undermining the benefit otherwise accruing from innovation clusters. Canada's geography is so vast that even the establishment of a significant number of institutes of applied research would leave some regions underserved. Accordingly, its innovation model is characterized by mobile individual experts, the Industrial Technology Advisers, who bring their knowledge and skills to the premises of the small and medium companies that they serve.

Innovation Clusters

Most of the institutes surveyed in the paper are located in innovation clusters of technology-intensive companies and other research organizations, including universities. ITRI is located in proximity to Hsinchu Science Park, and together with two universities and numerous high-tech companies, comprise one of the most famous and successful innovation clusters in the world. The Fraunhofer is responsible for establishing innovation clusters throughout Germany and is currently pursuing 19 cluster initiatives, each of which usually involves one or more Fraunhofer institutes, a local university, and an array of small, medium, and large industrial partners. Some Carnot institutes are responsible for governance of French Competitiveness Clusters (*pôles de compétitivité*) involving collaborations between research organizations, companies, and universities. Britain's High-Value Manufacturing Catapult is

---

[7] The UK's High Value Manufacturing Catapult operates sites at Bristol, Manchester, Sheffield, Coventry, and Glasgow. The Fraunhofer for Material and Beam Technology IWS operates facilities at Dresden, Dortmund, Wroclaw in Poland and East Lansing and Plymouth in Michigan.

TABLE 7-1 Governance Structures of Leading Research Organizations

	Fraunhofer	ITRI	IRAP	Catapult	Carnot
Direct Supervisory Authority	None	Ministry of Economic Affairs	National Research Council of Canada	Technology Strategy Board	National Agency for Research
Form of Entity	Private not-for-profit association	Government-owned research institute	Government program	Various private and public organizations	Public research institutions
Geographic Footprint	Widely distributed across Germany	One main site in Hsinchu, one beta site in Tainan	Widely distributed but heavily concentrated in Quebec and Ontario	Plans for distribution across the UK	Distributed across France
Prototype Development for Companies	Yes	Yes	No	?	Yes
Pilot Lines/ Simulation Platforms on Premises	Yes	Yes	No	Yes	Yes
Company Personnel Can Work Onsite & Use Laboratory Facilities	Yes	Yes	N/A	Yes	Yes
Spin-offs	Yes	Yes	No		Yes
Number of Institutes	60	1	18*	7	34
Staff	20,000	5728	4,000	Evolving	19,000
Patents	6,131	17,569	NA	NA	880/year
Annual "Core" Government Funding (Millions of Dollars)**	723	300	90	65	79

NOTES:
*IRAP integrated into 18 institutes of the National Research Council of Canada.
**Converted to dollars at prevailing rate, December 12, 2012.

comprised of five existing research organizations, each linked with a university and serving local industrial customers.

Government Funding

The research organizations surveyed here in Taiwan, Germany, the United Kingdom, and France receive government funding that they use to equip, staff, and operate research facilities that provide services to industry.[8] They also receive revenue from public organizations pursuant to research contracts that is calculated on the basis of the cost of each project. These organizations do not disburse government funds to companies and in fact charge companies and public bodies for research services. IRAP, by contrast, distributes government funds directly to companies for the development of technologies, to third-party innovation organizations that provide services for SMEs, and to companies to cover part of the cost of hiring graduates for R&D projects.

In addition to federal funding, Fraunhofer and Carnot institutes also receive substantial funding from the state and regional governments and the European Union. One objective of the Catapult initiative in the United Kingdom is to emulate the manner in which the continental organizations have succeeded in securing EU financial support.

A comparison of funding levels underscores the extent to which the volume of government core funding for Fraunhofer and ITRI dwarfs that being made available by the British government for core funding of the Catapult Centres. Moreover, the German figure would roughly double if annual revenues derived from by the Fraunhofer from contract research for government organizations were added to the core funding figures.[9]

Talent Shortage

All countries surveyed in this paper suffer significant shortages of engineers, scientists, and technicians, notwithstanding the existence of excellent education and training programs, some of which are operated by the research organizations covered by the survey. Taiwan has experienced a major talent drain of skilled professionals to mainland China and Southeast Asia, and government forecasts indicate that the island's educational system is not turning

[8]The Fraunhofer has also been receiving funding from the European Regional Development Fund and stimulus funding from the federal and *Länder* governments, which is being used for capital expenditures on plant and equipment.

[9]Within the research networks, the proportion of each institute's income that is attributable to government expenditures varies dramatically from institute to institute. For example, the Fraunhofer Institute for Silicon Technology ISIT receives 56 percent of its income from the private sector and 44 percent from public sources (Annual Report 2011). The Fraunhofer Institute for Wind Energy and Energy System Technology IWES derives only about 17 percent of its revenue from companies (Annual Report 2011/12).

out enough graduates to support local high-tech industries.[10] In any given month in 2011, Germany averaged 92,000 vacant engineering jobs, and the manpower shortage is seen as a major threat to the country's economic model.[11] Britain and France are hampered by the fact that not enough of their young people pursue careers in science and engineering.[12] A 2012 survey reported that 66 percent of Canada's employers have trouble finding the right people for specific jobs, and shortages are reported of mechanical engineers, mechanical designers, machinists, electro-mechanics, industrial mechanics, welders, and supervisors in industrial management.[13]

Given the realities of the manpower situation, one of the most important roles played by the research organizations surveyed is the training of significant numbers of young people in the practical application of science in an industrial setting. ITRI, Fraunhofer, Britain's High Value Manufacturing Catapult and many Carnot Institutes partner with nearby universities, and students from those institutions are drawn into part-time and sometimes full-time jobs at the research institutes. The Carnot institutes are staffed with 7,700 PhD students. Graduates taking positions in ITRI and the Fraunhofer institutes typically remain for six years before departing for positions in industry for which they are, at that point, highly qualified. IRAP's Youth Employment Program provides up to $30,000 in the form of nonrepayable contributions to companies to hire recent graduates for R&D projects, giving these young people specialized career-related work experience and skills.

Intellectual Property

The Fraunhofer and ITRI have amassed large patent portfolios that are a substantial source of income. It is the normal practice of the Fraunhofer to retain the patent and other intellectual property rights when a research contract is concluded; an industry partner may receive an exclusive license from the institute, but only for the particular application that was the target of the research. While ITRI also holds the intellectual property derived from its research, it licenses technology to Taiwanese companies on more favorable terms than they could secure from foreign sources. ITRI also periodically auctions off blocks of its patents. Although Britain's Catapults received numerous comments from interested parties during their formative period to the effect that they should not seek to establish large IP portfolios, the Technology

[10]*Central News Agency*, "Taiwan Faces Serious Brain Drain Crisis," April 18, 2011. *Taiwan Economic News*, "Taiwan's Hi-Tech Manufacturers Go to the U.S. to Solicit Talents," September 13, 2010.

[11]*Spiegel Online*, "Lack of Engineers Costs Germany Economy Billions,". April 16, 2012. *Deustche Welle*, "Labor Bottleneck Squeezes Germany's solar Sector," November 23, 2010.

[12]European Commission, 2011, *ERAWATCH Country Reports 2011: France*, p. 14. James Dyson, *Ingenious Britain: Making the UK the Leading High-Tech Exporter in Europe*, 2010.

[13]*Canada Newswire*, "Skill Shortage: Survey Says Canadian Employers Have Difficulty Finding the Right Talent," October 3, 2012.

Strategy Board has adopted IP guidelines similar to those followed by the Fraunhofer.

Transparency

The organizations surveyed here operate in a research space that is close to the market—some Fraunhofer projects have a commercial impact within six months of their inception. Information about recent and ongoing contract research for companies is therefore sensitive and the research institutions are reticent to disclose details of projects undertaken for industry. The Fraunhofer institutes publish extensive details of their contract research for public organizations, but very little information about their contract research for industry.[14] ITRI publicly displays product prototypes developed in its laboratories, but will not necessarily disclose the identities of the companies that will take them to market. While the importance of the Fraunhofer as a research resource for Germany's specialized middle-sized companies, the *Mittelstand*, is widely acknowledged, little public information exists to support a systematic independent verification of that fact. While the Carnot institute CETIM (Technological Institute of Mechanics), which supports manufacturing technology research, publishes an extensive collection of examples of its work for companies—which are identified—the information nevertheless is fragmentary, anecdotal, and dated.

Other aspects of the research organizations are to some degree opaque. It is difficult to secure a full accounting of the fate of spin-off companies from the institutes, which prefer to emphasize success stories. The Fraunhofer publishes comparatively detailed financial information in its annual reports, and some of its individual institutes do likewise, but with respect to the other research organizations, detailed financial information is difficult to locate.

Small Business and the "Missing Middle"

All of the organizations surveyed have a mandate to support small businesses, a recognition of the key role played by small firms to driving innovation. A theme that is growing in importance is the "M" part of the acronym "SME," e.g., the medium-sized businesses with several hundred to several thousand employees that have the scale and financial stamina to carry an innovation across the "valley of death" to commercialization. British, French, and Taiwanese policymakers are discovering that their industrial fabric "does not have enough medium-sized companies".[15] The phenomenon extends beyond

---

[14]The Fraunhofer Institute for Silicon Technology ISIT publishes a list of its industrial customers in its annual report, but most of the other institutes do not. Fraunhofer ISIT Annual Report 2011, pp. 22-24.

[15]Comment by Institut Carnot spokesman, October 2012. A recent study by the French Centre d'Analyse Strategique suggested that France's gap with the United States in R&D intensity was due

these countries—a 2012 Australian commentary lamented the fact that the country's economy was comprised mainly of a relatively few large companies and a "very high proportion of micro-businesses, the majority of which are non-employing."[16]

Germany's *Mittelstand*, medium-sized firms that specialize in and frequently dominate obscure niche industrial subsectors, are widely viewed as the key to the country's impressive ability to compete in export markets.[17] The *Mittelstand* are typically manufacturing and engineering firms based in small towns and rural areas with a deep sense of social obligation to their own workers and their local communities. They tend to resist offshoring and layoffs in economic downturns. The Fraunhofer institutes are cited as comprising "external, very well equipped research departments of the *Mittelstand* companies."[18] To the extent that this view is accurate, it validates the widely held view of the Fraunhofer-Gesellschaft as one of the most successful research organizations in the world. It does not necessarily follow, however, that programs that consciously seek to emulate the Fraunhofer model will be able to foster new communities of *Mittelstand*-like firms outside of Germany. The *Mittelstand's* main strengths arise less from size or methods than from core values and industrial traditions handed down over the generations that cannot readily be replicated by government programs and public expenditures alone.

BEST PRACTICES

The research organizations surveyed here frequently cite certain practices, policies, methods, and approaches as key to the way in which they operate. While these may or may not constitute "best practices," give the success of organizations like ITRI and Fraunhofer, their techniques command attention. The fact that some of these practices are found at more than one of the organizations studied is partially a reflection of "borrowing" from other innovation models.

---

to two main factors: (1) patterns of French industrial specialization, and (2) a shortage in France of R&D-intensive enterprises of intermediary size—Enterprises de Taille Intermediare (ETI)—with between 250 and 5,000 employees and either less than 1.5 billion Euro turnover or a balance sheet under 2 billion Euros. CAS, 2010, *R&D et Structure des Enterprises, une comparaison France/Etats-Unis.* Cited in European Commission, *Erawatch Country Reports 2011: France.* p. 7.

[16]*The Conversation* "The Missing Middle: What Australia Could Learn from Germany," May 19, 2012.

[17]*The Economist*, "German Business: A Machine Running Smoothly," February 3, 2011. *The Wall Street Journal*, "The Engine of Growth," June 26, 2011. Berad Venohr and Klaus E. Meyer, *The German Miracle Keeps on Running: How Germany's Hidden Champions Stay Ahead in the Global Economy*, Berlin: Berlin School of Economics, May 2007.

[18]Christian Hamburg, *Structure and Dynamic of the German Mittelstand.* Heidelberg and New York: Physica-Verlag, 1999, pp. 58-59.

Maintaining Mission Focus

The Fraunhofer and ITRI were founded decades ago to perform applied research of commercial value to industry, and they have remained relevant and successful by maintaining a disciplined focus on their original mission.[19] Within the scientific community, basic research is often viewed as more interesting than applied research, and examples exist where institutes set up to perform applied research for industry have drifted into other activities.[20] That has not happened in Germany or Taiwan. Both Fraunhofer and ITRI have developed a strong sense of institutional pride in the fact that they must continually demonstrate their own relevance by securing commercial research contracts from companies.

International Technology Induction

All of the countries surveyed share a common aspect—that is, that most of the basic scientific discovery that occurs globally takes place outside their own borders. Accordingly, while the research organizations have a mandate to serve as a bridge between the domestic research base and industry, their comparative effectiveness will increasingly turn on their ability to engage foreign research organizations and technology-intensive companies. The Fraunhofer Institute for Material and Beam Technology IWS, which maintains a relationship with the U.S.-based Fraunhofer Center for Coatings and Laser Applications, explains that this relationship enables it to identify emerging trends, and "the research and development work performed in the United States generates additional know-how and competencies, which benefit the project acquisition in the German and European markets."[21]

ITRI is "arguably the most capable institution of its kind in the world in scanning the global technological horizon for developments of interest in Taiwanese industry, and executing the steps required to import the technology—either under license or joint development—and then absorbing and adopting the

---

[19]The Fruanhofer ousted its first President, Nobel Prize winner Walther Gerlach, because he was a basic scientist who had little regard for basic research. Christine Egger, "Nachdenken im Auftrag: Eine Geschishte der Fraunhofer Gesellschaft,". *Aventinus Bavaria*, October 22, 2010.

[20]The Hauser Report noted (p. 10) that Korea's Electronics and Communications Research Institute (ETRI) had strayed away from applied and into fundamental research and was competing as well as collaborating with universities. Taiwan's Development Center for Biotechnology (DCB) was set up to function as an ITRI-style intermediary organization for applied research in the life sciences, but moved toward a focus on basic R&D, one factor underlying Taiwan's difficulty in establishing a viable biotechnology industry. Andrea Yung, "A Long Haul for Biotech," *Topics*,. October 2009. The UK's Faraday Centres were created in the early 1990s to perform a Fraunhofer-like bridging function between Britain's research base and its private sector, but by the time they were operational the first five pilot sites for the Faraday project had become "merely a new way of training postgraduate engineers." *New Scientist*, "German Lessons, Half Learnt," November 21, 1992.

[21]Fraunhofer IWS Annual Report, 2011, p. 117.

technology for Taiwanese firms to use."[22] ITRI has an extensive array of joint R&D projects with world-class companies and research organizations, including Microsoft, IBM, Intel, Lawrence Berkeley National Laboratory, Novartis, Corning, and Applied Materials. Technology acquired from these collaborations is transformed in ITRI's laboratories into prototype products and manufacturing processes that are transferred to Taiwanese companies for commercial application.

Some of ITRI's international technology transfers have had epochal effects. ITRI established an R&D partnership with RCA in 1975 pursuant to which RCA agreed to transfer technology and know-how to ITRI for CMOS semiconductor design, process, and testing technology and to train ITRI engineers. RCA supplied CMOS technology and trained a group of 37 Taiwanese students—the so-called RCA 37—in the fundamentals of semiconductor manufacturing. Out of the RCA 37, "virtually the entire senior echelons of the subsequent semiconductor industry in Taiwan [were] formed."[23] The ITRI trainees returned to Taiwan and built a pilot facility for manufacturing integrated circuits, serving as a practical confirmation for the trainees, whose "hidden knowledge" gained from RCA was passed on to other trainees that had not been chosen for the original mission. Ding-Yuan Yang, who led the RCA 37, recalled later that—

> *Now looking back, setting up an IC demonstration plant in a research institution was something truly unique in the world. Normally a private plant is set up, and they transfer technologies by themselves. Using the power of the government to establish a demonstration plant in a research institution, to plant it as a seed to spread it out, was a very unique method.*[24]

In retrospect, the technology induction from RCA to Taiwan was a Promethean moment that made possible not only the development of a world-class semiconductor industry in Taiwan but also follow-on industries such as displays, computers, photovoltaics, and telecommunications. The process continues today as ITRI acquires technology abroad and develops products and proves manufacturing processes on demonstration plants within the institute for dissemination to companies entering emerging industries such as flexible displays, green buildings, and cloud computing.

---

[22]John A. Matthews and Dong-Sung Chu, *Tiger Technology: Thee Creation of a Semiconductor Industry in East Asia*, Cambridge: Cambridge University Press, 2000.
[23]John A. Matthews, "The Hsinchu Model: Collective Efficiency, Increasing Returns, and Higher-Order Capabilities in the Hsinchu-Based Industry Park, Taiwan," Keynote Address, Chinese Society for the Management of Technology: 20th Anniversary Conference, December 10, 2010.
[24]Interview with Ding-Yuan Yang. 2011. "Taiwanese IT Pioneers". D.Y. (Ding-Yuan) Yang. Recorded February 23, 2011. Computer History Museum: 2011.

Spin-offs (ITRI, Fraunhofer, Carnot)

Many public research organizations around the world give rise to new companies as employees depart to commercialize technologies they have helped to develop. ITRI, Fraunhofer, and the Carnot institutes have formalized this process, endorsing the concept of spin-offs as a mechanism for technology transfer to industry and establishing institutional structures and programs to enhance their spin-offs' prospects for success in the marketplace. ITRI has established a venture capital subsidiary, the Industrial Technology Investment Corporation, which screens proposed spin-offs and takes small equity stakes in roughly one-third of them, with the expectation of selling that interest at some point in the future. The Fraunhofer manages spin-offs through an internal division, Fraunhofer Ventures, which takes an equity stake of up to 25 percent in about half of the spin-offs. ITRI, Fraunhofer, and the Carnot institutes provide incubation facilities and services (legal, accounting, business counseling, technology support) to their spin-offs.

ITRI spin-offs are noteworthy because of their scale and market impact. The United Microelectronics Corporation (UMC) involved a spinoff of 31 people from ITRI microelectronics organization as well as a substantial portion of its semiconductor production equipment.[25] In 1986-1987, ITRI created Taiwan Semiconductor Manufacturing Corporation (TSMC) involving spin-off of 130 professionals (including Chairman Morris Chang), semiconductor process technology, and a complete manufacturing plant for integrated circuits. Winbond Electronics Corporation, created in 1987, involved spin-off of about 200 professionals from ITRI. These spin-offs dwarf those of the other organizations studied here in scale, degree of risk, and actual impact in the market, which was to create a national semiconductor industry and to project Taiwan into the ranks of the world leaders in the field within the space of a single generation. Some of ITRI's smaller spin-offs have quickly developed into midsize successful firms. Phison Electronics Corp., a maker of flash memory systems, began as an ITRI spin-off with 12 employees in 2000 and by 2012 had become a company with 560 employees and annual revenues of over $1 billion.[26]

ITRI has helped its spin-offs integrate into global supply chains. ITRI and the Taiwanese government arranged a joint venture with Philips when they created TSMC in 1987, the world's first pure-play semiconductor foundry. This arrangement gave the new company one of the world's biggest semiconductor producers as its first customer, at a time when the greatest perceived risk facing TSMC was the uncertainty of demand for its services. The fact that Philips was sourcing from TSMC soon drew the attention of companies like Intel and Texas

---

[25]ITRI's microelectronics unit was the Electronic Research and Service Organization (ERSO), which now operates as the Electronics and Optoelectronic Research Laboratory.

[26]Phison, *Corporate Overview*, 2012.

Instruments, who became customers. Philips also deployed its IP umbrella over the new company, using a web of cross-licensing agreements transferred to the new company, protecting it from lawsuits.[27] More recently, ITRI introduced its spin-off Phison Electronics Corp to Toshiba and on the basis of ITRI's presentation, Toshiba invested in Phison and became an important customer.[28]

Creating Industry Chains

ITRI has established "technology integration centers," which coordinate and integrate research and technology from the institutes' specialized core laboratories. The original rationale for the centers was to promote silo breaking and multidisciplinary collaboration among ITRI's thematic core laboratories. But the centers have evolved into mission-oriented organizations that seek to establish indigenous capacity for the integrated production of advanced technologies throughout the creation of entire industry supply chains. The centers have their own budgets and can commission research by ITRI's core laboratories. They oversee the development of capabilities with respect to production equipment, materials, components, manufacturing processes, testing, and certification. The director of ITRI's Display Technology Center, Dr. John Chen, explained in 2012 that—

> *The biggest strength of ITRI is the multidisciplinary cooperation. We create a complete manufacturing supply chain in its early stages. That is the secret. Then you can scale it up, you can have a complete supply chain for the industry. So DTC does not just work with display companies, but also materials suppliers and equipment makers.*[29]

During the past three decades, as ITRI was creating domestic industry supply chains through spin-offs and technology transfer, Britain was witnessing the partial or complete disaggregation of many of its long-standing industrial chains as the application of orthodox-free market policies by successive Conservative and New Labour governments forced a punishing restructuring on the country's manufacturing sector. "When the big factories closed, the supporting infrastructure decayed. Import dependency is the legacy...British manufacturing has downsized into workshops as it loses its industrial districts...on the available evidence, it is no longer possible to construct a large, heavy engineering product with high British content."[30] The contrast with

---

[27]Tommy Shih, "Scrutinizing an Economic Model: The Taiwanese Semiconductor Industry Revisited," Uppsala University, 2009, p. 14.
[28]Interview with K. S. Pua, Phison Electronics Corp., Hsinchu, Taiwan, February 14, 2012.
[29]Interview with DTC Director John Chen, Hsinchu, Taiwan, February 14, 2012.
[30]Centre for Research on Socio-Cultural Change, *Rebalancing the Economy (or Buyer's Remorse)*, Working Paper No. 87, 2011, pp. 29-33.

Taiwan is stark. In 2011, Barry Lam, former head of Kinpo electronics and founder of Quanta Computer, observed the following about the Taiwanese IT sector:

> *The supply chain is very complete in Taiwan. We have semiconductor foundries here in Taiwan. We have good design houses. We have many good assembly houses. We also have many components, such as CD-ROM drive. We gave up on the hard drive business at the time, so we didn't catch the momentum when the industry was blooming. Assembly was done mostly in Thailand or Southeast Asia. We pretty much know how to make most components in mass production. So from components...that's why even when we make resistors and capacitors, our price is still very cheap. For components, from PCB boards to chips, we can make it all in Taiwan. So, we can even complete the design in Taipei. And, why? Because all the vendors are concentrated in Taipei. Taipei is not big, so it's easy to deal with everything in Taipei and everything can be done here. This is good, isn't it?*[31]

The Fraunhofer has contributed to the survival and flourishing of industrial chains in established German industries like autos, machinery, and chemicals. Many of its industrial customers are small and medium firms that produce intermediate industrial products for incorporation into end products like construction equipment and motor vehicles. The technological support of Fraunhofer institutes enables these SMEs to remain competitive with their Asian counterparts, ensuring that all or most of the industry chains remain in Germany or other European countries.

Engaging the Private Sector

In Britain and France, a long-standing challenge has been the difficulty experienced by companies, especially small ones, seeking to ascertain which research organization could assist with a specific technological concern. In 2011, Dr. Tim Bradshaw, the head of Enterprise and Innovation at the Confederation of British Industry, told Parliament that no one "really knows" what potential support was available from the country's welter of hundreds of technology intermediary organizations, and that "SMEs, in particular, do not know where the best facilities are that they could go and tap into."[32] In France,

---

[31]Interview with Barry Lam, "Taiwanese IT Pioneers: Barry (Pak-Lee) Lam," recorded March 2, 2011 (Computer History Museum, 2011), p. 29.

[32]Testimony of Dr. Tim Bradshaw, House of Commons, Science, and Technology Committee, Technology and Innovation Centre Inquiry, Oral Evidence, December 15, 2010, Q27.

prior to the Carnot initiative, companies confronted a "vast, ossified public science and research system...and a lack of synergy between public research and industry."[33] IRAP, Fraunhofer, and ITRI have been successful in forging close links with businesses, albeit through differing institutional mechanisms.

The key to IRAP's success is a small, elite cadre of about 230 Industrial Technology Advisors, which deliver IRAP services and funding to individual companies across Canada. ITAs typically are veteran R&D professionals with extensive industrial experience and strong educational backgrounds.[34] The ITAs provide companies with R&D project advice, competitive technical intelligence, funding, networking, and technology and business advice. In one instance involving Boreal Laser, Inc., a maker of laser-based trace gas analyzers that detect hazardous gases, support from an IRAP ITA enabled the company to double its sales, move from an analog to a digital platform (improving flexibility and profits), establish clients in 45 countries, grow from 7 to 15 employees, develop a relationship with ExxonMobil worth over $3.5 million and create a community of suppliers with the ability to develop laser technology for other companies.[35] Britain and France offer advisory services to small businesses that resemble IRAP's services, although they are not part of the Catapult and Carnot initiatives.[36]

Fraunhofer institutes link the university system with private industry in parallel structures of thematic "research units" and "business units." Contacts with industry are made by Fraunhofer business units, which play a role somewhat analogous to IRAP's ITAs as a bridge between companies and the research base. The business units have business expertise as well as knowledge of the potential applications and markets for the particular technologies in which they specialize. When a company asks a Fraunhofer institute for research assistance, the business unit is the point of contact and the entry point into the system. The business unit evaluates the potential cost of research, its business feasibility, and the terms pursuant to which the parties will collaborate. The business units determine which Fraunhofer research units should perform the developmental research with respect to materials, equipment, processes, and so on. The business unit remains the point of contact with the client, while research

---

[33]Emmanuel Muller, Andrea Zenker, and Jean-Marie Heraud, "France: Innovation System and Innovation Policy," in Fraunhfer ISI, GIGA and Georgia Tech New Challenges to Germany in the Innovation Competition, August 2008, p. 148.

[34]Eighty percent of ITAs have specialized industrial experience, 75 percent of ITAs have masters or PhD education, 45 percent have run their own R&D facility or occupied a leadership position in an R&D facility, 34 percent have been entrepreneurs, and 41 percent have worked in institutions of higher learning. IRAP, "The ITA Advantage," *<http://www.nrc-cnrc.gc.cn/eng/irap/uhwt/advisors.html>*.

[35]National Research Council of Canada, 2012, "Local, Digital, and Growth-Oriented," *<http://www.nrc-cnrc.gc.ca/eng/irap/success/2012/boreal_laser_inc_trace_gas_analyzer.html>*.

[36]In the UK, the Manufacturing Advisory Service (MAS) is a government program that offers business and technological advice to SMEs. In France, regional technology transfer centers (*Centres Regionaux d'Innovation et de Transfert de Technologie)* known by the acronym CRITT, supports innovation by SMEs through specific advice by *Conseilleurs en Developpement Technologique*.

unit or units are the point of contact for the work with respect to university faculty laboratories and students.

The Fraunhofer has created formidable internal communications structures that enable staff to ascertain what competencies exist and what projects are under way throughout the Fraunhofer system. As a result, when confronting a technological challenge posed by a company, various Fraunhofer units can be assembled from one or more Fraunhofer institutes on an ad hoc basis to attack the problem, if need be on a multidisciplinary basis. Thus, a company that makes a contact with any Fraunhofer institutes engages the entire network and an appropriate research team quickly assembled. This process is facilitated by the fact that German scientific fields are traditionally highly networked and scientists are trained to work in concert on a multidisciplinary basis to achieve a given objective. A Fraunhofer scientist recently observed that establishment of collaboration networks is extremely difficult, but that once established, networks constitute a "huge competitive advantage."[37]

ITRI's interface with companies is facilitated by the fact that many of its research customers are colocated with ITRI in the Hsinchu innovation cluster. In addition, many company executives are ITRI alumni, and "most of them graduated from the same university, took the same classes, were taught by the same professors, and had similar work experience at ITRI."[38] The interface with industry is enhanced by ITRI's "technology alliances," which convene interested companies on a thematic basis for technology development and diffusion.[39]

The primary purpose of the Carnot and Catapult initiatives is to improve the flow of technology from the national research base to private industry. The Carnot institutes typically enjoy research alliances with domestic universities and domestic and foreign research organizations. Like the Fraunhofer, in eight thematic areas, the Carnot institutes have formed "Carnot alliances" of research units with relevant multidisciplinary competencies that can be brought to bear on behalf of industrial customers. Britain's Catapult centers have absorbed a number of "Knowledge Transfer Networks" (KTNs), networks established to connect companies in need of technology with research organizations capable of delivering it.

---

[37]Interview with Fraunhofer Institute for Manufacturing Engineering, Stuttgart, Germany, June 14, 2012.

[38]Anna Lee Saxenian, "Taiwan's Hsinchu Region: Imitator and Partner for Silicon Valley," Stanford Institute for Economic Policy: Research Discussion Paper No. 00-44, June 16, 2001.

[39]In 2008, ITRI formed the "A+ Machine Tool Technology Alliance" with six domestic manufacturers of numerically controlled machine tools to introduce "advanced techniques from Germany," to improve the accuracy of Taiwanese machine tools down to the submicron level. *Taiwan Economic News*, "ITRI Forms A+ Machine Tool Technology Alliance,". July 8, 2008.

Core Funding

One of Hermann Hauser's key findings was that the foreign applied research organizations which he surveyed received between 10 percent and 70 percent of their funding in the form of "core" government funds not linked to the performance of research contracts. He observed that "core funding from the public sector appears to be most in need at start-up for infrastructure and capacity building, but studies have shown a need for continued core funding for three functions: strategic high risk research of medium to long term duration; competence development; and the acquisition and maintenance of large-scale facilities and specialist equipment."[40] The Fraunhofer's steepest growth trajectory occurred after its core funding was locked in by a federal-*Länder* pact in the early 1970s.[41] The Carnot and Catapult initiatives in France and the United Kingdom involve provision of core government funding to research organizations in a conscious effort to replicate the Fraunhofer model.

Proportional Matching of Industry Funding (Fraunhofer, Carnot)

The Fraunhofer institutes' headquarters allocates government core funding to individual institutes according to a formula that rewards each organization's success in contracting with the private sector for research—the more contract research revenue from industry, the more funding the organization receives. France's Carnot institutes operate according to a similar principle, with additional incentives which reward contract research for SMEs. The net effect of these arrangements is to direct government funds toward thematic areas regarded as important by industry. A potential shortcoming of this model is that because it is driven by the needs and demands of companies and industries that already exist, it is unlikely to foster transformational research in thematic areas where no companies yet exist.

Branding

In Germany, France, and Britain, a common branding of institutes of applied research is being used, unifying organizations that are somewhat disparate under a designation intended as a seal of excellence. For Fraunhofer, the emphasis on building a brand has worked very well, and its name is closely associated with the high quality of German products and engineering. Hermann Hauser observed in his 2010 report that "it is widely believed in Germany that the renewal and greater use of the Fraunhofer brand in recent years has made a difference and enabled the network to promote itself and compete effectively

---

[40]Hauser, *Technology and Innovation Centres*, 2010, op. cit. p. 11.

[41]Marksu Winnes and Uwe Schimack, *National Report: Federal Republic of Germany*, Institute for the Study of Societies, TSER Project No. SOE1-CT96-1036, May 1999, p. 14, 45.

both nationally and internationally."[42] The Carnot and Catapult initiatives represent attempts to establish comparably strong brands in France and Britain, respectively.

Dual System

The term "dual system" refers to a form of education practiced in Germany and some other European countries in which students concurrently receive vocational education and apprenticeships in companies. But the term could equally be used to characterize any curriculum that combines a course of theoretical study with practical experience and training in an industrial environment. In Germany, where institutions of higher learning have emphasized technical and scientific training as part of their curricula for many generations, the Fraunhofer embodies the dual system in practice at the pinnacle of the educational system, with PhD candidates and postdocs who are pursuing courses of scientific study also learning to function in highly sophisticated, real-world industrial settings. The thousands of Fraunhofer alumni comprise a scientific and engineering elite capable of applying their knowledge for the benefit of industry. While Britain and France have long bemoaned the disconnect between academia and industry, Germany has produced generations of scientists and engineers who are at home in both worlds.

External Evaluations (Carnot, Fraunhofer)

France has pioneered the practice of requiring periodic evaluations of all public research organizations (as well as programs, universities, and schemes) by independent committees. The evaluations are conducted under the auspices of the Agency for the Evaluation of Research and Higher Education. The visiting committees which conduct the evaluations are comprised of individuals with relevant expertise who have no connection to the institution and in a majority of cases are not French. Fraunhofer institutes are also subject to periodic performance audits by external experts drawn from the university and business communities.

Political Nonintervention

The most successful foreign applied research institutes, Fraunhofer and ITRI, have benefited substantially from a relative absence of political interference with their activities. Although the Fraunhofer has occasionally been assigned special missions by the German government, it has not confronted periodic demands from the government or Germany's main political parties that it be shut down, downsized, subjected to budget gyrations, or assigned a

---

[42]Hauser, *Technology and Innovation Centres*, 2010, op. cit. p. 13.

fundamentally different role in the German research ecosystem. Its government founding has remained on a stable basis since a pact on research spending was reached between the federal government and the *Länder* in the early 1970s.[43] In Taiwan, ITRI benefited from the longstanding tradition of the Kuomintang Party—which held power from the time of ITRI's funding until 2000—of relying on "scientific" government planning, according technocrats a large degree of freedom from party and military control.[44] In the United Kingdom, government funding of applied research has traditionally been a contentious issue, although at present both New Labour and the coalition government support the Catapult initiative.

LESSONS FOR MEP

An overview comparison of MEP with successful foreign programs promoting applied research, such as Germany's Fraunhofer-Gesellschaft and Taiwan's ITRI, raises the threshold question as to whether there are applicable lessons for MEP from the foreign efforts, given that foreign programs are so divergent from MEP in scale, level of public funding, and character.

The origins and operations of these programs, explored in detail in Appendix 1, suggests that in fact there are two distinct areas in which these programs have potential lessons for MEP:

1. The addition of innovation services. Recent MEP efforts to add innovation and globalization to MEP services have been noted in the Findings of this report, along with issues related to their implementation. These or similar initiatives are very likely to be part of the MEP portfolio in the future, and therefore insights drawn from programs with similar components can be relevant. Programs such as Fraunhofer and ITRI appear to be among the best global examples of government support for innovation in manufacturing. As MEP moves forward to help U.S. SMEs enhance their international competitiveness—including the creation of the requisite workforces, supply chains, global outreach, and innovation clusters supporting local manufacturing employment—examples and lessons drawn from leading programs internationally should be directly relevant.
2. Further improving efficiency services. While MEP has over 20 years of experience in deploying lean production techniques to meet the needs of small U.S. manufacturers, leading foreign programs like Fraunhofer and ITRI have provided similar services. Given the scale of their efforts

---

[43]Marksu Winnes and Uwe Schimack, *National Report: Federal Republic of Germany*, Institute for the Study of Societies, TSER Project No. SOE1-CT96-1036, May 1999, p. 14, 45.
[44]Robert Wade, *Governing the Market: Economic Theory and the Role of Government in East Asian Industrialization*, Princeton and Oxford: Princeton University Press, 1990, pp. 247-48.

and the long history of successful activity, their experiences and approach, including the scale of effort, are relevant to MEP.

The extent of the differences between MEP and leading international programs is underscored by consideration of different kinds of relevant capabilities (see Table 7-2). There are substantial differences in current capabilities between the foreign programs and MEP (we note that while MEP has been adding innovation-oriented services for only 6 years, Fraunhofer has operated for over 60 years, so their limitations at MEP are to some degree to be expected).

Although at present MEP does not possess the assets, budget, or expertise to engage in many of the activities that constitute core functions of organizations like the Fraunhofer and ITRI, those activities are nevertheless relevant points of reference for MEP. MEP's mission now includes promotion of innovation in manufacturing by SMEs, and the methods employed by the most successful organizations seeking to foster innovation cannot simply be dismissed. If MEP cannot replicate such functions in their entirety as they are applied abroad, it may be able to capture some aspects of those functions either through its own internal processes as through expanded collaborations with other government programs, universities, and research organizations—all of which are important aspects of ITRI's and Fraunhofer's operations. Finally, organizations like the Fraunhofer and ITRI are of abiding relevance to programs like MEP because they are indicative of what the best U.S. competitors are doing to establish and maintain a competitive edge in manufacturing—whether or not U.S. policymakers approve of, or are prepared to replicate, such efforts.

The MEP's traditional program of promoting lean manufacturing involves the application of long-established production protocols such as Six Sigma to existing production operations in order to foster greater efficiency and larger markets and higher return. Fraunhofer and ITRI also offer technical support, management training, and advice with respect to manufacturing processes as part of their larger function of promoting innovation. However, in addition, Fraunhofer and ITRI projects addressing manufacturing technology also emphasize the development, introduction, and diffusion of new processes and materials including proprietary technologies. Lean manufacturing programs may enable SMEs to close the efficiency gap or even catch up with their global competitors; the Fraunhofer and ITRI programs, when successful, propel their participating companies into positions of leadership. In other words, while the original MEP mission sought to help SMEs apply lessons learned from Toyota in the 1970s and 1980s, the Fraunhofer has sought to enable German companies to meet the challenge from Toyota for market share in the 21st century. IRAP has steadily moved in this direction as well, with an increasing emphasis on stimulating "wealth creation by supporting technological innovation." To push new technologies toward the market, IRAP provides not only technical advice but also seed funding for promising projects.

TABLE 7-2 Comparing MEP with ITRI and the Fraunhofer Gesellschaft

	MEP	Fraunhofer/ITRI
Mission	Help SMEs catch up with global efficiency levels in manufacturing	Help SMEs advance the global state-of-the-art in manufacturing *and* help SMEs reach global efficiency levels in manufacturing
University Partnerships	About one-third centers	Always
Foreign Networks	No	Yes (extensive)
Sophisticated Internal Networks	No	Yes
Large "Bank" of Proprietary R&D	No	Yes
Spin-offs	No	Yes (including ITRI-TSMC)
Internal Venture Capital Fund	No	Yes
Prototyping Products	In some cases, e.g., MAGNET	Yes
Demonstration/Pilot Facilities	No	Yes
Technology Focus	Generic (exceptions: MAGNET, TechSolve)	Thematic/Sectoral
Linkage with National Technology Strategy	TBD	Yes
Leadership Role in Industry Rationalization	No	Yes

MEP provides expertise and advice to individual companies with respect to manufacturing technologies based on observation and study of those companies' existing production operations. ITRI and the Fraunhofer perform similar services, but they also establish demonstration manufacturing operations on their own premises to develop and test processes, machines, and materials that have often not been used previously, for the benefit of multiple industrial collaborators. Success in an MEP project is measured in terms of improvements to the client's bottom line, increased sales, new products, and leveraged investment. Success in the foreign programs has been manifested in such terms and in the rapid dissemination of cutting-edge manufacturing processes throughout an entire domestic industry, or the creation of entirely new industries (Taiwan), or the maintenance of a global competitive edge in manufacturing in traditional industries (Germany).

The establishment of organic manufacturing demonstrator capability at MEP's centers may be beyond the reach of the program for the foreseeable future, given the current model and the existing resource constraints. However, the promotion of demonstrator projects and facilities through collaborations with universities and interested companies (such as the Timken-University of Akron collaboration) could benefit from MEP support.

MEP's Next Generation Strategy, launched in 2007, seeks to "help firms grow by becoming more competitive and innovative." In that sense, its mission has moved closer to the goals of ITRI, Fraunhofer, and IRAP, all of which focus on innovation rather than improvements in process efficiency. As a result, MEP is now grappling with some of the same challenges confronting the leading foreign organizations, which include connecting manufacturers' needs with technology sources, technology scouting, supply chain issues, and the development and commercialization of products. Accordingly, MEP could benefit by reference to some of the practices and strengths of the foreign programs.

SMEs, whether located in the United States, Germany, Taiwan, or Canada, typically do not have much organic R&D capability and rely on external sources for most of the technologies which they may seek to adapt for commercial purposes. The principal sources of relevant research are universities, public research organizations, and technology-intensive major companies. The well-established foreign research organizations surveyed here (ITRI, IRAP, Fraunhofer-Gesellschaft) are successful largely because they have developed institutional arrangements for systematically linking the research base (national and international) to domestic companies, especially their SMEs, on their own premises. In addition, organizations like Fraunhofer and ITRI possess their own internal research infrastructure, expertise, and intellectual property. "The research facilities of Fraunhofer serve as external, very well equipped research departments at the Mittelstand companies."

Close Ties to Universities

Although the United States arguably enjoys the world's best system of research universities, the linkages between ITRI and Fraunhofer and the universities in Taiwan and Germany, respectively, are far more pervasive and intimate than the ties between MEP and U.S. universities. Some 18 MEP centers have a partnership with a U.S. university, whereas all Fraunhofer institutes are paired with one and sometimes multiple academic research institutions. ITRI is virtually colocated with two of Taiwan's leading research universities, Chiao Tung and Tsing Hua, and the personal and professional ties between the three institutions are so close that the boundaries between them are sometimes indistinct. And with respect to U.S. universities, an interesting question is presented as to which organizations' research ties are closer—MEP, on the one hand, or ITRI-Fraunhofer, on the other hand, each of which has developed strong research collaborations with leading U.S. academic research institutions.

MEP's Innovation Engineering initiative seeks to move in the direction of Fraunhofer-ITRI practice, which emphasizes the development of new manufacturing processes and materials rather than training companies to apply successful, but traditional, methods. This approach involves significant challenges. For example, staff at MEP centers are usually trained in lean manufacturing, not innovation, leading some staff directors to seek to replace current staff with new staff with different skills. Moreover, innovation in manufacturing takes places over a larger time frame than implementation of lean manufacturing. The Fraunhofer and ITRI address such challenges by "importing" expertise as need from university research departments and foreign companies (firms like Corning and Kodak have shared technology and know-how with ITRI projects). The involvement of university graduate and postdoc students in research projects (some of whom contribute through pursuit of relevant thesis topics) eventually generates a flow of expert graduates both in the research institutes and to companies. Some MEP centers are already involved in executive education programs with respect to manufacturing management; the programs could arguably benefit by observing the close integration of academic research, engineering, and business skill sets that occurs with respect to manufacturing innovation at the Fraunhofer and ITRI.

To be specific about the degree of integration between Fraunhofer institutes and German research universities, consider the relationship between the Fraunhofer Institute for Laser Technology ILT and RWTH Aachen, one of Germany's leading research universities. Fraunhofer ILT works closely with four RWTH university chairs: Laser Technology, Technology of Optical Systems, Laser Physics, and Nonlinear Dynamics of Laser Processing. On the one hand, the Fraunhofer ILT's "knowledge of current industrial and scientific requirements in the optical technologies flow directly into the planning of the [RWTH] curriculum." On the other hand, undergraduates and postgraduates "can put their theoretical knowledge into practice through project work at the three chairs and the Fraunhofer ILT." University courses at RWTH are jointly drawn up by the university and Fraunhofer ILT. Professor Thomas Taubner, who oversees a research working group at Fraunhofer ILT studying nano-optic concepts using lasers, also holds a junior professorship at RWTH Aachen for "Nano-Optics and Metamaterials," where he supervises research by students in physics involving new imaging techniques with nanometric spatial resolution.

The director of each Fraunhofer institute is also a faculty member at the institute's university partner, and typically identifies the most promising students to recommend for positions at the Fraunhofer. One Fraunhofer scientist who also serves on a university faculty observes that "my marketing effort is my lecture. I convince students to work [at Fraunhofer] 15-20 hours per week. I'm positioned to find the best. They can stay until they get their Ph.D.s (5-8 years) and use our excellent network to find jobs."

International Outreach

Internationally, ITRI and Fraunhofer have developed multiple research collaborations with high-tech multinationals and public research organizations which generate a steady flow of cutting-edge technology to domestic companies in Taiwan and Germany. The benefits of these collaborations are sufficiently evident to governments in the two countries that public funds may be used to support research projects in other countries. MEP does not have a network of international research collaborations and is not in a comparable position to serve as a conduit for new foreign technologies to U.S. SMEs. A question is presented as to how MEP could leverage existing U.S. government programs and resources to enhance its access to international technologies relevant to small and medium U.S. manufacturers.

Most MEP centers report that they are focusing more on multiple projects with repeat clients, a practice which is inconsistent with the National Institute of Standards and Technology's current emphasis on the need for MEP centers to reach out to SMEs not currently served. It should be noted that in Canada, IRAP's clients are SME firms "with the potential to innovate," a different approach which emphasizes assistance to the most innovative companies rather than extending help to technological laggards. Fraunhofer and ITRI R&D programs are characterized by multiple projects with repeat clients, including large firms, a dynamic that has been associated, in the German case, with the spectacular success of the Mittelstand companies in world markets. A question is presented whether MEP should be encouraged to emphasize outreach to a broader base of SMEs as a way to concentrate its efforts on the most dynamic and innovation companies.

The Key Role of Networks

MEP's partner, the consultancy RTI International, has been grappling with the fundamental challenge of "helping clients find technologies that they need," an issue which continues to vex policymakers in other countries as well. Initiatives include workshops, creation of online clearinghouses, and other initiatives which resemble those elsewhere, such as Britain's Knowledge Transfer Centers and France's AiCarnot. However, such arrangements pale before the formal and semiformal technology networking arrangements in Germany, in which the Fraunhofer is perhaps the central player, and the informal networks linking Taiwan's ITRI and a vast community of ethnic Chinese in Taiwan and abroad, including technology centers in Silicon Valley and Singapore.

The Fraunhofer relentlessly emphasizes networking and collaboration in every aspect of its operations, reflecting the recognition that competitive innovation requires joint efforts by many actors with specialized competencies working in concert toward a common objective. Fraunhofer institutes are linked with each other in formal alliances that draw on differentiated but related skill sets. They also maintain vast formal and informal networks throughout the German university system and with other public research organizations such as the Max Planck Society, a system of public research institutes focused on basic research. Finally, the Fraunhofer institutes participate in a web of international technology alliances. The net effect of the Fraunhofer's deep networking is the ability in response to a request from a client (at least in the best case) to assemble rapidly a multidisciplinary team of experts well qualified to develop a solution to the particular technological problem at hand. With reference to this example, MEP centers will benefit by greater attention to the capabilities and best practices of other MEP centers as well as increased familiarity with the research facilities and competencies of U.S. universities and public research organizations.

Projects vs. Prototypes

The MEP program focuses on improvements in manufacturing processes, tools, and sustainability. Historically, it has not emphasized the development of commercial product prototypes by individual companies. By contrast, ITRI and Fraunhofer themselves develop product prototypes—unilaterally or, more commonly, in industrial collaborations—and work closely with industrial partners to scale up production and commercialize those products. IRAP currently sees its mission as assisting SME firms to "develop, adopt and adapt technologies and incorporate them into competitive products and services to be commercialized in the global marketplace." A question is presented as to the extent to which MEP can engage directly in activities such as proof of concept, prototyping, and pilot manufacturing of commercial products on behalf of clients in a manner similar to ITRI and Fraunhofer. Traditionally, activities on behalf of individual companies have been regarded as an inappropriate use of public resources. The same perspective is observable in some other countries, most notably the United Kingdom, although policy has shifted to support initiatives such as the Catapult program. However, public support of commercial product development—particularly at the R&D stage—is more the norm than the exception globally, and is generally not inconsistent with existing international trade rules.

The question of national attitudes toward the appropriate role of government with respect to bringing commercial products to market is not unrelated to the performance of the research organizations examined in this study. In Germany and Taiwan and to a lesser degree, Canada, there is a rough consensus that governments should play a role in bridging the gap between knowledge creation and its application in a commercial context. Reflecting this consensus in Germany and Taiwan, applied research organizations in these countries have enjoyed a substantial, stable flow of government funding, a coherent and clearly-defined mission, and close linkages with and support from other government initiatives such as cluster promotion and support for SMEs. In countries where partnerships among governments, industry, and universities is taken as a given, the economy has benefitted from organizational and fiscal continuity in applied research. The resulting scale, concentration of resources, and institutional development has not proven effective, at least in the past.

The leadership of the Fraunhofer tends to assess the prospects for success of government-backed research institutes in other countries with reference to the commitment of governments to sustain core funding at adequate levels over the long term. Chronic uncertainty over the availability of funding and/or the need to devote substantial operational resources to fundraising activities undermines the ability of research organizations to attract and hold managerial and researcher talent and to maintain a focus on the primary research mission. A key lesson for U.S. policymakers with respect to MEP or any other program promoting industrial innovation to be drawn from the foreign experience is the critical importance of sustained public funding.

Chapter 8

Findings and Recommendations

FINDINGS

1. **A strong domestic manufacturing base is integral to sustaining innovation and maintaining global competitiveness in advanced technologies.**[1]

    a. There is growing and authoritative concern that the erosion of America's manufacturing and high-technology base threatens to undermine U.S. leadership in next-generation technologies and the high value-added employment gains that would follow expanded U.S. high-technology production and exports.

    b. Moreover, some analysts argue that maintaining a competitive onshore manufacturing sector and the associated skilled labor and technical institutions are linked and essential for long-term national competitiveness. They note that once manufacturing activity moves overseas, so do the required skills, networks and supply chains; and once offshore they, and the learning they engender, are difficult to recover. These analysts therefore argue that it is important for U.S. policymakers to be concerned with the capabilities and composition of the economy, just as policymakers are elsewhere.

    c. The emergence of new technologies and other favorable developments, such as shifts in energy costs, open fresh windows of opportunity for manufacturing in the United States that can be exploited by new policies.

[1]See the analysis in Chapter 1.

2. **Spread across 50 states and 60 centers, the Manufacturing Extension Partnership (MEP) is the leading U.S. government program designed explicitly to provide support services to the nation's small and medium manufacturers.**[2]

   a. **The target constituency:** Small and medium manufacturers represent 98 percent of all manufacturing enterprises in the United States. They account for two-thirds of all manufacturing employment and contribute over half of the total value added by all U.S. manufacturers.
   b. **Limited market alternatives:** Given their small scale, limited resources, and scattered locations, many small and medium manufacturers report that private-sector alternatives to MEP are limited at best. While there are now more consulting services in most states than there were when MEP was originally established, small firms report that they cannot afford the fees of private consultancies and, in many cases, find that they do not adequately provide the type of services extended by the MEP system.[3] In addition, other government programs do not focus as directly on providing technical and management advice to these small firms.
   c. **Reach of MEP:** Since its establishment in 1989, the MEP system has continued to grow to serve U.S. small and medium manufacturers. It now reaches out to some 7,000 manufacturers a year, providing a variety of services, from short-term cost reduction through lean manufacturing to longer-term growth initiatives.
   d. **Diversity of the program:** Given the engagement of a large number of state partners and the varying conditions in different parts of the country, there is considerable diversity in methods of operation and delivery of MEP services.
   e. **Focus of MEP services:** MEP is focused on providing services with proven technologies and methods to existing manufacturing firms. By contrast, many of the large-scale programs to support manufacturing around the world (described in Appendix A of this report) are designed to move new technologies forward from concept to prototype to commercial production.
   f. **Net impact:** Multiple assessments of MEP find that the program has a positive net impact, although it is important to note that not all MEP projects generate measurable returns.[4]

---

[2]See Chapter 2 for a description of the MEP system.
[3]NRC staff interviews with center directors in Pennsylvania, California, and Ohio, October-November 2012.
[4]See Appendix B of this report for a survey of the literature on MEP evaluation.

g. **Federal-State Funding Match:** The total MEP budget is about $300 million. One-third of this is provided by the federal government. Centers are expected to match federal funds with funding from other sources, notably fees for services to clients and funding from state governments. The federal contribution in FY 2013 was $123 million, with more than three-quarters going directly to the centers.
h. **Program Limits:** While effective in meeting its specific mission needs, MEP is not designed to address the broader spectrum of competitive opportunities—from additive manufacturing to flexible electronics—on its own. The program, however, can be an increasingly important element in the nation's portfolio of programs to support manufacturing and the jobs it brings.

3. **Overall, MEP's support for lean manufacturing shows evidence of success.**[5]

   a. Evidence of this overall success is found in academic reviews, noted above, as well as in the analysis, case studies, and interviews of company staff conducted for this study.[6]
   b. It is important to note that not all programs at all MEP centers are successful all the time, nor should that be expected. In addition to the quality and timeliness of the advice provided by the centers, successful outcomes depend on factors internal to the company and on market conditions.

4. **MEP's annual budget is relatively modest, given the importance of SME manufacturing in the United States and in comparison to similar foreign programs.**

   a. The total system budget of the MEP is about $326 million annually, comprising $128 million in federal support (in 2012) and

---

[5]See Chapter 3 for a review of MEP and lean manufacturing.
[6]See Chapters 4 and 5 for a review of MEP metrics and performance outcomes. Appendix B includes a survey of the evaluation literature, and Appendix C-2 documents the written responses of a number of center directors. The study also included a number of field visits by the committee to MEP centers, including centers in California, Georgia, Ohio, and Pennsylvania. The committee also held two workshops in Georgia and Ohio that brought together representatives from universities, state governments, innovation intermediaries, and small and large manufacturers. A major workshop launching this study also drew viewpoints from NIST MEP, MEP centers, and program users. See National Research Council, *Strengthening American Manufacturing: The Role of the Manufacturing Extension Partnership—Summary of a Symposium*, C. Wessner, Rapporteur, Washington, DC: The National Academies Press, 2013.

$198 million in state and local matches, fee income, and in-kind contributions.[7]

b. Despite a positive record of accomplishment, the National Institute of Standards and Technology (NIST) MEP's budget is not commensurate with the importance of SME manufacturing to the U.S. economy. As a point of comparison, Canada's IRAP program, which has a similar mission as MEP in terms of promoting and facilitating innovation, has been funded at approximately the same level as MEP and, recently, had its federal contribution doubled to $220 million a year.[8] Canada's population and economy are about one-tenth that of the United States.

5. **MEP centers vary in terms of structure, services, and business models. This commendable variety makes the assessment and management of the overall program challenging. At the same time, this variety provides fertile ground for experimentation where individual centers can adapt to the needs of manufacturers in their particular region. However, the MEP system does not capitalize on this diversity of organizational experience as much as it might, and learning across the MEP system could be enhanced.**

   a. The centers that make up MEP's nationwide network vary substantially in organizational form, business model, and regional scope.[9]
   b. The role of NIST MEP headquarters is to provide strategic direction and to assess and certify the regional MEP centers' operations based on established metrics, while providing one-third of the budgets of the regional centers on a matching basis.
   c. NIST MEP influences individual centers by evaluating them on terms designed by NIST MEP, with the implicit sanction of the withdrawal of funding. Yet, centers can only be influenced up t o a point. They have other stakeholders locally—often including the state in which they operate—and they have an over-riding imperative to maintain a business that is sustainable over the long run. While NIST MEP does sometimes decertify a center, leading to a recompetition for the region, in fact, recompetition as a

[7]As of this writing, the President has proposed in his FY 2014 budget an additional $25 million to this program.
[8]See the review of the Canadian IRAP program in Appendix A-1.
[9]See, for example, descriptions of California's CMTC, Enterprise Minnesota, Ohio's MAGNET, and Pennsylvania's Catalyst Connection in the National Academies 2011 conference on "Strengthening American Manufacturing: The Role of the MEP." The committee has also conducted additional studies of MEP centers, including the Georgia MEP center. For a description of the overall MEP system, see the NIST MEP homepage at <*http://www.nist.gov/mep/*>.

sanction is rarely applied. Most recompetitions occur as a result of changes in state policy or an institution's inability to meet matching requirements. While, in some cases, informal pressure is sometimes applied through direct contact by NIST MEP management, this appears to be neither systematic nor transparent.[10]

d. There are large differences in MEP center performance on a variety of metrics. While some centers perform well on particular metrics, there currently appear to be minimal structured efforts to distill and adopt best practices across the MEP system. This learning across the MEP system is especially important at a time when centers have been asked to make substantial changes in their operations and when financial pressures facing centers are in many cases acute.

6. **The current fixed funding formula, where NIST MEP uses federal funds to provide one-third of the funding, is outmoded and limits the adaptability of the MEP system.**

   a. **Fixed Matching Funding Formula:** Currently, MEP centers raise funds from state governments and client fees, as well as from other sources, such as philanthropic organizations. These funds are matched by NIST MEP at a ratio of one dollar for every two dollars the centers bring in.
   b. **Impact of matching requirements**:

      i. **Limited leverage:** NIST's matching contribution is critical to the ability of MEP centers to attract state resources and private revenue generation. Yet, NIST MEP's influence over the management of the individual MEP centers is limited.[11] The rigidity of the current matching formula limits, for example, NIST MEP's ability to provide additional incentives to encourage state MEP centers to learn from and adopt best practices from across the system.
      ii. **Declining resources:** For the MEP centers, the federal matching requirement also amplifies the cuts in funding from other sources. Thus, when a major matching partner—such as the state government—reduces or eliminates its funding contribution, even successful MEP centers can become

[10]See Chapter 2 on the relationship between NIST MEP and MEP centers.

[11]See the related analysis in Chapter 1. Complete program control by NIST MEP, in any case, would not be desirable given the program's broad geographic dispersion and the unique features of many regions. There can be, on a case-by-case basis, informal means of influence from NIST MEP headquarters.

trapped in a cycle of declining resources.[12] This can result in a suboptimal level of coverage and service in relevant manufacturing locations.

iii. **A greater focus on client revenue:** To sustain the federal match, the MEP centers currently must overcome declines in state funding through greater reliance on client revenue.[13] This matching structure incentivizes the MEP centers to focus on short-term client income rather than on the overall MEP mission. It is not clear how these short-term measures will mesh with MEP's new long-term strategy (described below) to drive manufacturing growth and innovation.

iv. **Impact on MEP mission:** The need to meet the federal match through fees and the reliance on in-kind contributions is pushing the program away from its presumptive target group—i.e., away from smaller companies and rural firms that should be a part of MEP's public vocation. MEP centers that work with the target audience of SMEs also tend to have smaller jobs impacts. Centers in markets with a greater percentage of smaller companies, or centers that choose to focus on the SMEs tend to be at a disadvantage in the calculation of results.

7. MEP efforts to add support for innovation and growth through its national network are commendable.

a. **MEP focus:** The committee commends efforts to enhance the innovation capacity and growth of small manufacturers under the new strategic orientation. This strategy is in line with recent academic and policy analyses that call for U.S. manufacturers to become more innovative and more focused on growth and international competitiveness.[14] MEP—with its nationwide network of centers and mission to support small manufacturers—is well placed to support this effort.[15]

b. **Mitigating risks:** The successful implementation of the Next Generation Strategy (NGS) will, however, require mitigating the risks faced by MEP centers (see Finding 8), developing a robust roadmap to expand new service capacities (see Finding 9), and

---

[12]See the related analysis in Chapter 1. Some centers (e.g., Minnesota and California) now operate entirely without funding from their state governments.

[13]Interviews with staff and clients at MANEX and CMTC (MEP centers in California).

[14]See Chapter 1 for a review of the related academic analysis and policy initiatives.

[15]Interviews with staff and clients at MRC and DIVRC (MEP centers in in Pennsylvania), and with Magnet (Ohio).

fostering the exchange of best practices among MEP centers (see Finding 11).

8. **Implementation of the Next Generation Strategy (NGS) poses risks and challenges for MEP centers.**

   a. **Revenue risks for centers:** MEP centers encounter significant risks as they seek to transition from a tight focus on lean production to a much wider range of services that require new clients, new contacts, new kinds of client conversations, new services, and new toolsets and capabilities. For some centers, this will require the addition of qualified staff and other resources. Centers must make this transition to NGS while maintaining revenue levels in the absence of direct financial assistance from NIST MEP.
   b. **Demand Risks for New Services:** While the new services offered under the Next Generation Strategy are designed to support growth and innovation among the nation's manufacturers, this supply-side push needs to be matched by sustained demand for these services for this strategy to be economically viable for the MEP centers. [16] At present, some centers see sufficient demand for the new services, but others do not.[17] There is the additional possibility that this focus on innovation-related initiatives could simply result in the reclassification of existing activities. MEP centers face the need to provide services that clients are willing to pay for, in order to ensure that they meet matching funds requirements.

9. **NIST MEP introduction of innovation and growth-related services are too narrowly based.**

   a. **Early identification of services**: Currently, the introduction of new services such as "Innovation Engineering" and "Technology Acceleration," are largely experimental, with much of the experiment being supported and funded by NIST MEP. Bearing this in mind, NIST MEP may have committed too early and too fully to a particular set of services that may or may not turn out to be successful and cost-effective across the MEP centers.
   b. **Role of single providers**: NIST MEP has selected one contractor for each type of growth service. These contractors are to provide training for the MEP centers so that they can rapidly become self-sufficient in delivering new services. NIST MEP has awarded large

[16]See Chapter 6.
[17]See comments by MEP center directors, compiled in Appendix C-2.

contracts to single providers for each strategic sector. Such a single provider approach carries inherent risks.

c. **Need for clear milestones**: NIST MEP has a strong history of developing metrics for its activities. It is therefore surprising and a matter of concern that NIST MEP appears to have no clear milestones for the consultants associated with the provision of new strategy services. The largest and most expensive component of the new strategy is an innovation-engineering program, provided to MEP centers through the Innovation Engineering Leadership Institute (IELI). Despite repeated requests, the committee has not been provided with systematic and independent evidence about the take-up or impact of the innovation-engineering program. Two other NIST MEP initiatives—a technology scouting program offered by RTI International and the ExporTech program to help companies develop international marketing strategies—have both provided some evidence about outcomes, but not about agreed milestones with NIST MEP.

d. **NIST seems not to have drawn on expertise from the best centers within the system.** Some of the substantial funding for contractors might have been more effectively allocated to centers with strong track records on innovation and growth, perhaps for the explicit purpose of supporting change at other centers (see Finding 11).

10. Successful implementation of the Next Generation Strategy depends on the strength of the individual MEP center: [18]

a. **Capital infusion**: Among the MEP centers reviewed for this study, those best adapting the new services appear to have received additional state or other investments. These resources supported their development of and then transition to growth services.

b. **Leadership:** Center leadership and the composition of each center's advisory board are important factors in their effectiveness and adaptability; both of these factors appear to vary substantially among centers, suggesting that effective take-up of NGS across the MEP system will be uneven.

c. **Local and regional networks:** MEP centers that successfully add new strategy services appear to be especially well connected to the local manufacturing community and other stakeholders (such as local universities). They have also developed mechanisms for

[18]Interviews with the leadership of several MEP centers, including those of Georgia Tech, Magnet Ohio, and MRC and DVIRC in Pennsylvania. The committee also received written submissions of comments from 42 MEP centers.

enhancing these connections, such as executive education and networking projects.

11. **Adopting best new strategy practices within MEP.** Development and rollout of new strategy services are both financially risky and technically demanding. It is therefore important that the experience of individual MEP centers be shared effectively across the network, so that follower centers can benefit from the experience of path-breakers and avoid potential pitfalls. Most centers do not have the resources to afford false steps. Hence, the gaps identified below could have significant long-term consequences both for individual centers and for the program as a whole.

    a. **Peer-to-peer activity:** While some centers do reach out to other centers, there is no easy way to determine which centers have the most experience with specific issues and approaches. NIST MEP does not provide an effective online forum for peer-to-peer interaction between center director and other center staff.
    b. **Role of the MEP Annual Conference:** A number of center directors observed that the MEP annual conference provided the primary path for informal and formal knowledge exchanges.[19] Currently, no annual conferences are planned for 2013 or 2014.
    c. **Identifying best practices:** Currently, NIST MEP does not formally identify best practices, nor does it circulate the results of such assessments.[20]
    d. **Cross-center training:** While some centers do provide training and direct client services to other centers (for example, Pennsylvania's DVIRC has provided some new strategy services to the New Jersey MEP center), MEP CORE metrics apparently do not record and recognize these initiatives.

12. **NIST MEP evaluations and programs are evolving from a focus on quantified metrics derived from a survey of clients and center reporting to a new emphasis on qualitative metrics.**

    a. **The previous evaluation system, Minimally Acceptable Impact Measures (MAIM), had several flaws:** Notably, the system collected extensive information of uncertain value in time frames that were arguably too short. The centers were evaluated on the percentage and character of survey responses they generated.

[19]Interviews with MEP center staff at DVIRC and MRC in Pennsylvania, and MANEX and CMTC in California.
[20]For a review of external assessments of the MEP program, see Appendix B.

Moreover, there appears to have been insufficient separation between the respondents and the MEP centers despite the use of a third party to administer the survey.

i. **Self-reported data:** MEP evaluations of center activities have traditionally depended on data reported by participating manufacturers and on center reports related to center processes, e.g., those related to the acquisition of distribution of center funding, and the use of center expertise. While MEP's reliance on self-reported data is in line with the evaluation efforts of other small business support programs, some aspects of this approach may negatively affect the quality of information and related assessments.

ii. **Timing of surveys:** Deployed only six months after MEP centers submit client information, the NIST MEP survey is arguably premature, and thus unable to capture the impacts of longer-term MEP interventions, especially those resulting from innovation-oriented services.

iii. **Use of data:** Collected data appears not to have been systematically utilized by NIST MEP to drive improvements in center operations.

b. **New CORE metrics provide a more nuanced view of center activities:** In 2012, NIST MEP introduced a new CORE (Center Operations Reporting and Evaluation) metrics system. The new metrics are a positive development, as they reduce reliance on the client survey and seek a balanced scorecard that includes attention to center innovation and growth services. Further, the effort to include qualitative indicators is welcome, as is the emphasis on utilizing the metrics to help centers manage better.[21] In addition, the survey is now deployed on a more flexible schedule, at least in principle.

c. **Concerns about the new CORE metrics remain:** Although these metrics were only recently introduced (in 2012), there are several issues:

i. **Survey burden**: The client survey has not changed, and concerns remain that it places a substantial burden on recipients. In addition, the NIST MEP survey requires an extremely high response rate, which heavily burdens the centers and may well unintentionally tend to bias the results. As a consequence, while reliance on the survey is reduced in

[21]NIST MEP Webinar for NRC staff, July 23, 2012.

the development of the overall assessment, the survey itself retains its previous flaws.

ii. **Complexity of survey:** At the most basic level, some firms find the survey difficult to complete, with too many complex calculations and requests for data that many small firms simply do not compile. [22]

iii. **Potential for bias:** Given that the center's certification and funding depends on positive responses from their clients, MEP center staff are incentivized to encourage their clients to respond actively to survey requests and to ensure that responses are as positive as possible.[23]

iv. **Survey timing:** While MEP has added survey windows to help MEP clients better report the long-term impact of MEP center services, these may still be insufficient especially in long-cycle industries, while it appears that a large majority of clients will continue to be surveyed on the standard six-month cycle.

v. **Transparency:** The new qualitative metrics may require a more transparent assessment process, especially if they become important for the allocation of NIST MEP funding.

13. Major U.S. trading partners have initiated substantial programs to support manufacturing.

a. A key element of the committee's task was to examine programs to support manufacturing in other countries. As documented in Appendix A of this report, many other countries intervene with substantial resources and well-equipped facilities early in the product development cycle. Drawing on a portfolio of programs and policies, they provide shared-use facilities, expertise garnered from working on multiple projects over long periods of time, and connections to other parts of the innovation ecosystem, be it suppliers, customers, or sources of finance, as well as a ready and low-cost source of technical support.

b. Major U.S. trading partners understand that a robust domestic industrial base that can produce advanced products in high volumes, and the high-skilled jobs that this productive activity generates, are integral to maintaining global competitiveness. [24] They believe that these shared facilities and capabilities also

---

[22]See Chapter 4.

[23]Interviews with MEP center directors, Pennsylvania, California, and Ohio, and Georgia, as well as during the NRC MEP workshop in Washington, DC, November 29, 2012, and December 6-7, 2012.

[24]See the analysis of foreign policies and programs to support manufacturing in Chapter 7.

increase their chances of successfully advancing to next-generation technologies.[25]

14. **Leading national programs to support manufacturing include Canada's Industrial Research Assistance Program (IRAP), Germany's Fraunhofer Institutes, Taiwan's Industrial Technology Research Institute (ITRI), and more recently, France's Carnot institutes and the United Kingdom's Catapult Program.**[26] **The experience of these manufacturing support organizations show that they can have positive technological, industrial, and educational effects in a given country or region.**[27] **These positive effects include**:

    a. **Incremental Improvements:** Continuous incremental technological improvement in established local industries, enabling them to compete successfully with new entrants.
    b. **Research and development (R&D) services:** The provision of R&D services and facilities to innovative small and medium enterprises which could not otherwise readily access such resources.
    c. **Supply chains:** The creation, augmentation, and continual upgrading of onshore industry supply chains.
    d. **Support for start-ups:** The provision of high-quality support for start-ups, including but not limited to spin-offs, to enable them to commercialize innovative products and processes.
    e. **New industries:** The creation of new technology-intensive products, services, and industries in emerging sectors.
    f. **Technology diffusion:** The diffusion of advanced domestic and international leading-edge technologies throughout local industries.
    g. **Practical training:** The provision of practical training and skills to young people pursuing academic studies in the fields of science, mathematics, and engineering.
    h. **Practical application of knowledge:** The efficient mobilization of knowledge from the research base to practical applications benefitting local companies and industry sectors.

---

[25]For a review of leading national programs to advance innovation in manufacturing in Canada, Germany, France, the United Kingdom, and Taiwan, see Appendix A of this report.

[26]Among these, Canada's IRAP is the most similar to MEP in concept. Appendix A of this report provides detailed descriptions of the scope and structure of leading national efforts to support small, medium, and large manufacturers.

[27]For a review of some of the leading applied research organizations, see Appendix A of this report, which provides an overview of leading programs in Canada, Germany, France, the United Kingdom, and Taiwan.

i. **Innovation clusters:** Collectively, these activities contribute to the development of successful innovation clusters and regional economic development.

15. **While the United States has numerous public organizations engaged in applied research, notably in fields such as medicine, agriculture, energy, and defense, large segments of the U.S. manufacturing sector are often underserved with respect to technological support from the national research base.**

a. **Fragmented U.S. programs:** Existing federal programs to facilitate applied research for manufacturing are much more limited in funding, focus, and scope as compared to similar programs in other technologically advanced countries.[28] They do not have the same broad impact as do these foreign programs.[29]
b. **Limited commercialization:** Promising new technologies developed from basic research conducted in the United States, often with public support, are increasingly commercialized outside the United States with little effect on domestic value added or jobs.
c. **Supply chain gaps:** Significant gaps exist in some domestic industry chains in technology-intensive U.S. industries.
d. **Limited manufacturing skills:** Some industry leaders have argued that the United States needs to enhance the training of scientists, engineers, and technicians with the skills necessary to staff manufacturing facilities in technology-intensive industries.
e. **Limited private technology transfer**: While the private sector (particularly in larger firms and in consultancy) engages in technology transfer, it is not fully able to address the needs identified here.
f. **Awareness:** In many cases, small and medium businesses often do not know where or how to pursue technological support from the research base.[30]
g. **MEP role:** The role of the MEP in linking with proposed U.S. manufacturing institutes and other major manufacturing initiatives has not been made explicit.

16. **The most successful foreign applied research programs around the world share elements of the following characteristics.**

---

[28]See Chapter 1 for a partial list of federal programs to assist manufacturing.
[29]See Chapter 7 and Appendix A of this report for a review of leading national manufacturing programs.
[30]See Appendix B.

a. **Substantial and sustained funding over a long-time horizon**, to provide first-class equipment and infrastructure and retain qualified staff to maintain focus on core missions and client base.
b. **A high degree of autonomy** in establishing technology strategies, deploying resources, and developing relationships with the private sector.
c. **Funding from contract research:** Foreign programs often derive a significant fraction of their income from contract research, primarily from the private sector, and receive public funding according to a formula rewarding the volume of industry contract work and support for SMEs.
d. **Institutional support for start-ups:** They establish internal structures, including venture capital funds and incubators, and provide legal assistance with respect to intellectual property rights, to support the start-up and spin-off of new companies.
e. **Links to local clusters:** They contribute directly to the development of local innovation clusters.
f. **Partnering with universities:** They collaborate with universities, often drawing a significant proportion of total professional staff from the ranks of students pursuing advanced degrees in the science, mathematics, and engineering fields.
g. **Training:** They contribute directly to the training of technical staff.
h. **Strategic priorities:** Some establish as a strategic priority the creation and augmentation of onshore industry chain.
i. **Collaboration encouraged:** They encourage companies to collaborate in onsite research and demonstration projects, including the operation or simulation of entire industrial production processes.
j. **Networks:** They develop organizational networks linking sister institutions, universities, and internal departments in a manner which permits the deployment of a combination of research competencies to address specific technological problems raised by individual companies.
k. **Centers with business expertise:** They develop internal business expertise, including an understanding of the potential commercial applications of new technologies, with respect to relevant industry sectors and subsectors.
l. **Regular review:** Centers are subject to periodic assessments by independent reviewers on a longer-term basis.

RECOMMENDATIONS

While the committee finds that the MEP program provides valuable help to small manufacturers, with the enhancements recommended here, the program will be an increasingly important element in the nation's portfolio of programs to support manufacturing and the jobs it brings.

1. **NIST MEP should focus more on driving the overall improvement of MEP centers rather than focusing on the outcomes of individual projects.**

    a. **Develop performance incentives:** To improve overall center performance, NIST MEP management should be tasked with developing incentives to drive center performance.

    b. **Promulgate best practices.** NIST MEP should identify and help promulgate best practices among centers. The need for much better communication of best practices is strongly underlined by the need to share tools and techniques that work in implementing the Next Generation Strategy. Many centers have limited resources, which constrains their ability to undertake new and higher-risk strategies. They need access to relevant experience and best practices to help ensure success.

        i. **Peer-to-peer communication:** NIST MEP should encourage centers to provide help and support to each other: A peer-to-peer model may work better than standard vertical communication between individual centers and NIST MEP in this regard.

        ii. **Pilot programs**: NIST should provide additional funding for pilot programs that assess the performance of MEP centers against relevant comparison groups.

        iii. **Centers of excellence.** NIST MEP should consider fostering centers of excellence amongst the centers. This may be challenging in such a distributed environment, but fostering these centers of excellence can have important systemwide impacts.

        iv. **Annual conference.** NIST MEP should consider reinstating the MEP annual conference as soon as possible.

        v. **Center directors' forum.** NIST MEP should consider providing—or encouraging the independent development of—an online forum in which center directors and other staff can exchange information, focused on the transfer of best practices among them and the efficient solution of operational problems and questions.

c. **Address revenue stability for the centers.** The proposed shift in the business model has potentially significant financial implications for the MEP centers, given that many of them rely on the revenue stream from current clients to remain in operation and to meet the matching requirements (see below).

2. **The Manufacturing Extension Partnership should use its resources to leverage maximum beneficial outcomes for the U.S. manufacturing sector rather than focus on reaching the maximum number of manufacturers.**

   a. With its current resources, MEP should focus its analysis on identifying and supporting manufacturing firms that are most able to benefit from MEP services. As with any service provider, spreading MEP services too thinly may well result in suboptimal results overall.
   b. MEP should identify those markets and clients where the impact of program resources would be greatest—particularly in terms of long-term productivity, sustainability and innovation performance—that will in turn, anchor good-quality jobs. This strongly suggests that MEP centers should focus on selective client acquisition. Market penetration metrics should focus on generating a sustainable client base.
   c. In addition, MEP should consider metrics that measure market penetration and innovation over time rather than a single year, as manufacturers may not need MEP services every year. Adopting a longer, multiyear time frame for analyzing center and program performance would provide a better assessment.

3. **MEP should continue to encourage lean manufacturing.**

   a. Given that lean manufacturing remains at the core of a successful manufacturing support program, MEP should encourage centers to maintain the current capacity in this area and integrate lean manufacturing into new initiatives including those related to innovation.
   b. NIST MEP should adjust its CORE metrics to better reflect the continuing importance of lean manufacturing for the centers.

4. **MEP should continue efforts to enhance growth, increase innovation, and improve sustainability though its Next Generation Strategy. At the same time, NIST MEP needs to address the challenges inherent in this transition.**

a. **Recognize risk to centers:** NIST MEP must recognize that maintaining its revenue base is necessarily a primary center objective, and that growth strategies must be adapted accordingly with variation across centers accepted (see Recommendation 5).
b. **Analyze demand for growth services:** NIST MEP should help centers analyze the demand for growth services, and review patterns of demand across centers. This should include a review of centers with leading edge services. It should also identify firm populations where growth services are likely to be effective.
c. **Benchmark service providers:** NIST MEP should develop appropriate benchmarks for all major service providers and review them regularly. Single provider contracts should be used sparingly, should require a detailed justification, and should have clear deliverables and metrics, and be closely monitored. Competitions for contractors should be conducted wherever possible.
d. **Eliminate duplication of services:** Based on a regional survey of effective alternative private- and public-sector service providers, NIST MEP should encourage centers to avoid the direct provision of services already provided effectively to manufacturers, including NGS services.
e. **Identify success factors:** NIST MEP should identify emerging best practices in implementing the Next Generation Strategy, reward them, and encourage their adoption, where relevant, across the system.
f. **Develop networks:** MEPs might consider developing networks connecting SMEs to the local value chain, including original equipment manufacturers, universities, and national laboratories. These networks could be regional, sector based, or cross-sectoral.
g. **Improve administration:** Achieving the goals of the new strategy while maintaining services related to lean manufacturing creates new requirements of people, mechanics, and execution. This calls for the recruitment and training of staff to address new challenges.
h. **Advisory boards:** Effective leadership, supported by rigorous advisory boards, or their equivalent, will be essential for NIST MEP and MEP centers in their strategy and operations. The makeup of advisory boards comprising diverse expertise including manufacturing savvy can accelerate the progress and performance of MEP's service to manufacturing firms.

5. **NIST MEP should significantly improve its collection and analysis of performance data.**

   a. **Develop analytic capacity:** MEP should develop both in-house and external data and the analytic capacity to provide ongoing and independent evaluations of its new initiatives. This will aid MEP to identify programs and initiatives worth expanding, those that should be adjusted, and those that are a poor use of resources. MEP should also assess cost savings and earnings gains against public support for the program.

b. **Consult stakeholders:** The new strategy should include metrics that are both clear and take more fully into account the informational demands placed on participating small firms. The selection of subjective performance criteria should also include input from state and other stakeholders.

c. **Independent assessment:** MEP initiatives, and in particular the specifics of the new strategy, should be periodically assessed by an expert and impartial third party. As per best practice, outcomes data should be collected by independent survey organizations. Contractors conducting surveys should use a standardly-accepted, rigorous methodology to ensure the validity of survey responses and adequate response rates.

d. **Response rates:** Benchmarks for response rates should be developed over time. MEP survey evaluations should be run independently of MEP centers, and the contractor running them should be incentivized to develop and seek an appropriate response rate with high-quality data.

e. **Survey Timing:** As MEP moves to support projects with much longer time horizons—for example focused on innovation—it will be necessary to adjust the timeframe of outcome surveys to match this new approach. With the new innovation-oriented interventions, six months is too short a timeframe to gather meaningful data.

f. **Identify center best practices:** MEP data and analyses should identify best practices, and measure the relative accomplishments of individual initiatives and center activities. The centers should be given considerable latitude and be viewed as laboratories for new initiatives, where rapid and effective assessment in turn leads to the adoption and transfer of best practices.

g. **Develop milestones:** The effectiveness of support programs should be examined and, if they are to be continued, an integrated set of metrics and related milestones should be developed to gauge their deployment. The use of these ancillary programs should be linked to the adequacy of the MEP funding.

h. **Data analysis as a service:** Analysis of the data could be more effective if considered as a service that NIST provides to centers for internal and state-level management of the center.

6. With the adoption of the improvements recommended here, Federal funding for the MEP program should be at a level

commensurate with its mission, and take into account relevant international benchmarks.

a. Funding for MEP should be commensurate with the importance of manufacturing to the growth of the economy and the program's proven ability to contribute to improved firm performance and adapt to the changing needs of the manufacturing sector.

   i. MEP's annual federal budget of $123 million (in 2012) is modest, compared to the target population and to the efforts of our competitors in the global economy. As a point of comparison, Canada's IRAP program, which has a similar mission as MEP in terms of promoting and facilitating innovation, has been funded at approximately the same level as MEP and, recently, had its budget essentially doubled to $220 million a year. Canada's population and economy are about one-tenth that of the United States.
   ii. The current level of funding is not adequate to maintain the program's focus on small firms, build new services around the Next General Strategy, and provide the resources required to drive the improvements recommended by this assessment.

b. The committee believes these additional funds can be used effectively by the program, particularly with the more flexible approach recommended above.

   i. The program needs to expand in order to support its new strategy while maintaining the integrity of the current portfolio of services that have proven effective.
   ii. Additional funding for MEP centers should, in principle be distributed to the centers, but *not* be entirely on a pro rata basis. For example, NIST MEP should be encouraged to offer centers additional matching funds for pilot projects that support new ways to accelerate and enhance movement toward the objectives of the NGS.
   iii. NIST MEP should direct some of the additional funding recommended by this committee to develop innovative services, encourage diffusion and adoption of best practices, and fund the data collection, analysis, and evaluation needed.

7. **NIST MEP should be more flexible in the management of the funding of MEP centers.** The fixed federal match of one-to-two from the states and centers should be changed to a match approach with more flexibility for NIST management and the centers.

a. The current MEP funding model, on its own, can result in perverse impacts on a center's viability, as a loss of state funding translates into a loss of federal funding. Declines in center revenues (even short-term declines) can result in a decline in federal revenues.
b. The matching requirement for MEP is outmoded: Significant risks exist in the current funding levels due to the existing problems with the match and the concern about putting the current funding model at risk due to the adoption of the Next Generation Strategy—NIST's new strategy for the centers. By changing the match and recommending additional program funds, the committee seeks to address these concerns and strengthen MEP's financial stability.
c. To enhance program flexibility and adaptability to new circumstances, the NIST MEP director should be permitted to waive the two-to-one match, as circumstances require. Secondly, MEP centers should not be required to match new funds at a one-to-one rate. Consideration should be given to moving the entire system to a lower match that is more consistent with other Department of Commerce programs and with due attention to maintaining state contributions. [31]

8. **Any effort to establish programs to further support manufacturing should thoroughly assess existing U.S. resources, organizations, and institutions already engaged in applied research and should take into account lessons from U.S. and international best practice.** These best practice lessons might include, inter alia:

   a. **Focus:** These new organizations should not be considered a replacement for the MEP program, nor should they replicate existing public and private technology transfer and applied research organizations. Where possible, existing institutions with sufficient capability and that are well-anchored in the local and regional innovation system should be given priority over the creation of greenfield facilities. Creation of new organizations should be targeted to (a) filling gaps that are identified in the existing U.S. ecosystem for applied research and (b) pursuit of promising new technologies in thematic areas in which the market, and the policies of competitors, are unlikely to produce a satisfactory outcome for the U.S. economy.
   b. **Draw on best practice:** The organizations and the enabling legislation should draw on international best practices. Likewise,

[31]For a discussion of this point, see Chapter 1.

the MEP program should seek to learn from foreign programs through the exchange of best practices.

c. **Branding:** Initiatives to support manufacturing might bear a designation drawing on the name of an American innovational pioneer (e.g., Edison, Noyce) to establish a brand synonymous with excellence.
d. **Encourage public-private partnerships:** The most successful foreign applied research institutes involve public-private partnerships, a structure which leverages public financial support, research excellence, and private-sector participation to ensure a mission focus on research with commercial and industrial relevance.
e. **Dual mandate:** The organizations should have a dual mandate: (1) to support incremental and continuous technological improvements in established industries and (2) to facilitate the development and testing of new products, either for existing innovative companies or to help develop new companies in emerging technological areas.
f. **Integration with field services:** If significant numbers of small manufacturers are to be engaged with new manufacturing support organizations, it is vital to provide pathways and assistance so that small manufacturers can access and use these new capabilities and opportunities. In the United States, the MEP already offers a mechanism to achieve this. Ensuring linkages with new manufacturing support organizations would leverage the MEP's existing public and private investment and expertise in outreach, services, and tools to contact small firms and enhance innovation upgrading.
g. **Invest in research infrastructure:** Public investments should be focused on the plant, equipment, and human capital for applied research, whether at universities or other institutions. Companies should be incentivized to furnish equipment and/or prototypes. These resources should be made available to industrial partners for precompetitive R&D, prototyping and process R&D.
h. **Incentives for firm development and training:** The organizations should be incentivized to prioritize support for small businesses, while also providing support for medium and large companies on a case-by-case basis. Incentives should also be provided for each organization to develop apprenticeship programs in advanced manufacturing. Cooperation with local institutions, e.g., community colleges, should be encouraged and incentivized.
i. **Provide stable public funding:** While public funding for research organizations should not be open-ended, it should be substantial and sustained; i.e., it should be provided for a sufficiently long-time horizon to enable institutions to maintain focus on their core

mission (rather than fundraising), acquire necessary equipment and facilities, and retain key personnel.

j. **Encourage start-ups:** Research organizations receiving public funding should be tasked with fostering start-ups to commercialize promising technologies through spin-offs from the organizations and provision of incubation services that include access to the organizations' research infrastructure.

k. **Foster collaboration and networking:** Research organizations receiving public funding should incentivize internal collaborations and team building by their own personnel and external collaboration with other institutions and industry consortia and associations.

l. **Foster entrepreneurialism:** Research organizations receiving public funding should include staff with business training that are tasked with linking onsite research to specific commercial applications, including the establishment and strengthening of supply chains, improvement of manufacturing efficiency, and the commercialization of new products.

m. **Conduct periodic arms-length reviews:** Each research organization receiving public funding should be subject to periodic audits by external objective experts with backgrounds relevant to the work of the organization to assess effectiveness and cost-benefit balance.

APPENDIXES

Appendix A1

Canada's Industrial Research Assistance Program (IRAP)

The National Research Council of Canada's Industrial Research Assistance Program (NRC-IRAP) currently supports over 8,500 Canadian SMEs in their R&D and innovation-related activities.[1] IRAP has a growing international reputation, and has been identified as a best practice in providing a broad suite of innovation support services and funding to Canadian SMEs.[2] This chapter examines the key features of the IRAP program: the origin and history of the program; its organization and governance; the resources dedicated to it; the services provided; the characteristics of its clients; the benefits derived from the program, the program's strengths and weaknesses, and lastly suggestions on how IRAP might be improved.

HISTORY

The Industrial Research Assistance Program (IRAP) was initiated sixty years ago in 1962, under the management of the National Research Council. It was designed to provide industrial firms with assistance to support research and was a response to the post-war belief that Canadian industrial research, and government support for such research, was insufficient to maintain Canada as a major industrial nation.

Today, the program's mandate is to stimulate wealth creation by supporting technological innovation. This mandate is achieved through two strategic objectives. The first is to provide support to small and medium-sized

[1]John McDougall (President of NRC), "National Research Council Canada," Presentation at the 10th Annual Re$earch Money Conference on May 11, 2011.

[2]Council of Canadian Academies, "The State of Science & Technology in Canada," (2006). <*http://www.scienceadvice.ca/en/assessments/completed/science-technology.aspx*>.

> The original objective of IRAP was to "stimulate a build-up of competent research teams in industry" through the funding of "relatively long-term applied research projects in science and engineering."
>
> NRC 1969, *Summary Report on IRAP for Study Group reviewing Federal Government R&D Incentives Program.*)

enterprises (SMEs are defined as firms with 500 employees or less) for the development and commercialization of their technologies. As the current Director General of IRAP has recently stated, "IRAP assists SME firms to develop, adopt and adapt technologies and incorporate them into competitive products and services to be commercialized in the global marketplace."[3] The second is to collaborate with, and in some cases to fund regional and national organizations that support the development and commercialization of technologies.

In summary, IRAP's objectives and corresponding services are situated in a key space within the innovation continuum, as noted in Figure APP-A1-1.

This focus on SME firms is important since Canada is a nation of SMEs: for example of a total of 1.14 million business establishments in Canada that have employees only 2,708 are large businesses (defined as having 500 or more employees)—the remaining 1,136,053 are SMEs (99.8 percent) with an average size of 6 to 7 people. As a result, IRAP is challenged to deliver services to such a large client group dispersed across Canada's wide geographic landscape.[4]

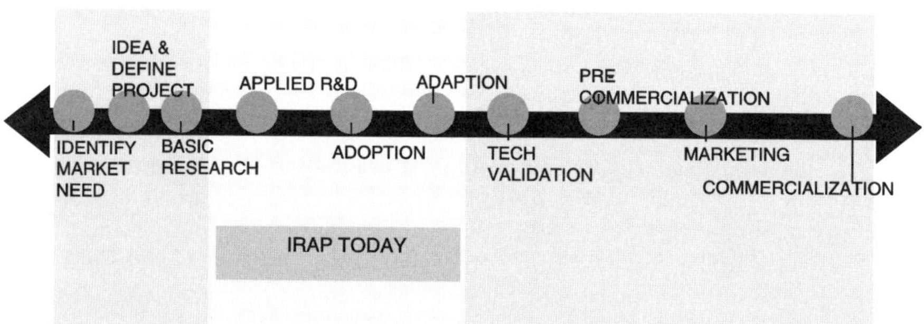

FIGURE APP-A1-1 The innovation continuum.
SOURCE: Modified from a Presentation by Dr. Tony Rahilly (Director General), "Our Program Overview," on October 5, 2010.

[3]Bogdan Ciobanu (Director General of IRAP), "Supporting SME Growth Through Innovation" on May 8, 2012 at the National Academies of Science conference in Toronto, Ontario: "Canadian Approaches in Promoting Innovation in Manufacturing."

[4]Industry Canada, "Key Small Business Statistics," July 2010. <*http://www.ic.gc.ca/eic/site/sbrp-rppe.nsf/vwapj/KSBS-PSRPE_July-Juillet2010_eng.pdf/$FILE/KSBS-PSRPE_July-Juillet2010_eng.pdf*>.

How Has the Program Changed Over Time?

"Throughout its history, IRAP has gone through a complex evolution of expanding functions and coverage, as well as numerous reorganizations" of its administrative structure. Most of this evolution appears to reflect a need to respond to altered economic conditions.[5] The timeline can be briefly summarized as follows:

- ✓ 1962-1978: (later known as IRAP-P) **Contributions to Large Projects**—IRAP began as a single program providing long-term technical and financial assistance, lasting up to five years per project.

- ✓ 1978-1984: The program expanded and diversified in response to a growing realization of the importance of SMEs. Three new program elements were added:

    1. IRAP-M: **Contributions to Small Projects** which increased assistance to small- and medium-sized enterprises (SMEs) having little or no in-house R&D capability and experience.
    2. IRAP-L: **Contributions to Laboratory Investigations** covered up to 75 percent of total costs (not to exceed $6,000 in 1983) of small firms subcontracting to "laboratories, institutes, or consulting firms" with the task of solving specific technical problems or fulfilling 'testing requirements with their products or processes.'[6]
    3. IRAP-H: **Contributions to Firms Employing Undergraduates.**

- ✓ 1984-1988: IRAP placed a new emphasis on technology diffusion and decentralization through:

    4. IRAP-R: **Contributions to Major Projects Involving Technology Transfer** focused on complex collaborative R&D projects.
    5. IRAP-S: **International Technology Services** were created to promote co-operative R&D between foreign owners of the technology and Canadian companies.

- ✓ 1991-1999: as outlined in the NRC's *Strategic Plan* (1991), IRAP began to focus on "increased decentralization, increased selectivity with an emphasis on 'more complex projects with greater technical impact,' the consolidation of the program's five elements into 'two,

---

[5]Richard G. Lipsey and Kenneth Carlaw, "A Structuralist Assessment of Technology Policies—Taking Schumpeter Seriously on Policy," Industry Canada Research Publication: Working Paper Number 25, October 1998: pp. 87-108.
[6]Lipsey and Carlaw 1998, p. 88.

more flexible elements,' and continued efforts to build and strengthen the IRAP network through 'increased linkages with complementary programs and services.'"[7] The two consolidated programs are noted below.

1. **Technology Enhancement (TE)** allowed support for small scale initiatives enabling SMEs to demonstrate and improve their technical competence as a basis for more substantive investments.

    - Up to 75 percent of eligible project costs were covered under TE.
    - This includes the salary costs of subcontractors and consultants, and travel-related expenses up to a maximum of $15,000.

2. **Research, Development and Adaptation (RDA)** focused on more complex projects involving applied R&D (unproven) as well as the adaptation of technology. RDA sought to meet the needs of firms that lacked specific technical expertise or a crucial technology to undertake a project.

    - Funding of between $15,000 and $350,000 was provided covering up to 50 percent of project costs.

✓ 2000-2005: Throughout the first half of this decade, IRAP's program scope expanded and diversified.

1. IRAP organized 3 **Technology Missions** to Asian countries resulting in the signing of 20 MOUs between Canadian and Asian companies. Thus, there was an increased international focus on getting SMEs into the global marketplace.
2. IRAP also fostered integration of sustainable development practices into the innovation processes of Canadian SMEs through a joint initiative (Eco-Efficiency Innovation initiative) with the **Ontario Centre for Environmental Technology Advancement** (OCETA).[8]

---

[7]NRC (1991), *Strategic Direction for NRC: A Council Discussion.* Ottawa.

[8]NRC (2000-01), *Performance Report.* Ottawa. <*http://www.collectionscanada.gc.ca/webarchives/20060117225755/http://www.tbs-sct.gc.ca/rma/dpr/00-01/nrc00dpr/nrc00dpre.pdf*>, NRC (2001-2002), *Performance Report.* Ottawa. <*http://www.collectionscanada.gc.ca/webarchives/20060120051523/http://www.tbs-sct.gc.ca/rma/dpr/01-02/nrc/nrc0102dpr_e.pdf*>, NRC (2002-2003), *Performance Report.* Ottawa. <*http://www.collectionscanada.gc.ca/webarchives/20060120100152/http://www.tbs-sct.gc.ca/rma/dpr/02-03/nrc-cnrc/nrc-cnrc03d01_e.asp#S2_2*>.

3. The **Canadian Technology Network (CTN)** and IRAP collaborated to develop regional initiatives and a ***Technology Cluster Strategy*** focused on linking existing local skills and strengths and opportunities in emerging sectors to core NRC R&D priorities.[9]
4. IRAP joined with **Technology Partnership Canada (TPC)** to support and fund SMEs with projects at the pre-commercialization stage.

✓ 2005-2009: IRAP developed a new strategic plan emphasizing the needs of medium-sized businesses and on helping more businesses grow from small to medium size.[10] To achieve this objective, the total number of funded projects decreased but the size and number of larger projects increased. In addition, because of the success of the IRAP program, new roles and responsibilities were directed toward IRAP including:

1. IRAP partnered with the NRC's **Canadian Institute for Scientific and Technical Information (CISTI)** in a pilot program to provide Competitive Technology Intelligence (CTI) to firms through NRC-IRAP ITAs.
2. IRAP provided Technology assessments for DFAIT's **International Science and Technology Partnership Program** (ISTPP).
3. IRAP provided application, technology, and commercialization validation for PWGSC's **Canada Innovation Commercialization Program** (CICP).
4. IRAP provided regular support for Department of Foreign Affairs and International Trade (DFAIT)'s **Trade Missions.**
5. IRAP provided support for SMEs in communities in Southern Ontario through the disbursement of $45 million from the **Federal Economic Development Agency for Southern Ontario** (FedDev).

---

[9]For more information on NRC Cluster Strategy, see *<http://www.nrc-cnrc.gc.ca/eng/clusters/index.html>*.

[10]NRC (2003-2004), *Performance Report.* Ottawa. *<http://www.collectionscanada.gc.ca/webarchives/20060120074932/http://www.tbs-sct.gc.ca/rma/dpr/03-04/nrc-cnrc/nrc-cnrcd3401_e.asp#_Toc81043592>*, NRC (2004-2005), *Performance Report.* Ottawa. *<http://www.collectionscanada.gc.ca/webarchives/20060120062945/http://www.tbs-sct.gc.ca/rma/dpr1/04-05/nrc-cnrc/nrc-cnrcd45_e.pdf>*, NRC (2004-2005), *Performance Report.* Ottawa. *<http://www.collectionscanada.gc.ca/webarchives/20071207170610/http://www.tbs-sct.gc.ca/dpr-rmr/0506/nrc-cnrc/nrc-cnrc_e.pdf>*.

- ✓ 2009-2011: In response to the 2008 global financial crisis and recession, the Government of Canada developed a comprehensive stimulus and economic recovery plan called "Canada's Economic Action Plan". As part of this plan, IRAP received a monumental investment in Budget 2009 of $200 million over two years. The funding included $170 million to double IRAP's contributions to firms and $30 million to help hire more than 1,000 new post-secondary graduates via its Youth Employment Program.[11] NRC-IRAP estimated that this would enable it to provide larger contributions to firms per project and allow it to support approximately 1,400 additional SMEs (beyond its annual base) over the two year period of the stimulus package, see Table APP-A1-1.[12] This increased flow of resources is discussed further in section 3.

  IRAP was also given responsibility to deliver the **Digital Technology Adoption Pilot Program (DTAPP** is an $80 million pilot program that began in October 2011 and that will run until March 31, 2014).[13] It is delivered by IRAP ITAs who offer SMEs advice on ICT systems and technologies, and provide funding to improve business performance. It is a component of the Government of Canada's overall Digital Economy Strategy to accelerate digital technology adoption and investment in Canada in order to increase SME productivity.

- ✓ 2012: In the 2012 federal Budget (March 2012), IRAP received a $110 million increase in funding per year, effectively doubling its budget. It

TABLE APP-A1-1 The Impact of Budget 2009 Over Two Years

Purpose	2009-2011	Projected Impact
Contributions to firms	$170 M	1,400 firms
Graduates through YEP	$30 M	1,000 post-secondary graduates
With existing organizational structures and existing operating dollars		

NOTE: NRC, "NRC-Industrial Research Assistance Program (NRC-IRAP)," March 2010 *<http://acamp.ca/alberta-micro-nano/images/docs-conventional-energy/Finance/Generic%20IRAP%20PPT_FINAL%20ENG_March-2-2010.pdf>*.

---

[11]Government of Canada, "Canada Economic Action Plan: Budget 2009," Tabled in the House of Commons by the Honorable James M. Flaherty, Minister of Finance on January 27, 2009 *<http://www.budget.gc.ca/2009/pdf/budget-planbugetaire-eng.pdf>*.

[12]NRC, "National Research Council Canada Backgrounder," *<http://www.dehnel.com/images/D-Pace-IRAP_backgrounder_PCO_e.pdf>* and Government of Canada, "Canada Economic Action Plan: Budget 2009," Tabled in the House of Commons by the Honorable James M. Flaherty, Minister of Finance on January 27, 2009 *<http://www.budget.gc.ca/2009/pdf/budget-planbugetaire-eng.pdf>*.

[13]NRC, "Digital Technology Adoption Pilot Program (DTAPP)," *<http://www.nrc-cnrc.gc.ca/eng/ibp/irap/digital-technology-adoption/dtapp-index.html>*.

was also asked to develop and deliver a new "concierge" service to SMEs across the country in order to provide SMEs with one-stop shopping to access the range of other federal innovation support programs. This influx of funding was in response to the Jenkins Report (a comprehensive review of support for business R&D in Canada with the objective of optimizing the contributions of the Government of Canada to improve business innovation and competitiveness). In particular, recommendation two of the Jenkins Report advised the Government to redeploy funds from "Indirect" tax credit programs like the Scientific Research and Experimental Development (SR&ED) program to a more complete set of "direct" support business through programs like IRAP.[14]

ORGANIZATION AND GOVERNANCE

The IRAP program is administered through the National Research Council of Canada (NRC). The NRC is Canada's largest federal research and technology organization with a budget in 2010-11 of $749 million, employing over 4,000 staff in most areas of science and engineering research and includes 1,500 visiting workers.[15] The NRC is a Government of Canada statutory organization with its mandate outlined in the *National Research Council Act*. NRC's powers include:[16]

1. Undertaking, assisting or promoting **scientific and industrial research** in different fields of importance to Canada;
2. Investigating **standards** and **methods of measurement**;
3. Working on the **standardization and certification** of scientific and technical apparatus and instruments and materials used or usable by Canadian industry;
4. Administering NRC's **research and development activities**; and
5. Providing vital **scientific and technological services** to the research and industrial communities (IRAP derives its authority from this statutory power).

The IRAP Program is overseen by the NRC's Senior Executive Committee composed of a President, five (5) Vice Presidents, the Secretary General, the Director General of Human Resources Branch, the Director General of IRAP and the Chief Financial officer. Services are delivered on a

[14]Review of Federal Support to Research and Development—Expert Panel Report, "Innovation Canada: A
Call to Action" <*http://rd-review.ca/eic/site/033.nsf/vwapj/R-D_InnovationCanada_Final-eng.pdf/$FILE/R-D_InnovationCanada_Final-eng.pdf*>.

[15]John McDougall (President of NRC), "National Research Council Canada," Presentation at the 10th Annual Re$earch Money Conference on May 11, 2011.

[16]National Research Council Act, R.S.C., 1985, c. N-15, <*http://laws.justice.gc.ca/PDF/N-15.pdf*>.

decentralized basis across Canada primarily by the ITAs. In addition, the NRC-IRAP program has an Advisory Board in place to provide an external perspective on the strategic direction and management of the program. The 10 to 12 advisors are appointed by the President of the NRC, and represent a cross-section of national industry associations, provincial organizations, universities and government.[17]

As shown in Figure APP-A1-2, the ***Director General*** of NRC-IRAP reports directly to NRC's ***President*** and is responsible for managing the national delivery of the program and demonstrating the program's alignment with its strategic objectives. The position is supported by six ***Executive Directors*** located at the National Office and in each of the five regions. The Executive Directors are supported by ***Directors*** who are responsible for directing the ***Industrial Technology Advisors*** (ITAs) in their responsibilities for providing technological and funding assistance to small and medium size firms.

Innovation Network Advisors (INAs) in the region, who report to the Executive Directors, focus on building effective regional innovation system relationships and, where warranted, work with innovation support organizations to provide expanded or new innovation assistance to SMEs. They also work closely with the Directors and the ITAs in each region and help identify gaps in regional innovation assistance available to respond to SME needs and work with organizations to provide resources to address these unmet needs.

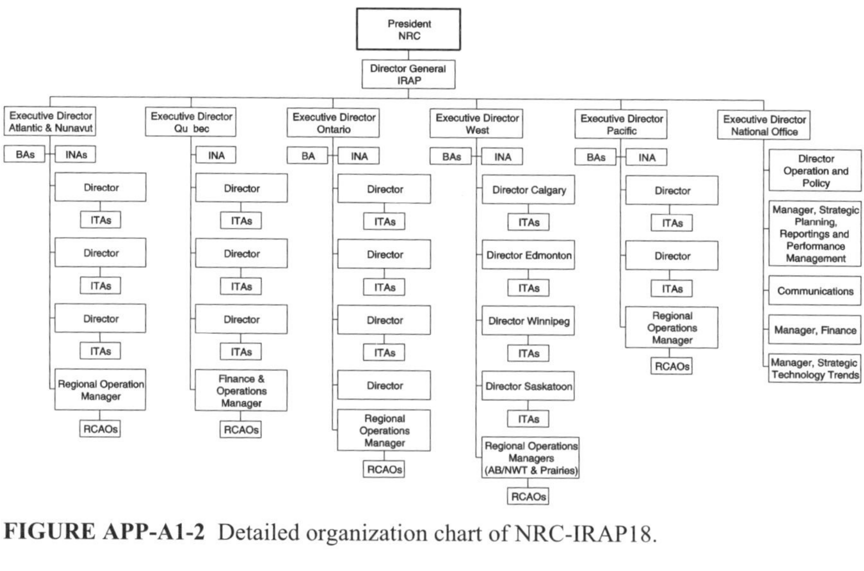

FIGURE APP-A1-2 Detailed organization chart of NRC-IRAP18.

---

[17]NRC, "NRC-IRAP Advisory Board," <*http://www.nrc-cnrc.gc.ca/eng/ibp/irap/about/advisory-board.html*>.

[18]NRC, "Internal Audit," September 2007 <*http://www.nrc-cnrc.gc.ca/obj/nrc-cnrc/doc/audit-nrc-irap.pdf*>.

Business Analysts (BAs) who are located in all regions except Québec, manage a portfolio of clients in collaboration with ITAs, and conduct due diligence on client companies and their projects to ensure that the basic business functions are planned for and provide support and advice to clients throughout the process.

Each region has either a ***Regional Operations Manager*** (ROM) or a ***Finance and Operations Manager*** (FOM) reporting to the Executive Director, responsible for ensuring day-to-day operations of the program in the areas of resource management, quality assurance and performance management. Finally, ***Regional Contribution Agreement Officers*** (RCAOs) who work under the supervision of the ROM / FOM are responsible for a wide range of activities undertaken in support of the delivery of the program with regard to financial contributions. More specifically they administer contribution agreements with firms and organizations, including assisting the ITA with preparing and reviewing agreements and amendments, reviewing claims and processing payments in accordance with the terms of the contribution agreement, and advising ITAs, INAs, signing authorities and clients on appropriate modifications to agreement clauses.[19]

Regional Distribution

IRAP field staff is located in 147 offices in 100 communities across Canada in regional economic development organizations and networks, business associations, universities, and colleges. In a 2007 survey of ITAs, 80 ITAs (37 percent) indicated that they are co-located at NRC facilities.[20] With the construction of new facilities in support of technology clusters, previously dispersed business lines of NRC (including IRAP ITAs) are now housed under a single roof in some cities. Examples include Chicoutimi, Québec's Aluminum Technology Centre; Edmonton, Alberta's National Institute for Nanotechnology and Fredericton, New Brunswick's Institute for Information Technology.[21]

IRAP RESOURCES AND BUDGETS

The NRC-IRAP budget allocations for FY2008-2009 to FY 2011-12 are shown in Table APP-A1-2. Over this period of time, IRAP's A-Base budget (the normal amount it receives every year) has averaged about \$89 million per year in contribution payments to SMEs, as well as \$48 million per year in wages and operating costs—for a total IRAP budget of \$137 million per year.

IRAP's B-Base funding is also identified in Table APP-A1-2 (B-Base funding is one-off incremental funding for a limited time and for a specific

---

[19]NRC, "NRC-IRAP Advisory Board," *<http://www.nrc-cnrc.gc.ca/eng/ibp/irap/about/advisory-board.html>*.

[20]Impact Evaluation 2007, p. 45.

[21]Impact Evaluation 2007, p. 45.

purpose). This B-Base funding is primarily the result of the Government of Canada's comprehensive stimulus and economic recovery plan called "Canada's Economic Action Plan" in response to the 2008 global financial crisis and economic recession. As part of this plan, IRAP received an investment in Budget 2009 of $200 million over two years. The funding included $170 million to double IRAP's contributions to firms and $30 million to help hire more than 1,000 new post-secondary graduates via its Youth Employment Program. As a result, note then that contribution payments in FY 2009-2010 and 2010-2011 (at a level of $235 million) are unusually high for each year—and then in 2011-12 contribution funding reverts to the more normal level of $89 million.

TABLE APP-A1-2 NRC-IRAP Budget (FY2008-2009 to FY2011-2012)

Programs ($000)	2008-2009	2009-2010	2010-2011 budget as at October 1, 2010	2011-2012
A-Base Contributions				
Contribution to firms	71,002	73,601	72,384	72,714
Contribution to Organizations	11,518	11,371	11,763	11,380
Youth Program	5,000	5,000	5,000	5,000
Total A-Base Contributions	87,520	89,972	89,147	89,094
B-Base Contributions				
Canada's Economic Action Plan (CEAP—Firms)	0	90,000	80,000	
Canada's Economic Action Plan (CEAP—Youth)	0	10,000	20,000	0
Youth Career Focus Program Funding	0	262	3,624	0
FedDev Ontario (CAF)	0	17,500	26,600	
FedDev Ontario (SODP)	0	27,500	16,150	0
Total B-Base Contributions	0	145,262	146,374	0
Total Contributions	87,520	235,234	235,521	89,094
Wages	33,904	32,550	37,742	35,208
Operating	14,781	12,176	8,741	12,292
FedDev Ontario (CAP and SODP) Operating Budget (Salary and Ops)	0	2,118	2,001	
Total Wages and Operating	48,685	46,844	48,484	47,500
Total Budget	136,205	282,078	284,005	136,594

SOURCE: NRC-IRAP, "Program Overview," January 2011.

This table clearly illustrates the one-off incremental funding that was provided to IRAP during 2008-2009 and 2010-2011. This B-Base funding in effect doubled the NRC-IRAP's national budget and quadrupled NRC-IRAP Ontario's budget.[22] Note however that the amount spent on wages and operations remained constant at $48 million. Thus, while additional funding to the IRAP program for contribution to SMEs is beneficial, it also placed enormous pressure on IRAP staff to deliver the program with no new administrative or operating resources.

In addition, during this two year period as part of the Government's stimulus and economic recovery plan, IRAP was asked to deliver two support programs (involving $88M in payments) for the Federal Economic Development Agency for Southern Ontario (FedDev Ontario). These programs were the Southern Ontario Development Program (SODP) aimed at stimulating local economies and enhancing the growth and competitiveness of local businesses and communities,[23] and the Community Adjustment Fund (CAF) aimed at quickly minimizing the impacts of the global economic downturn by creating employment opportunities, and assisting communities with populations of less than 250,000 to adjust and restructure their economies.[24]

Figure APP-A1-4 shows the key components of IRAP's budget for a normal year, in this case 2011-12. Note that just over half of IRAP's 2011-2012 budget (53 percent) is allocated to payments to firms and another 8 percent to regional organizations that support business innovation. Note also that the portion of total IRAP funding spent on wages is high at 26 percent. This is a reflection of the quality and experience of the ITAs and the large number of ITAs across the country. It also reflects the fact that like the rural doctor, ITAs make house calls (i.e. they regularly visit each client SME at their location). Also note that the portion spent on the Youth Program (employing recent technology graduates) is small at 4 percent. Arguably, considering the high employment rate for youth in Canada at 13.9 percent in early 2012 (at nearly twice the overall unemployment rate of 7.3 percent), more funding could be allocated to this program.[25]

Figure APP-A1-5, identifies the types of organizations that receive funds to support private sector innovation (8 percent of the 2011-2012 budget).[26] These recipients represent a broad cross section of innovation intermediaries

---

[22]NRC (2009-2010), "Departmental Performance Report," *<http://www.tbs-sct.gc.ca/dpr-rmr/2009-2010/inst/nrc/nrc-eng.pdf>*, p. 22.

[23]Canada's Economic Action Plan, "Southern Ontario Development Program," *<http://www.feddevontario.gc.ca/eic/site/723.nsf/eng/h_00097.html>*.

[24]Canada's Economic Action Plan, "Community Adjustment Fund (Budget 2009 and Budget 2010)," *<http://www.actionplan.gc.ca/initiatives/eng/index.asp?mode=7&initiativeID=126>*.

[25]The Star, "Youth jobless rates remain high globally, report says," on May 22, 2012 at *<http://www.thestar.com/business/article/1182249--youth-jobless-rates-remain-high-globally-report-says>*.

[26]Impact Evaluation 2007, p. 33.

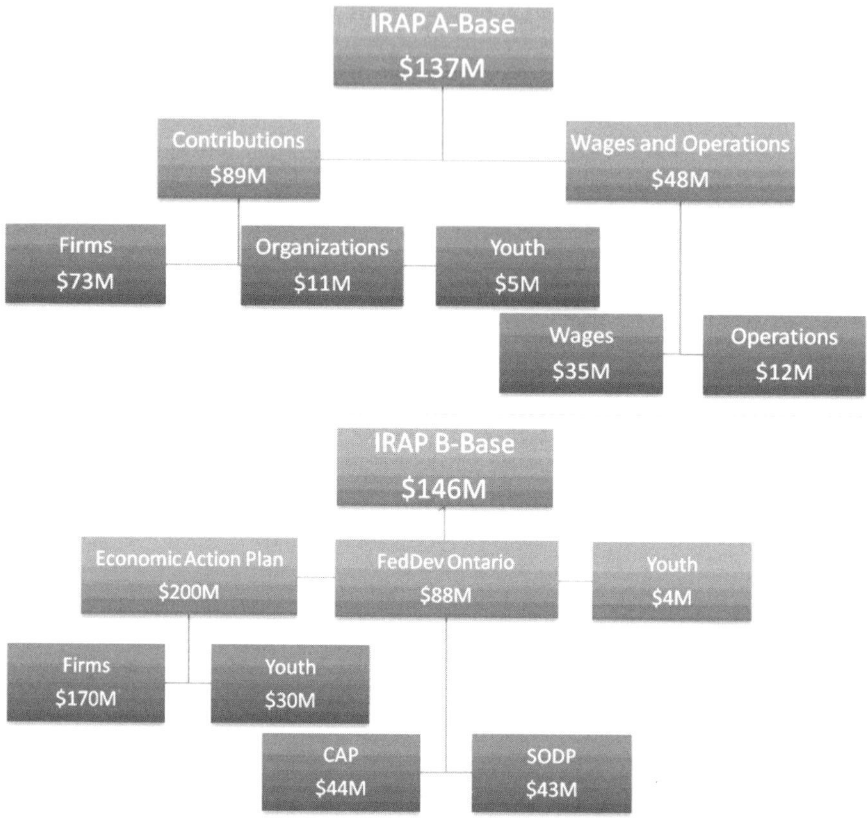

FIGURE APP-A1-3 A summary of NRC-IRAP's A and B base funding during 2009-2010 and 2010-2011 stimulus (Millions of Dollars).

that offer innovation services to SMEs. For example, to pick one IRAP region, the types of organizations in Western Canada that receive funding include: Alberta Innovates, Arctic Energy Alliance, BioAccess Commercialization Centre, Biomedical Commercialization Canada, Calgary Technologies Inc., Canadian Environmental, Technology Advancement, Corporation (CETAC) WEST, Communities of Tomorrow, Composites Innovation Centre, Enterprise Saskatchewan, Entrepreneurial Foundation of Saskatchewan, Government of the Northwest Territories, Industrial Technology Centre, Manitoba Innovation, Energy and Mines, Springboard West Innovations, TEC Edmonton, Western Economic Diversification Canada, and local universities, colleges and technical institutes.

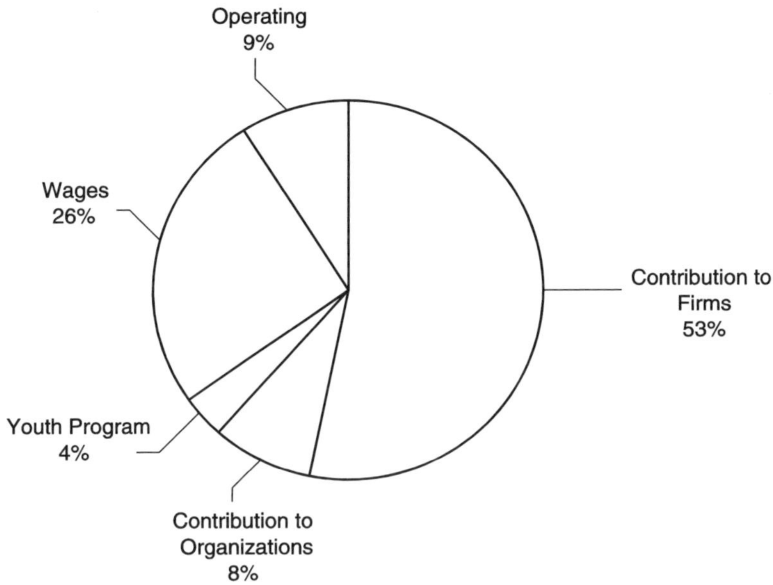

FIGURE APP-A1-4 Budget breakdown.

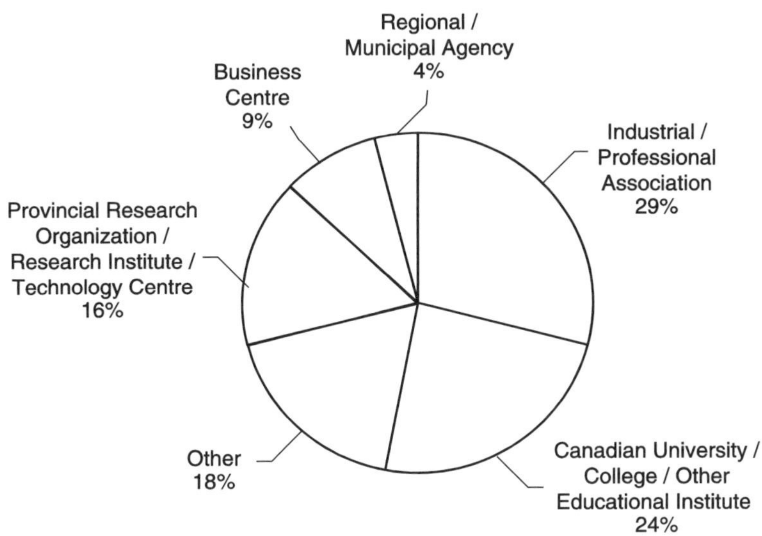

FIGURE APP-A1-5 Types of organizations.

SERVICES

The NRC-IRAP program provides support to Canadian SMEs for the development and commercialization of their technologies through the following two main groups of services:

1. **Technical and Business Advisory Services are** provided by over 230 Industrial Technology Advisors (ITAs) to mentor clients' projects through all stages of the innovation and commercialization process. ITAs are viewed as a trusted advisor, which often results in long-term relationships and partnerships formed between NRC-IRAP and Canadian SMEs. ITAs can also recommend project funding as noted below. They also provide **Networking and Linkage Services** through their extensive networks which can connect SMEs with industry experts, providers of risk capital, potential commercial partners and other investors. Access to the ITAs' commercial networks opens up opportunities for SMEs to connect to knowledgeable individuals and organizations about local sources of financing, regional R&D initiatives, technology brokers and technology transfer centers.

2. **Financial Assistance** is provided to support innovation projects, graduates, and organizations. For example, in the 2009-2010 fiscal year, IRAP contributions supported 2,597 firms, 2,947 projects, and the creation of 11,921 jobs.

   a. **Financial contributions to firms to develop technologies** provide up to 75 percent or $1,000,000 per year of the total eligible activities (e.g. eligible expenditures include up to 80 percent of labor costs and 50 percent of sub-contractor costs; but exclude material and capital expenditures). Note that an eligible firm must be an SME incorporated in Canada that is committed to growing its business and generating profits, through the development and commercialization of innovative, technology-driven new or improved products, services, or processes.
   b. **Financial contributions to third party innovation organizations that provide services for SMEs.**
   c. **Financial assistance to hire new graduates under the Youth Employment Program.** The program provides up to $30,000 in a non-repayable contribution to help clients hire graduates for an R&D project. This provides post-secondary graduates (aged 15 to 30) with tailored, career-related work experience. SMEs benefit from the knowledge of new graduates while graduates gain

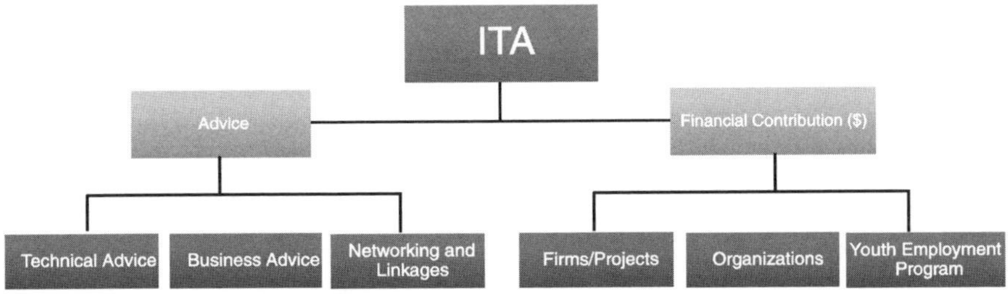

FIGURE APP-A1-6 NRC-IRAP services.

valuable work experience that will assist their career choices. IRAP supported 781 recent graduates in 2009-2010.[27]

In summary as shown in Figure APP-A1-6, the ITA is the main delivery agent of the IRAP program. ITAs deliver either advice, or financial contributions. The advice can be technical or business advice, and also extends the network of the SME to include individuals and organizations knowledgeable about local sources of financing, research and development institutions, technology brokers, and technology transfer centers. ITAs can also provide financial assistance in the form of a grant to projects with merit, or funding to regional organizations that offer other innovation services to SMEs, or funding to hire post-secondary graduates.

The Key IRAP Resource: The Skills and Role of the ITA

IRAP employs over 230 Industrial Technology Advisors (ITAs) which account for approximately two thirds of IRAP staff. They deliver the core IRAP services to SME firms. Each work with about 42 clients per year and 11 of their clients will receive funding.[28] They provide R&D project advice, technology and business advisory services, competitive technical intelligence, networking and linkages, funding, and investment readiness counseling.

The ITA's goals are to help SME's develop/improve new products and processes, improve efficiency and productivity, adopt and adapt technologies, and increase sales and add jobs.

In this regard, the experience of the ITA field staff sets them apart from most other federal and provincial government service providers. The profile of the ITA advisors is outlined below and in Figure APP-A1-7:[29]

- 75 percent have masters or PhD degrees.
- 80 percent have previous specialized industrial experience.

[27]NRC (2009-2010), "Departmental Performance Report," <*http://www.tbs-sct.gc.ca/dpr-rmr/2009-2010/inst/nrc/nrc-eng.pdf*>.
[28]NRC-IRAP, "Program Overview," January 2011.
[29]NRC-IRAP, "Program Overview," January 2011.

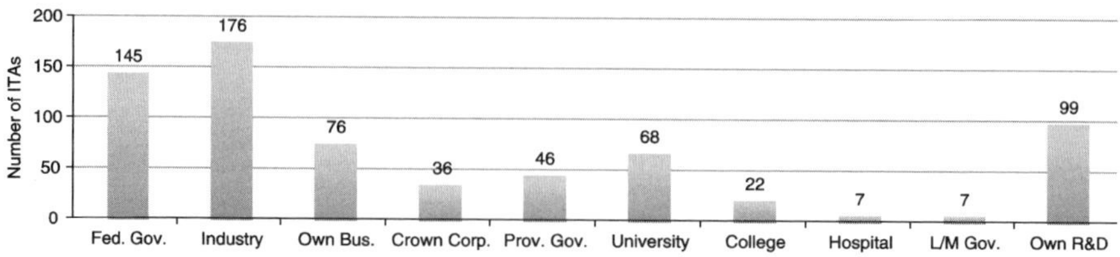

FIGURE APP-A1-7 ITA experience.
SOURCE: NRC-IRAP, "Program Overview," January 2011.
NOTE: ITAs can have more than one employment experience.

- 55 percent have worked in another Federal government laboratory or department.
- 41 percent have worked in a college, cégep or university.
- 45 percent have run their own R&D facility and/or been a leader within a R&D facility.
- 34 percent have been entrepreneurs.
- 16 percent have experience in some type of Crown corporation.
- 24 percent have some experience in working either in the provincial or municipal government level.

Also see Table APP-A1-3 for a summary of how ITAs spend their time on the job (and how they feel they should spend their time). In summary, the ITAs feel they should spend more time developing funded projects and managing contribution agreements and less time on administrative tasks.

Table APP-A1-4 identifies the top four advisory services (out of 22 services) requested by SME clients from ITAs. SME clients are not charged for this advice. The top four are: **management** advice, **financial** advice, **technical** advice and **marketing** advice. This summary illustrates the broad range of business advice offered by ITAs. However note that there are considerable differences in ranking between the advice actually requested and what clients might be willing to pay for. For example, of the top four ranked services, no client firms were particularly willing to pay for these services. Instead, the client firms indicated they were more willing to pay for: access to legal advice, partnering, competitive technical intelligence, and patent/literature searches.

The Role of the ITA in Funding a Project

A good SME project proposal to IRAP will include clear commercial goals and objectives (identifying challenges, risks and opportunities), a work plan, and a budget estimating costs, and demonstrate a good business case to pursue the project. The proposal should identify potential applications and

TABLE APP-A1-3 Summary of How ITAs Spend their Time

ITA Survey—How do ITAs spend their time?	Percent 4 years ago	2006-2007	How ITAs feel they should spend their time
Developing funded projects and monitoring/managing contribution agreements	38.6	34.3	38.4
Providing technology information and advice	23.2	17.8	26.6
Engaging in linkage and networking activities (e.g. presentations; building, maintaining and expanding contacts; attending seminars and courses)	15.5	12.1	17.6
Completing administrative tasks (e.g. information tracking, reporting, due diligence)	17.1	29.5	11.8
Other	5.5	6.3	5.6
TOTAL	100	100	100

SOURCE: Impact Evaluation 2007, p. 58.

markets, potential impacts and benefits to the company and to Canada (in the form of growth, job creation, wealth generation and improved quality of life), and must advance the financial position of the company in increasing market share, profits and sustainability.

IRAP payments (see step 5 in Figure APP-A1-8) begin after the project is approved. The client will then start to implement and manage the project, submitting status reports and monthly claims according to the costs incurred and dispersed (note that good record keeping is mandatory and the potential for audits is high).

In Figure APP-A1-8, note that step 7 identifies the mandatory requirement to report yearly on **Benefits to Canada.** This includes reporting on the following tangible outcomes:

1. New or improved products.
2. New or improved processes.
3. Increased capabilities and competitiveness.
4. Improved efficiency and increased productivity.
5. Introduction of new technologies.
6. Acquiring new patents.

TABLE APP-A1-4 NRC-IRAP Advisory Services Accessed by Client Firms (2002 to 2007)

Advisory Services Accessed by Clients	Percentage of Clients Accessing Service*	Ranking based on Access	Ranking Based on Willingness to Pay for Service
Management Advice	24.2	1	5
Financial Advice	23.3	2	7
Technical Advice	22.7	3	6
Marketing Advice	21.8	4	10
Patent Search and Literature Search	17.5	5	4
Mentoring and Coaching	17.2	6	13
Expertise Linkages	16.9	7	18
Technical Assessment	16.9	8	11
Referral to Other Organizations (including non-governmental)	15.9	9	22
Project Development	15.7	10	12
Market Assessment	15.2	11	9
Project Management	15.0	12	15
Market Linkages	14.6	13	16
Competitive Technical Intelligence	12.0	14	3
Business Strategies	12.0	15	8
Linkages and Referrals to Other Innovation Systems	11.9	16	21
Identified Business Opportunities	10.5	17	14
Promotion and Trade Shows	9.9	18	20
Partnering	8.0	19	2
Access to Legal Advice	7.6	20	1
Export Development	5.9	21	17
International Trade Missions	5.8	22	19

SOURCE: Impact Evaluation 2007, p. 23.
NOTE: (*) Percentages sum to more than 100 percent since clients can identify accessing more than one advisory service.

7. Attracting and retaining the best technical employees.
8. Increasing sales, revenue, productivity, market share, new markets and creating new jobs.

Regarding contribution funding, an important distinction should be made between funding "firms" and funding "projects." Firms are eligible to receive funding for a project that will support the development of their innovative technology. (The project must lead to: increased skills, knowledge and technical competencies, improved management capabilities, or new or

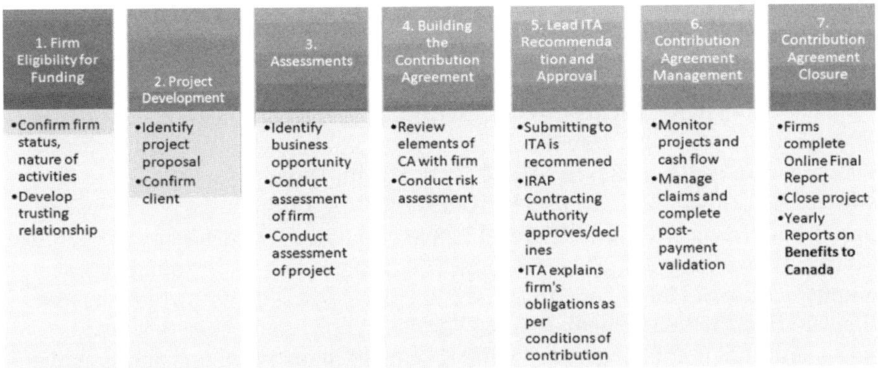

FIGURE APP-A1-8 How does funding work?

improved products, processes and services). Note that firms may enter into several project agreements with IRAP over time. There is no limit on the number of projects, or any program graduation required. However an overall limit on the level of contributions that may be made to a given firm, in any single year, is set at $1,000,000.[30]

Not surprisingly as a result, there is a higher level of project activity than firm activity because of variability in project length as well as the potential for successive projects in a given fiscal year. For example, between 2002-03 and 2006-07, IRAP started and funded approximately 5,800 discrete projects, but only 4,100 firms were funded.

Note in Table APP-A1-5 that **the number of funded client firms has decreased by 21 percent between 2002-03 and 2006-07 while the total contribution funding values have declined only very slightly (0.8 percent decline)**. As a result, the average payments to firms have **increased by 24 percent.**

TABLE APP-A1-5 Number of Firms Actively Funded by NRC-IRAP

Firms/Fiscal Year	2002-2003	2003-2004	2004-2005	2005-2006	2006-2007	Percent Change
Total Firms Actively Funded	1746	1745	1609	1498	1393	-21
Total Contribution Funding (in millions $s)	62.2	62.4	60.6	68.3	61.7	-0.8
Average Contribution ($)	35,624	35,759	37,663	45,594	44,293	24

SOURCE: Impact Evaluation 2007, p. 11.

[30]Impact Evaluation 2007, p. 13.

IRAP'S CLIENTS

NRC-IRAP clients can be described as Canadian SME firms with under 500 employees with the potential to innovate. The most recent statistics from 2009-2010 indicates that NRC-IRAP worked with a total of 8,578 SMEs of which 2,871 (about 33 percent) received some form of funding.[31] Note that 84 percent of clients have fewer than 50 employees, and 67 percent have fewer than 20 (recall that the average size of an SME in Canada is 6 to 7 people). These companies are not large enough to have extensive R&D departments—instead they are faced with technical, financial or managerial challenges that are obstacles to their advancement.[32] Clients span all industry sectors and Canadian geography; see the detailed regional client breakdown and description in Table APP-A1-6.

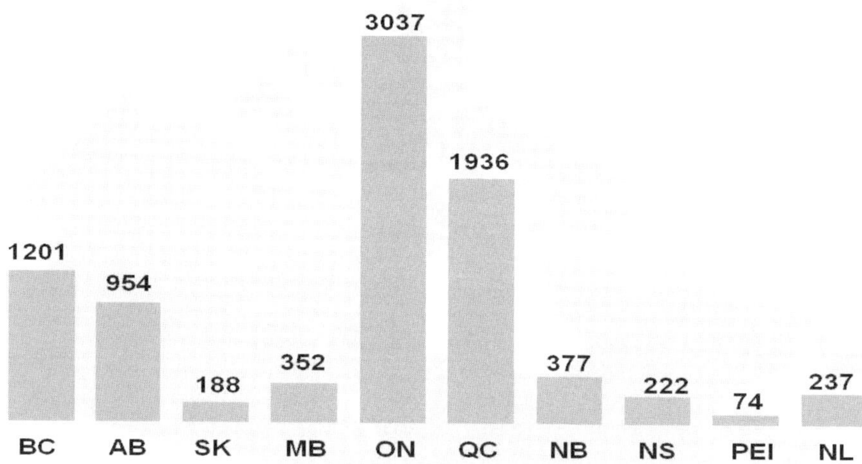

FIGURE A1-9 National overview NRC-IRAP clients by location (2009-2010).

[31]John McDougall (President of NRC), "National Research Council Canada," Presentation at the 10th Annual Re$earch Money Conference on May 11, 2011.
[32]NRC-IRAP, "Program Overview," January 2011.

A Snapshot of Client Firms

Various characteristics of client firms are discussed below, including:

1. Size of Firm (employee size).
2. Age of Firm.
3. Industry Sector of Firm.
4. Location of Firm.
5. Goods and Services.
6. Project Duration.
7. Amount Funded.

Size of Firm. Between 2002-03 and 2006-07 the average number of employees in the firms that received funding was 32 employees.[33] As mentioned earlier, over this five year period, about 67 percent of IRAP funded firms had fewer than 20 employees (23 percent had 1 to 4 employees, 25 percent had 5 to 9 employees and 19 percent had between 10 and 19 employees). More recently, again as of 2011, about 67 percent of IRAP funded firms had fewer than 20 employees.[34] While this suggests that IRAP is targeting smaller SMEs, it raises some issues about the adequacy of support for medium-sized SMEs.

Age of Firm. Note from Table APP-A1-7 that firms have been incorporated for about 11 years on average when they signed their first agreement with IRAP (with a median age of 7). This implies that most firms assisted by IRAP cannot be considered 'startups.'

Firm Sector. The IRAP program does not target specific sectors. Funding and advice is available to any firm that meets eligibility criteria, regardless of sector. However, the most recent data indicates that the largest served sector is ICT at 37 percent of all clients funded, with Manufacturing and Materials clients following at 14 percent, Health and Life Sciences at 11 percent and Energy and Environmental firms at 9 percent (see Figure APP-A1-10).[35]

Location of Firms. The data sets provided in the *Public Accounts of Canada* show all the firms (614) that received *over* $100,000 of IRAP funding (totaling $216.3 million in 2010-2011) and plot them geographically. The average amount transferred to industry for these 614 firms was $199,839 with a median of $160,000. In addition, there were a total of 2,481 recipients of transfer payments *under* $100,000 totaling $93.4 million which were not listed in detail.[36]

---

[33]Impact Evaluation 2007, p. 13.

[34]NRC-IRAP, "Program Overview," January 2011.

[35]Impact Evaluation 2007, p. 12.

[36]The live maps (<*http://batchgeo.com/map/42c2079a5ae7425be116dbd9f8d37dd2*>) include a web link to the firm's website, the location, and the amount of money received through the IRAP Program.

TABLE A1-6 IRAP Clients Across Canada

Region	Clients in key sectors and technologies	Partners	Success Stories
Pacific Region: British Columbia and Yukon **(33 ITAs)**	Forestry, mining, electronics, software and hardware, advanced materials, industrial engineering, food technology, biotechnology, construction, aerospace, fuel cells, electrochemistry, information technologies, engineering and physics	British Columbia Innovation Council, LifeSciences BC, BC Technology Industry Association, Okanagan Research and Innovation Centre, Victoria Advanced Technology Council, Western Economic Diversification Canada	e.g. Xenon-Pharmaceuticals, Ostara Nutrient Recovery Technologies Inc., Profile Composites Inc., Simon Fraser University, Green Steam Inc.[37]
West Region: Alberta, Saskatchewan, Manitoba, Northwest Territories **(47 ITAs)**	Advanced Chemistry and Materials, Advanced Manufacturing, Biotechnology and Bioprocessing, Construction, Energy, Environmental Technologies, Food and Agricultural sciences, Health-care and Pharmaceuticals, Information Technology, Electronics and Telecommunications, Transportation	Alberta Innovates, Arctic Energy Alliance, BioAccess Commercialization Centre, Biomedical Commercialization Canada, Calgary Technologies Inc., Canadian Environmental, Technology Advancement, Corporation (CETAC) WEST, Communities of Tomorrow, Composites Innovation Centre, Enterprise Saskatchewan, Entrepreneurial Foundation of Saskatchewan, Government of the Northwest Territories, Industrial Technology Centre, Manitoba Innovation, Energy	e.g. Boreal Laser Inc., Manitoba Harvest, Userful Corporation, Quantiam Technologies Inc., Quantum Technical Services.[38]

[37]NRC, "NRC-IRAP—On the leading edge of the mighty Pacific," <*http://www.nrc-cnrc.gc.ca/eng/ibp/irap/about/bc-yukon.html*>.
[38]NRC, "West: Alberta, Saskatchewan, Manitoba and the Northwest Territories," <*http://www.nrc-cnrc.gc.ca/eng/ibp/irap/about/alberta-nwt.html*>.

		and Mines, Springboard West Innovations, TEC Edmonton, Western Economic Diversification Canada, Local universities, colleges and technical institutes	
Ontario Region (70 ITAs)	Electronica, software, photonics, cleantech, manufacturing, life sciences, medical devices, construction, and aerospace	Federal Government: FedDev Ontario, Business Development Bank of Canada, Natural Sciences and Engineering Research Council of Canada, Foreign Affairs and International Trade Canada, Export Development Canada; Provincial Agencies: Ministry of Research and Innovation, Ontario Centres of Excellence, Ministry of Economic Development and Trade; NFP: ventureLAB, MaRS, Communitech, Sault Ste. Marie Innovation Centre, Northern Centre for Advanced Technology Inc., Ottawa Centre for Research and Innovation	e.g. Wolf Steel Inc., WinMagic Inc., Axiom Group Inc., Hendrick Seeds Inc.[39]

[39]NRC, "Ontario So Much to Discover," *<http://www.nrc-cnrc.gc.ca/eng/ibp/irap/about/ontario.html>*.

TABLE A1-6 ***Continued***

Region	Clients in key sectors and technologies	Partners	Success Stories
Quebec Region (50 ITAs)	Communications, Environment, and Manufacturing	Business Development Bank of Canada, Canada Economic Development for Quebec Regions, Centres locaux de developpement, Reseau Trans-tech	e.g. Motion Composites Inc., Technologies HumanWare Inc. Developpement Effenco Inc., Medicago Inc., Muridal Incorporated
Atlantic Region: NB, Newfoundland and Labrador, PEI, NS Nunavut **(Over 30 ITAs)**	Agriculture and aquaculture, wood products, advanced manufacturing, computer science and telecommunications, nutraceuticals, food products and biotechnology	Atlantic Canada Opportunities Agency, Fisheries and Marine Institute of Memorial University, Dalhousie University, New Brunswick Community College, PEI BioAlliance	e.g. Quark Engineering and Development Inc., Soricimed Biopaharma Inc., Acadian Seaplants Limited, Allens Fisheries

Quebec is the location of the largest number of funded firms (*over* $100,000) by the NRC-IRAP program at 178 firms (29 percent of the total) with Ontario a close second at 143 firms (24 percent) and British Columbia coming in third with 84 firms (14 percent). Not surprisingly, in Western Canada the numbers of funded firms are clustered around more densely populated cities (i.e. Vancouver 76 firms, Calgary 40 firms, Edmonton 26 firms, Saskatoon 17 firms, and Winnipeg 21 firms). In Central Canada, Toronto and the surrounding area have the highest number of funded firms at 119, followed by Montreal at 70 firms and closely followed by Ottawa with 69 firms. Eastern Canada has significantly fewer funded firms, given the smaller population, with the largest number of recipients centered on the three key cities of Halifax with 23 firms, St. John's with 22 firms, and Fredericton with 14 funded firms.

TABLE APP-A1-7 Age of Firms

	Age of Firm (Years)					
Percentage of Firms / Age of Firms	+/-1 to 2	3-5	6-10	11-15	16-20	Over20
Percentage of NRC-IRAP Firms	22	21	22	11	9	15

SOURCE: Impact Evaluation 2007, p. 13.

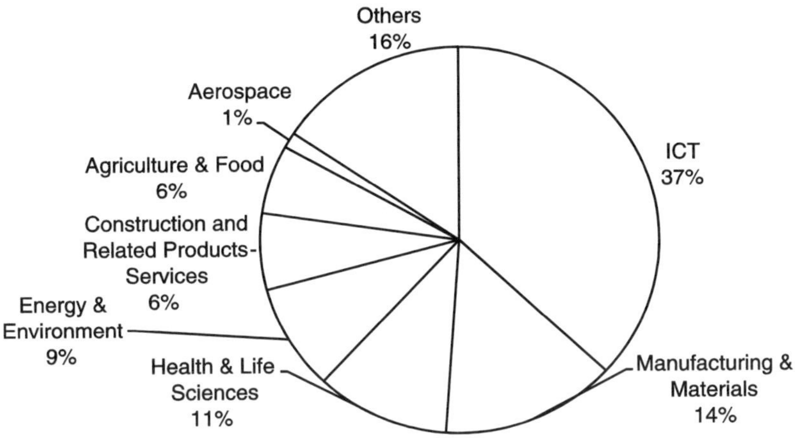

FIGURE APP-A1-10 IRAP Clients by industry sector, 2010-2011.
SOURCE: Bogdan Ciobanu, 2012.

Goods and Services. Figure APP-A1-11 illustrates that a higher proportion of the funded firms have been in the "goods" producing sector with the remaining categorized as "services" producing. This is not surprising given the presence of the manufacturing industry in the goods producing sector. Note however that the number of firms in the goods producing sector has been declining steadily since 2003-04.

Project Duration. The duration of a typical IRAP funded project

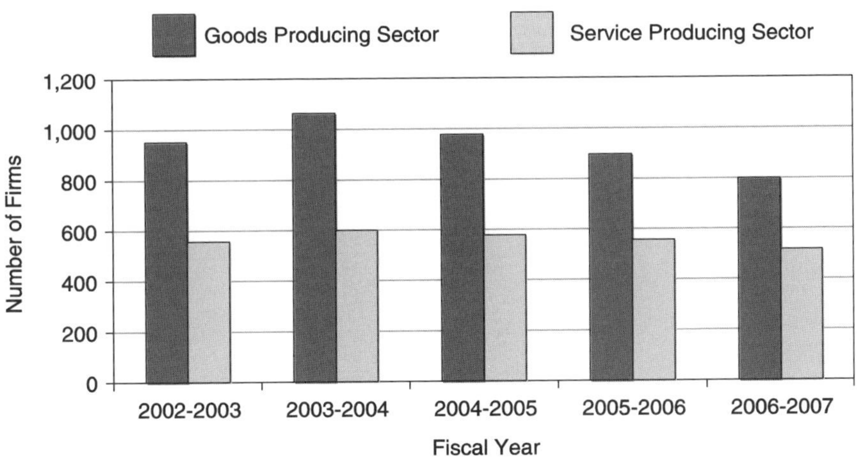

FIGURE APP-A1-11 Distribution of the NRC-IRAP funded firms.
SOURCE: Impact Evaluation 2007, p. 16.

averaged 9.5 months, with the shortest project duration at 2 days, and the longest *project lasting about 5 years. Again, this data suggests that IRAP funds relatively small projects that have shorter durations.*[40]

Number of Projects per Client Firm. As noted in Table APP-A1-8, a significant majority of IRAP clients (72 percent) receive funding for only one project. The remaining 28 percent receive funding for multiple projects with 18 percent receiving funding for two projects. This raises the issue of whether or not IRAP should be providing more sustainable funding to higher growth medium-sized SMEs in order to help them build critical mass to enter global markets and expand.

Amount Funded. The average total amount contributed to projects (i.e. dollars actually received by firms over the five years from 2002-03 to 2006-07) is approximately $56,000 per project.[41] Note however that the median amount per project over the five years was $20,000. Furthermore, while the median contribution per project grew from $15,000 to $36,750 over the five year period (an increase of 145 percent) these are still relatively small project investments to develop technologies for global markets.

BENEFITS

The IRAP Impact Evaluation Report of 2007 (a mandatory evaluation of the IRAP program conducted every five years—with the next evaluation due later this year) concluded that NRC-IRAP produces significant economic impact, and generates a higher SME capacity for R&D, knowledge creation and commercialization. The total wealth creation benefits of the Program were identified in the range of $2.3 billion to $6.5 billion (and estimated to be about $4.2 billion) over the 5 year period (2002-03 to 2006-07)—i.e. benefits exceed the costs of the program by a factor of 4 to 12 (estimated at 8). In fact, according to the Impact Evaluation Report:

- There is a strong correlation between NRC-IRAP assistance and firm growth—for example, for each 1 percent increase in both contribution agreement funding and advisory services, firms show an 11 percent increase in sales and a 14 percent increase in employment.[42]

TABLE APP-A1-8 Number of NRC-IRAP Funded Projects per Firm (by Fiscal Year)

Number of Projects per Funded Firm	Percentage of Total Funded Firms
1 Project	72.40
2 Projects	18.48
3 Projects	6.04
4 + Projects	3.08

SOURCE: Impact Evaluation 2007, p. 16.

40 Impact Evaluation 2007, p. 15.
41 Impact Evaluation 2007, p. 14.
42 Impact Evaluation 2007, p. 28.

TABLE APP-A1-9 Median Total Contribution Amount Received by Firms over the Life of NRC-IRAP Funded Projects (by fiscal year)

Median CA / Fiscal Year	2002-2003	2003-2004	2004-2005	2005-2006	2006-2007	Median Total CA Across 5 Years
National Median Total CA ($s)	15,000	15,000	20,000	25,500	36,750	20,001

- There is a positive and significant impact on productivity—a 1 percent increase in NRC-IRAP assistance (both funding and advisory) lead to a 12 percent increase in firm productivity, while a 1 percent increase in funding led to a 3 percent to 5 percent decrease in production costs.[43]
- There is a positive and significant relationship between NRC-IRAP assistance and R&D capacity and capabilities—a 1 percent increase in financial funding led to a 13 percent increase in R&D spending by the firm and a 3 percent increase in R&D staff.[44]

In addition, based on a review of the 437 final reports from NRC-IRAP clients, 94 percent of clients reported an increase in their technical knowledge and capabilities, 70 percent reported an increase in their ability to conduct R&D, and 93 percent reported that they anticipate commercial benefits to accrue as a result of IRAP assistance.[45]

Most recently, the Director General of IRAP, Bogdan Ciobanu, shared the following impact numbers in a presentation (May 2012). He stated that every $1 invested by IRAP generates an overall return on investment of $12 in socio-economic benefits to Canada in the form of:

- ✓ $10.44 in new sales.
- ✓ Increased employment.
- ✓ $2.5 invested by industry.

In short, he stated that the program continues to produce positive and significant benefits from its investments.

There are also a variety of third party sources who rate IRAP positively. For example, *The State of Science and Technology in Canada* report by the Canadian Council of Academies in September 2006 stated: "NRC-IRAP was rated the number one program in terms of support for commercialization of

---

[43]Impact Evaluation 2007, p. 28.

[44]Impact Evaluation 2007, p. 28.

[45]NRC (2009-2010), "Departmental Performance Report," <*http://www.tbs-sct.gc.ca/dpr-rmr/2009-2010/inst/nrc/nrc-eng.pdf*>, p. 22.

science and technology."[46] From an international perspective, IRAP is also gaining a reputation as a best practice. Some recent references include:[47]

- Thailand-based ITAP program recognized as best practice program, modeled after NRC-IRAP.
- 2007 Australian Department of Industry, Tourism and Resources "Study of the Role of Intermediaries in Support of Innovation."
- Columbia -- Bogotá's emulation of NRC-IRAP in discussion.
- Chile adoption of NRC-IRAP's Contribution to Organizations Program.
- IRAP highly regarded and referred to at a recent Technopolicy Conference on Creating World Class Regions in Germany (September 2010).
- ProInno Europe recently tabled IRAP as the "Good Practice Identification—Advice for transferability Innovation of Internationalization of SMEs Policies."

From the firms' perspective, the Program is also viewed very positively:

- For example, Ernie Davidson, an NRC-IRAP ITA, urged Mike Lazaridis (founder of RIM) to apply for NRC-IRAP funding to investigate Surface Mount Technology. As a result, sales to Sutherland-Schultz were so strong that RIM's annual revenue went over the $1M mark in 1990 for the first time in the company's history.
    - *"That was the breakthrough that got me the Sutherland-Schultz contract because we were probably the only company in town that even knew what surface mount technology was."*

      Mike Lazaridis, Founder and co-CEO, Research In Motion (RIM)

- *"There should be another billion dollars into IRAP because it is working."*

  Terry Matthews, Chairman and CEO, March Networks; Chairman, Mitel; Chairman, Wesley Clover International

[46]Council of Canadian Academies, "The State of Science & Technology in Canada," (2006). <*http://www.scienceadvice.ca/en/assessments/completed/science-technology.aspx*>.

[47]Program Overview Presentation by Dr. Tony Rahilly, Director General on October 5, 2010.

- *"IRAP is critical to building Canada's innovation economy...The popularity of the IRAP program is a testament to the strength of IRAP's reputation as a valuable resource in the science and tech community."*

Avvey Peters, Executive Director,
Government and Communications,
Communitech

What Metrics are Used to Measure These Benefits?

Director General Ciobanu articulated the following revised IRAP Metrics Framework (2012) in Figure APP-A1-12 to measure impact and benefits. This framework focuses on measuring five key outputs: increased revenues from goods and services developed by the project; total new employees; total new technical employees; new R&D spending; and equity raised as a result of the project.

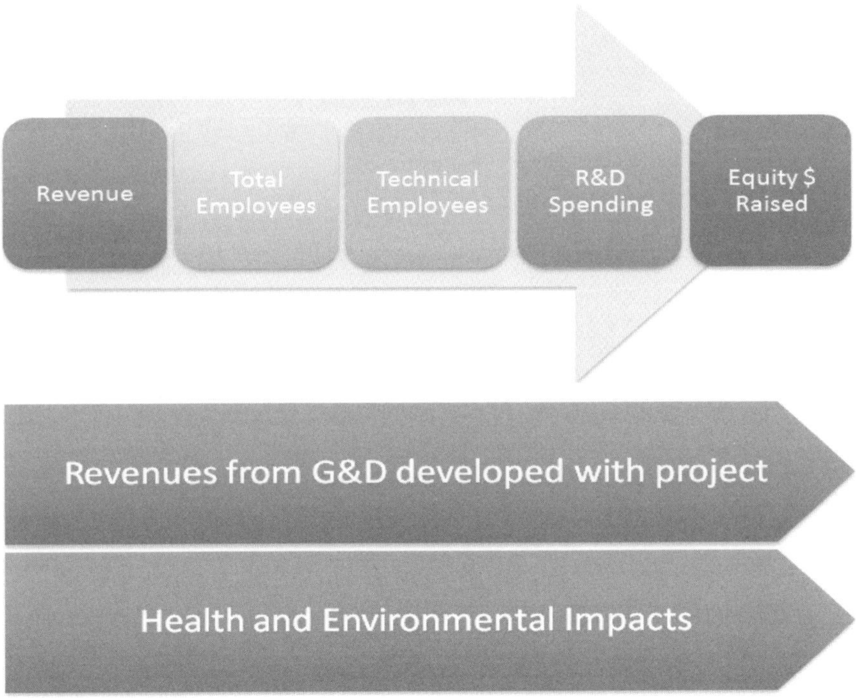

FIGURE APP-A1-12 Metrics framework.

As a result, the desired impacts from IRAP investments are increased revenue, new and improved products and services, increased sales and high quality jobs. Client firms are responsible for most of this information tracking and input. This improves the line of sight between managing the project investment and ensuring the monitoring of the return on investment to Canada. In developing the metrics for SME funded projects, consideration is also given to:[48]

- ✓ Information collected by the Canadian Revenue Agency, Scientific Research and Experimental Development (SR&ED) Tax Incentive Program, and Statistics Canada to ensure alignment.
- ✓ Information available from SME clients' fiscal year end statements to facilitate ease of completion.
- ✓ IRAP's reporting and accountability requirements, including those of formal evaluations.

Are There Particular Success Stories?

The Federal Budget is the government's most important annual policy statement reflecting through its announced spending, tax, and regulatory initiatives, the government's policy priorities for the year. Most recently, three brief success stories from the NRC-IRAP Program were highlighted in the Federal Budget 2012:

The National Research Council's Industrial Research Assistance Program (NRC-IRAP) has a strong track record of providing innovative small and medium-sized companies with financial resources, customized services and access to highly skilled people. Recent success stories include:[49]

1. **Wolf Steel Inc.** (Barrie, Ontario)—The firm has been able to design, manufacture and market a new high-efficiency furnace—the only gas furnace manufactured in Canada—with assistance and advisory services from NRC-IRAP.
2. **Quark Engineering and Development Inc.** (Halifax, Nova Scotia)—The firm has transformed their TetherBerry networking technology into a multi-platform and Bluetooth-enabled device, leading to better market penetration, new business opportunities and increased revenues, with the assistance of NRC-IRAP.
3. **Motion Composites Inc.** (Saint-Roch-de-l'Achigan, Quebec)—The company designed and produced a wheelchair that is lighter, more

---

[48]Shannon Townsend, Margaret McKay, Julia Rylands, Melissa Littau, "Measuring Results: From Actions to Impact," at the 15th Annual PPX Symposium on May 25-26, 2011.

[49]Government of Canada, "Jobs, Growth, and Long-Term Prosperity: Economic Action Plan 2012," tabled in the House of Commons by the Honourable James M. Flaherty, P.C., M.P. Minister of Finance on March 29, 2012. <*http://www.budget.gc.ca/2012/plan/pdf/Plan2012-eng.pdf*>.

durable and more affordable than a traditional wheelchair, with financial support from NRC-IRAP.

IRAP STRENGTHS AND CHALLENGES

There are many strong features of the IRAP Program. The following five give a sense of their diversity.

1. The IRAP program operates in every major city and community and every Province and Territory across Canada, **connected by a common culture and communications backbone**. This permits it to serve SMEs wherever their location and bring the suite of IRAP advisory and financial services to them. It also permits an IRAP ITA in one part of Canada to use the common platform and bring in ITA expertise from other parts of the country to meet a specific SME's needs, wherever they are located.
2. The IRAP program is **integrated into the research and technology expertise of the National Research Council** and its 18 Institutes across Canada, and many ITAs are co-located at these Institutes. This permits the ITAs to obtain privileged and easy access to NRC researchers and knowledge in virtually all areas of science and technology, and to obtain their advice on specific SME research and technology challenges and opportunities.
3. The background and experience of the ITAs is quite unique in the Canadian innovation program delivery system. As the Director General of IRAP characterized the ITA capability in a recent interview (June 2012), **"the ITAs understand technology in a business context."** He explained that while the ITAs have strong backgrounds in technology, they have even stronger backgrounds and experience in building and managing small businesses. He believes it is this combined skill set that makes their advice so relevant for SMEs.
4. The ITAs try to **establish long term relationships** with each SME client—reflecting the culture change that the program has gone through from funding "projects" per se—to funding "firms". In this regard, the ITAs develop a personal sense of responsibility for the performance of their portfolio of SMEs and seek to advise each firm on its business growth over the longer term.
5. The ITAs are **"embedded" in the communities** in which they operate and develop extensive knowledge of the regional innovation ecosystem in which their SMEs perform. In this regard, the ITAs can also fund 3rd party innovation service providers in the local area who may offer SMEs services not available through IRAP or the NRC. In this way the ITA can strengthen the entire regional innovation cluster and ecosystem.

As highlighted in the benefits section above, IRAP is generally commended and is a highly regarded program. Nevertheless, here are several key weaknesses:

1. The Jenkins Report while generally very laudatory in its assessment of the program "IRAP was widely praised as an effective, well run program that provides industry with non-repayable contributions, mentorship and technical business advice," also highlighted several weaknesses of the program, stating: "The main criticism is the **exhaustion of funds** very early in the fiscal year, but this is also one indication of the high level of demand for the program. Some believe that the amounts of IRAP **funding awards are too small** to be effective and that the **application process is excessively difficult** for first-time applicants." Our assessment of IRAP's performance suggests these weaknesses as areas for improvement.
2. IRAP has not yet articulated a strategy to deal with the tension inside the program between focusing on assisting smaller SMEs and start-ups (these are riskier investments for an ITA) versus focusing on providing multiple rounds of investments for more mature SMEs who have demonstrated a capacity to perform (these are less risky investments for the ITA). In what proportions should funding be allocated between these two competing client groups?
3. The inherent skill sets and background expertise that each ITA brings to the program to support the client is a potential concern. Specifically, many ITAs come to IRAP from Information and Communication Technology (ICT) backgrounds, and the impact on the program has been a somewhat unbalanced level of support being targeted towards ICT firms and projects. As IRAP expands how will it ensure its skill sets encompass most key areas of emerging science and technology? Alternatively, should IRAP fund only firms in a few priority areas of science and technology?
4. How will IRAP managers cope with a rapid expansion of funding (an incremental $110M per year) and a rapid expansion in responsibility to deliver other federal programs (such as the Digital Technology Adoption Pilot Program) or to provide services to other federal program managers (such as performing technology and business assessments for the rapidly expanding Canada Innovation Commercialization Program)? The IRAP program has not yet demonstrated the capability to manage such extraordinary increases in both permanent resourcing and in expanded responsibility.
5. In 2009-10 IRAP provided assistance to its portfolio of 8,578 firms, out of a total population of 1.14 million SMEs in Canada. Of the total of 1.14M SMEs, the Director General of IRAP has identified approximately 75,000 SMEs that are technology-based and could potentially require some form of IRAP advisory or financial assistance. From these he suggested that about 25,000 technology-based SMEs

would be currently eligible for some form of assistance. If these rough estimates are accurate, then he noted that IRAP in its current structure and business model may be reaching only 1/3rd to 1/10th of its potential market. This raises issues in a decentralized federation of how IRAP could collaborate more closely with other federal and provincial programs (there are over 250 federal and provincial innovation support programs) to deliver relevant innovation services to a larger market of eligible SMEs.

Appendix A2

Fraunhofer-Gesellschaft: The German Model of Applied Research

The Fraunhofer-Gesellschaft (Fraunhofer Society) is a network of German institutes for applied research. The Fraunhofer's primary mission is to perform contract research for German industry, particularly small and medium enterprises (SMEs), which translate basic research from universities and non-university research organizations into commercial products and industrial processes. The 60 Fraunhofer institutes in Germany, with an average staff of 400, perform research and test industrial processes on their premises which enjoy state of the art equipment and deep human competencies. Fraunhofer's funding is derived from diverse sources, including federal, state, and European Union public funding, fees from contract research for industry and public organizations and foundations, and licensing fees for intellectual property. It has created roughly150 spin-off companies, with a very high rate of success.

Each Fraunhofer institute specializes in a particular technology or sector, and a Fraunhofer institute exists for virtually every sector of significance to a modern industrial economy, ranging from renewable energy, aerospace, automotive manufacturing, microelectronics and information technology. Each Fraunhofer institute is paired with a German university and can utilize the most promising students as part-time researchers, thereby giving students practical experience in commercially-oriented research and manufacturing environments. The institutes generate technology for commercial products and processes, enable companies to test equipment and industrial processes on pilot manufacturing lines, and foster a continual flow of trained engineers and technicians to the private sector.

The German model of innovation is widely credited as a major factor underlying Germany's strong international competitive performance since the onset of the global financial crisis in 2008. German exports are at an all-time high and GDP growth outpaces the rest of the Euro zone and the United States. German competitive strength are often attributed to factors such as the country's highly-skilled work force; a longstanding "dual system" of education/apprenticeship, which marries academic learning and practical skills; the export prowess of Germany's small and medium enterprise (*Mittelstand*), which dominates hundreds of world markets for niche technologies; a web of mutually-supporting networks of companies, associations, and research organizations that can address complex technological challenges and diffuse technology through collaboration; and the excellence of German engineering and manufacturing technology in companies of all sizes. The Fraunhofer plays a central role in supporting all of these aspects of German competitive strength.

Reflecting Germany's international competitive performance, the Fraunhofer is probably the world's most thoroughly studied applied research organization, with other countries examining the question of whether they could apply the "Fraunhofer model" wholly or partially and emulate Germany's success. Of particular interest is Fraunhofer's role in Germany's abiding strength in "traditional" industries which have eroded in other developed economies, such as automobiles and machinery. Germany companies have remained competitive in these sectors through a continual process of incremental improvement in existing products and processes, an approach that may not yield technological breakthroughs but builds a competitive edge over a long timeframe. The Fraunhofer emphasizes performance of a large number of short-term research projects that have near-term commercial impact, an approach which reinforces the German method of incremental innovation. It continually diffuses competitively relevant innovation technology throughout the German economy and trains highly-skilled engineers, technicians and managers. Public funding of the Fraunhofer reflects a longstanding German national consensus that public investments must be made in research infrastructure that includes infrastructure for applied research with commercial relevance. [1]

It is important, however, not to overrate the impact and the importance of Fraunhofer Gesellschaft for the German research and innovation system. While Fraunhofer Gesellschaft have contributed in significant ways to the recent success of German industry, this success has also rested on factors such as the high demand for capital goods/machinery by emerging economies and the match of this demand with Germany's industry profile. Other factors supporting German competitiveness include a competitive currency, worker training programs, policies to subsidize the retention of skilled workers in economic downturns, and a dense network of supporting institutions, including localized banks with long-term relationships with Mittlestand firms.[2]

---

[1]Dirk-Meints Polter, a former Fraunhofer deputy director, addressed this point in 1992, characterizing "subsidy," as a misleading word: "we see ourselves as part of a scientific and technological infrastructure that is provided by the government, just as phones and roads are provided." *New Scientist.* 1992. "German Innovation, British Imitation." November 21.

[2]Additional instruments include innovation vouchers and Industrial Collective Research institutes. "In numerous Länder, including North Rhine-Westphalia, Baden-Württemberg and Bavaria,

INTRODUCTION

The Fraunhofer has an enviable reputation throughout Germany, Europe, and the world. A report by the U.K. House of Commons Science and Technology Committee commented in 2011 that "the name Fraunhofer resonates across the world and is widely associated with an impressive network of German technology and innovation centres."[3] In 2010, the United Kingdom's Minister of State for Universities and Science, David Willetts, testified that the Fraunhofer Institutes "have been a key part of Germany's success in advanced manufacturing and high grade engineering.[4] In 2010, Fraunhofer was Germany's second most popular employer for graduates in the natural sciences, fourth most popular for information and communications technology graduates, and seventh most popular for engineering graduates[5]. The amount of contract research business with industry—an accepted benchmark of Fraunhofer's effectiveness in promoting industrial innovation—is at an all-time high and increased by 15 percent between 2010 and 2011.[6]

Recent German Economic Performance

A recent resurgence of interest in the Fraunhofer reflects the fact that Germany has weathered the financial crisis which began in 2008 far more

---

innovation voucher programmes have been launched. Normally aimed at SMEs that conduct no R&D of their own, the programmes subsidise part of companies' costs for consulting or external R&D services. The Federal Ministry of Economics and Technology (BMWi) launched a similar programme in 2010" <*http://www.e-fi.de/fileadmin/EFI_2011_en_final.pdf*>. The Industrielle Gemeinschaftsforschung, (IGF) is a "a mechanism enabling businesses to solve shared problems through shared projects. This kind of pre-competitive research closes the gap between basic research and industrial application. The results are available for everyone interested and the basis for individual adaptations within enterprises" <*http://www.aif.de/en/collective-research.html*>.

[3]House of Commons. 2011. Second Report: Technology and Innovation Centres. Committee on Science and Technology. February 9. p. 41.

[4]Ibid. pp. 7. German industrial leaders attribute their abiding success in export markets, in substantial part to Fraunhofer and other similar German research organizations. In 2008, Der Spiegel interviewed Ekkehard Shutz, the CEO of Germany's biggest steel firm, ThyssenKrupp. Commenting on his firm's success in a global market increasingly affected by low cost manufacturers based in Asia, he observed that despite Germany's higher cost structure, it remained a "good place for production. You can tell by the fact that the lion's share of our investment still flows into our German plants. We have exceedingly well-trained employees here, a good research environment, close cooperation with our customers and suppliers, our six partner universities, and non-university research institutions like the Fraunhofer Society and the Max Planck Institute." Spiegel Online International. 2008. "We are not driving the Price Hike." July 16.

[5]Reid, Benjamin et al. 2010. Technology Innovation Centres: Applying the Fraunhofer Model to Create an Effective Innovation Ecosystem in the U.K. Submission to the House of Commons Science and Technology Committee, December.

[6]Fraunhofer Magazine. 2012. "Boost in Earnings." February.

successfully than most industrialized nations.[7] German manufacturing as a share of GDP stood at 20.7 percent, versus 12.7 percent for the United States and 11.5 percent for the U.K.[8] "[D]espite despite the onset of the financial and economic crisis in 2008, German research-intensive industry was able to develop a leading position among the established economies. Sectors which particularly profited from this development include the manufacture of road vehicles, machinery, electrical machinery, and chemicals."[9]

This performance poses the obvious question of how a wealthy, high-wage country with a strong currency could compete so successfully in a global market where the terms of competition are increasingly driven by low-cost Asian firms.[10]

German Model of Innovation

The German model of innovation applies science, technology, and engineering to drive incremental but constant improvements in processes and technologies and aims at niche areas where competition is less intense than in large commodity markets.[11] Although Germany has a vast array of research organizations capable of supporting such a strategy, the Fraunhofer is the foremost, and is most often cited as the driver of German competitiveness in export markets. The perception of Fraunhofer as a success story underpinning German competitive achievements in global markets has led other countries to study its methods and structure to determine whether part or all of the

---

[7]Spiegel Online. 2011. "Stronger than Expected Growth: German Economy Defies Crisis." November 15. German GDP grew 3.6 percent in 2010, with the rest of the Euro zone growing at 1 percent or less and the United States at 2.9 percent. German exports in 2011 reached 1 trillion Euros, an 11.4 percent increase over 2010 and the largest volume in German history. Spiegel Online. 2012. "News of Germany's Strong Growth isn't Welcome Everywhere." February 9.

[8]Manufacturers Alliance for Productivity and Innovation.

[9]DIW Economic Bulletin. 2011. "German Manufacturing Withstands Rise of Emerging Economies." June. The *Christian Science Monitor* observed in 2011 that Germans have looked around lately to find they have the preeminent world-class export economy in Europe. No one else comes close. German precision tools are coveted in Asia and Russia like Fabergé eggs. Germany is building much of the Summer Olympic and World Cup facilities in Brazil. The next generation of Eurostar trains linking the Continent and Britain will be made by Siemens of Germany, not, as they traditionally have been, by Alstom of France—a blow to French pride. *The Christian Science Monitor*. 2011. "Germany—the new mini-superpower." January 30.

[10]Xinhua. 2012. "What's Behind the Success Story of German Manufacturing Industry?" February 23; Spiegel Online. 2012. "How the German Economy Became a Model." March 21.

[11]"Mass production has never been the norm in [Germany] to the degree that it has been in the United States (and Japan)…as a consequence, Germany's factory work force shows a much stronger element of craftsmanship…this is also the result of the basic education that apprentices get in addition to their on-the-job training." Junne, Gerd. 1989. "Competitiveness and the Impact of Change: Applications of 'High Technologies'." In Katzenstein, Peter J. (ed.) Industry and Politics in West Germany: Toward the Third Republic. Ithaca: Cornell University Press. pp. 256.

"Fraunhofer model" could be adopted domestically.[12] The Fraunhofer model was extensively reviewed in 2010 and 2011 by the House of Commons Science and Technology Committee, and aspects of that model are now being adopted in the United Kingdom.[13] In the United States, the Fraunhofer was the subject of studies during the 1990s by the Council on Competitiveness and the Office of Technology Policy in the Department of Commerce.[14] In 2005, French Research Minister Francois d'Aubert commented that his country needed to develop "Partnership research…on the model of the German Fraunhofer Institutes."[15] Other countries have demonstrated similar interest.[16]

Non-German observers tend to cite the Fraunhofer's funding model as its strongest positive feature. That model, based on a formula of one-third core public funding, one third private contract research and one third publicly-funded contract research, ensures stable long-term funding. The model is particularly attractive to Anglo-American observers whose research institutions have suffered from erratic funding and a partisan political environment in which government support for science is controversial and uncertain. Fraunhofer staff acknowledge the success of the institute's funding approach, but emphasize its relational aspects as the real basis for its success. Fraunhofer institutes operate in vast, multiple overlapping human and institutional networks embracing universities, companies, research organizations, trade associations, and foundations, organized by scientific field and areas of interest. Relevant units of these networks can be brought to bear on research projects, consortia, and development alliances to address specific tasks based on their particular competencies. Any private or public entity which enters into a research relationship with Fraunhofer gains entrée to these networks.

Fraunhofer is more than a networking organization. It possesses deep and broad organic competencies and institutional scientific memory, reflecting

---

[12]In the run-up to the United Kingdom's 1992 general election, each of Britain's leading political parties vied to convince voters that their plan to apply the Fraunhofer model in the UK was the superior choice. *Physics World.* 1992. "Fraunhofer Fever Hits the UK." March. *New Scientist.* 1992. "British Innovation, German Style." March 21. *New Scientist.* 1992. "German Innovation, British Imitation." November 21.

[13]Hauser, Hermann. The Current and Future Role of Technology and Innovation Centres in the U.K. Report for Lord Mandelson. The Daily Telegraph. 2011. "Follow the German Model and be Patient for Manufacturing to Thrive." August 25. The Sunday Telegraph. 2010. "Welcome to Berlin, Peter. Is it the Future?" February 7.

[14]Mitchell, A. Duff. 1998. The Fraunhofer Society: A Unique German Contract Research Organization Comes to America. Burton, Daniel F. and Kathleen M. Hansen. 1993. German Technology Policy: Incentive for Industrial Innovation. Council on Competitiveness Occasional Paper, Challenge, January-February.

[15]Le Monde. 2005. "Research: The Key Points of the Law." January 17.

[16]In 2008, the Dubai Institute of Technology (DIT) signed an agreement with Fraunhofer pursuant to which the institute agreed to help DIT develop a "comprehensive R&D and innovation model." MENA Business Reports. 2008. "DIT, Fraunhofer Sign Agreement on Development of R&D Model." September 22; Irish Times. "Bavaria or Bust." June 8.

its permanent staff of scientists, technicians, and managers. Its institutes are extremely well-equipped and most of them operate multiple pilot manufacturing lines and other demonstration facilities.[17] The Fraunhofer is a beneficiary of the "power and generosity of the…German machine tool industry," which permits its labs to be equipped with state-of-the-art machines loaned on generous terms.[18] The institute holds a massive patent portfolio which can be deployed on behalf of clients seeking to license cutting-edge technology. It is relentlessly focused on practical applications of technology.

Supporting the Mittelstand

The Fraunhofer is frequently cited in connection with the export strength of Germany's small and medium enterprises. The best of these firms, the so-called Mittelstand, are typically family-owned, highly specialized, based in small cities and towns, and "build products that dominate obscure industrial subsectors". "There is no doubt that the German Mittelstand were one of the main factors that generated economic success and prosperity in Germany after World War II."[19] Observing the slogan, "don't dance where the elephants play," the Mittelstand pursue niche markets, particularly at the high end of the product spectrum, and make continuous incremental improvements in their products to maintain leadership over their competitors.[20] "It might be that their products will not be the cheapest, but they definitely have the best quality."[21] A 2007 study

---

[17]Typically the Fraunhofer Institute for Molecular Biology and Applied Ecology IME in Aachen states in its 2011/2012 Annual Report that "The R&D activities in the various IME business areas involve certain platform technologies that need sophisticated apparatus and infrastructure as well as highly trained staff…The services provided include sequencing, chip technologies, proteomics, metabolomics, recombinant protein production, protein purification, protection structural and functional analysis, antibody manufacturing, and high throughput imaging technologies and are available to the working groups within the IME as well as the external clients." Fraunhofer IME. 2012, Jahresbericht/Annual Report: 2011-2012. p. 15.

[18]The director of the Automation Department at the Fraunhofer Institute for Production Systems and Mechanical Constructions Technologies IPK, Gerare Dulen, commented in 1987 that in his department, "There are DM 50 million of machine tools in place. Most of them have been lent to us by industrialists. But we do developmental work for their account. Often, they turn prototypes over to us even before commercializing them. We then complete the development and programming of these prototypes. Afterwards they leave the machines with us at no charge for several months. The advantage of this arrangement is that we always have the most modern machines." Industries et Techniques. 1987. "Applied Research: The Fraunhofer Method." October 20. JPRS-ELS-88-006.

[19]Hamburg, Christian. 1999. Structure and Dynamic of the German Mittelstand. Heidelberg and New York: Physica-Verlag. pp. 1. Volker Treier, chief economist of the German Chamber of Commerce, characterizes the Mittelstand as the backbone of the German economy. Treier, Volker. 2011. "The Engine of Growth. Wall Street Journal. June 26. "If a particular job can be best done by a machine, then the chances are that the machine in question was built in a small town in Germany." The Economist. 2011. "German Business: A Machine Running Smoothly." February 3.

[20]Software Magazine. 2011. "Finding Hidden Gems in the German Mittelstand." October.

[21]Boeing, Phillip. 2012. "What's Behind the Success Story of German Manufacturing Industry?" Xinhua. February 23.

found that over 1,130 small and medium-sized German companies occupied the number one or two position in the world market for their products, or the number one position in the European market.[22] According to one source, the Fraunhofer's "main focus [is] on the Mittelstand companies," and "the research facilities of Fraunhofer serve as external, very well equipped research departments of the Mittelstand companies."[23]

The Skilled German Workforce

The Fraunhofer is cited in connection with another aspect of the German model, the country's highly trained, technologically-adept work force, which has been a subject of foreign admiration and study for over a hundred years.[24] To some extent this phenomenon is cultural—German blue collar workers draw on a tradition of craftsmanship with the attitude that "excellence on the shop floor is every bit as important as in the Nobel Prize caliber laboratory."[25] The pursuit of excellence in manufacturing is widely attributed to Germany's longstanding "dual system" of vocational training, pursuant to which a course of academic study of a practical discipline is undertaken in parallel with apprenticeship at a company or a public vocational institute.[26] The Fraunhofer applies the principle of the dual system at the highest educational levels, with masters and PhD candidates and postdocs simultaneously pursuing courses of study in science or engineering while performing work and acquiring practical experience in a Fraunhofer institute.[27]

---

[22]The study was co-authored by Professor Bernd Venohr, a German management consultant generally regarded as one of the world's leading experts on the Mittelstand and built on findings in earlier work by another expert, Hermann Simon, who coined the term "Hidden Champions" to describe the best of these companies. Venohr, Bernd and Klaus E. Meyer. 2007. The German Miracle Keeps Running: How Germany's Hidden Champions Stay Ahead in the Global Economy. Berlin: Berlin School of Economics, May.

[23]Hamburg, Christian. 1999. Structure and Dynamic of the German Mittelstand. Heidelberg and New York: Physica-Verlag. pp. 58-59.

[24] Thelen, Kathleen, 2004. How Institutions Evolve: The Political Economy of Skills in Germany, Britain, the United States, and Japan. Cambridge: Cambridge University Press. pp. 39.

[25]"This has much to do with the fact that the pursuit of excellence in Germany has not been limited to university-trained elites, but has occurred within each layer of German society...One relevant consequence is that managers trained in engineering or the sciences often feel they should be able to prove the mettle of their own skills in front of workers who have high standards against which to measure performance." Beyerchen, Alan D. 1990. "Trends in the Twentieth Century German Enterprise." In *The Academic Research Enterprise within the Industrialized Nations: Comparative Perspectives*. Washington, DC: The National Academies Press. pp. 80.

[26]In 2008, roughly 58 percent of Germany's upper-secondary school students were also enrolled in vocation or technical programs, and many students combine apprenticeships with the pursuit of graduate degrees, entering the work force in their mid-20s with highly developed practical skills. Helper, Susan, Timothy Krueger, and Howard Wial. 2012. "Why Does Manufacturing Matter? Which Manufacturing Matters? A Policy Framework." Brookings. February. p. 27.

[27]There is increasing overlap between the demand by industry for workers trained in vocational schools and universities. Professor Kathleen Thelen of Northwestern University, who has

Weaknesses of the German System

The German innovation system is by no means perfect. German innovation tends to build on established structures rather than creating entirely new ones; product and process improvements are slow, albeit continuous. "The resulting pattern of innovation is one that is more likely to generate improvements of existing products of existing firms and sectors than to give rise to new ones."[28] The postwar German system has not produced "radical, path-breaking innovations found, for example, in America."[29] Some observers believe that Germany's "heavy concentration of R&D and innovation activities in the automotive sector…may lead to an unbalanced innovation system."[30] "Germany never came close to a computer industry that was able to compete with its U.S. counterpart."[31] Start-ups which become spectacular successes are rare, and would-be startups confront obstacles such as inadequate availability of venture and bank capital and legal and societal factors that punish market failure.[32] As a

---

extensively studied the German training system, notes that "firms (even small, medium-sized handicraft firms) are increasingly inclined to recruit new workers from among applicants holding university degrees…[there is] increasing competition between job applicants from the traditional vocational training track and those with technical college and university degrees. Thelen, Kathleen. 2004. How Institutions Evolve: The Political Economy of Skills in Germany, Britain, the United States, and Japan. Cambridge: Cambridge University Press. pp. 274.

[28]Streeck, Wolfgang. 1995. "German Capitalism: Does it Exist? Can it Survive? In Colin Crouch and Wolfgang Streeck, eds. Modern Capitalism or Modern Capitalisms? London: Francis Pinter. p. 14.

[29]Harding, Rebecca. 2000. "Resilience in German Technology Policy: Innovation Through Institutional Symbiotic Tension." Industry and Innovation. December. p. 223.

[30]Astrom, Tomas, Marie Louis Eriksson and Lars Niklasson. 2008. International Comparison of Five Institute Systems. Copenhagen: Forsknings-og Innovationsstyreisen. December 23. p. 108. Numerous Fraunhofer institutes pursuing research themes with little apparent nexus with the automotive sector are in fact active in that sector. The Fraunhofer Institute for Experimental Software Engineering IESE is working with Audi on the development of software for electric vehicles, with BMW on a model for estimating the cost of software projects, and with "companies from the automotive domain" on software testing techniques. Fraunhofer IESE, "Automotive and Transportation Systems." <*http://www.iese.fraunhofer.de/en/customers_industries/automotive/referenzprojekt_mbtech.html*>. The Fraunhofer Institute for Communications Systems ESK has a large "Business Unit Automotive" working on automotive software engineering, external communications from motor vehicles, and internal transmission and processing of data in vehicles. <*http://www.esk.Fraunhofer.de/en/automotive.html*>. The Fraunhofer Institute for Chemical Technology ICT is collaborating with Faurecia to develop advanced industrial processes for composites for application in the automotive field. Fraunhofer ICT, "Faurecia Signs R&D Agreement with the Fraunhofer ICT for Advanced Industrial Processes in the Field of Composites." <*http://www.ict.fraunhofer.de/en/press_media/pressemitteilungen/2012/release4.html*>.

[31]Wieland, Thomas. 2006. "Innovation and Culture, Technology Policy and the Uses of History." Munich Centre for the History of Science and Technology. p. 7.

[32]Uwe Waltz, a professor at the Center for Financial Studies at the University of Frankfurt observed in 2011 that "entrepreneurship is not valued in Germany as it is in the United States. On top of that, U.S. business men are likely to go on the market at the first opportunity, while Germans like to perfect the final technical details before going into business." Deutsche Welle. 2011. "Germany's Venture Capital Market Starts to Take Off." June 1.

result, with the notable exception of SAP, there is no German counterpart to Intel, Apple, Google, or Facebook. The former area of East Germany absorbs a substantial part of the federal research spending but generates comparatively little innovation.[33]

Dr. Stefan Kulhman, Director of the Fraunhofer Institute for Systems Innovation Research in Karlsruhe, told a National Academies symposium in 2007 that Germany's education system, long considered quite strong, had declined to a point of crisis and required "and expensive restructuring."[34]The German university system is widely criticized and may prove incapable of supplying enough graduates with the skills required by German industry.[35] In 2011, on any given month, Germany averaged 92,000 vacant engineering jobs. The director of the Köln Institute for Economic Research IW, Hans-Peter Kloes, commented that "if we cannot close the engineering gap, the continuing shortage of qualified employees will become a threat to the German business model."[36] Even the vaunted vocational training system is under strain; a 2010 study by Germany's education ministry found that while there was an unmet need in German industry for 9-10,000 apprenticeships, one in five apprentices "did not stick out the apprenticeship," the number of students entering apprenticeships was declining, and "businesses are facing a huge drop-off in apprentice numbers."[37]

The Fraunhofer is implicated in at least some of these systemic weaknesses:

- Because the Fraunhofer business model is based on demand for research from industries that already exist, it has little economic

---

[33]"East Germany and Berlin together eat up nearly one-quarter of the…federal research budget while employing only 11 percent of the country's R&D personnel and accounting for 6 percent of its patent production." Kuhlman, Stefan. 2007. "The Record and Challenge in Germany." In National Research Council. *Innovation Policies for the 21st Century*. Charles W. Wessner (ed.) Washington, DC: The National Press. p. 70.

[34]Ibid. p. 70.

[35]National Research Council, *Rising to the Challenge: U.S. Innovation Policy for the Global Economy*. Wessner, Charles W and Alan Wm Wolff (eds.). Washington, DC: The National Academies Press. 2012. pp. 274. The Economist. 2009. "On Shaky Foundations." June 25. The Times Higher Education World University Rankings 2011-12 place four German universities in the top 100 institutions worldwide. The highest ranked German university was Ludwig Maximillian University in Munich. The Economist. 2011. "German Universities: Mediocre but at least They're Free." June 30.

[36]Spiegel Online. 2012. "VDI Study—Lack of Engineers Costs German Economy Billions." April 26. Bernd Rau, one of the founders of the Mittelstand company Roth and Rau, commented in 2010 that his company's search for skilled workers was becoming acute. "It is more and more difficult, especially for the IT people, and also in engineering and mechanical engineering." Deutsche Welle. 2010. "Labor Bottleneck Squeezes Germany's Solar Sector." November 23.

[37]Deutsche Welle. 2010. "Germany faces Youth Employability Crisis, Says New Report." April 28.

incentive to incur the costs and risks associated with creating entirely new industries.

- The Fraunhofer and other non-university research institutions compete with German universities for funding and personnel, and may have contributed to the erosion over time of university-based scientific research in Germany.
- The Fraunhofer's emphasis on research with relatively low risk, short time horizons, and early commercial payoff undermines its ability to support sectors such as biotechnology in which risks are substantial and the path to market is often many years long.

Although Fraunhofer centers have been established in the United States and other countries, they have not enjoyed the degree of success of their affiliates in Germany and it is questionable whether the Fraunhofer model is adaptable in its main aspects to a U.S. industrial innovation context:

- ***The full extent to which the Fraunhofer relies on government funding is frequently not appreciated.*** A widespread misimpression exists that the Fraunhofer's primary funding source is the private sector. In fact, one-third of Fraunhofer's funding consists of "core" money provided by the German federal and state governments, roughly another third comes from research contracts with government entities, and one third is provided through research contracts with the private sector—which are frequently supported by government grants and other financial assistance. While Germany as a nation tends to regard such public expenditures as a necessary form of infrastructural investment, it is unclear that such a high-level of government spending on commercially-related research could be reconciled with U.S. economic ideology or fiscal realities.
- ***A system of Fraunhofer-like research institutes could weaken U.S. university-based innovation.*** The U.S. Bayh-Dole Act of 1980 provided that U.S. universities conducting federally-funded research could own the patents for the technologies they developed, a policy shift that fostered an explosive growth of new companies commercializing university-based R&D. The Fraunhofer model interposes an intermediary organization between universities and industry, at that intermediary—not universities or companies—ends up as the owner of most of the intellectual property rights derived from government-funded research. While the Fraunhofer actively licenses its IP to industry, it is not clear that the German model is as dynamic as the current U.S. university-based system which—in contrast to the German—has given rise to entirely new technology-intensive industries. In addition, an intermediate research organization will

inevitably compete with universities for public and private funding as well as personnel.[38]

- ***The Fraunhofer model derives much of its strength from other elements of the German innovation system which are not present to the same degree in the U.S.*** The United States has no close parallels to the German system of stable government funding for commercially-relevant research; to the *Mittelstand*; to the "dual system" of vocational and training; or to the German tradition that competition should be subordinated to cooperative arrangements. These aspects of the German system reflect complex, multigenerational cultural and institutional evolution and cannot easily be replicated elsewhere. In this connection, an exchange which occurred in Britain's House of Commons Science and Technology Committee in 2012 is worth noting. MP Sir Peter Williams commended the Fraunhofer model to his colleagues and urged that the country learn from it:

> ***Williams:*** *Down the decades everybody, from John Fairclough onwards, has said to the Government 'Look at the Fraunhofer-Gesellschaft; there are lessons to be learned'. It is a fact that they do not have to debate the issues that we are debating today because they have been through the valley of death as a nation, as it were, and they prosper by valuing and backing engineering with their Lander, federal government, financial institutions, Mittelstand families, scientists, engineers, and business folk playing like a team, which is why they are going to win the European Championship as well...*
>
> ***MP Jim Dowd:*** *So we could have all these benefits as well if all we did was change every single piece of our social structure.*[39]

The Fraunhofer and the German innovation model are nevertheless of interest from an Anglo-American perspective because of Germany's success in holding its leadership position in traditional manufacturing industries and in retaining a major manufacturing presence in Germany itself, with the implications which that holds for employment and regional economic development. Germany remains strong in industries like automobiles,

---

[38]In the British Parliament's recent deliberations over the establishment of intermediate research organization patterned on the Fraunhofer, concern was expressed by the Institute of Physics that the result "may well be to put further pressure on the universities which use research contracts as alternative sources of funding." House of Commons Science and Technology Committee. *Technology and Innovation Centres.* Second Report of Session 2010-11. Volume I, Report, p. 27.
[39]House of Commons, Science and Technology Committee. "Bridging the Valley of Death: Improving the Commercialization of Research." June 20 2012. Transcript of Oral Evidence.

machinery, metals, and chemicals—industries that have eroded, shed jobs, and begun to move offshore in the United States and the United Kingdom. For this reason, the German innovation system warrants study.

THE INSTITUTIONAL SETTING

The German innovation system is complex, featuring multiple levels of governance and interlocking responsibilities between different ministries and between authorities at different levels of the political system.[40] Significantly, "in contrast to many other countries, [in Germany], there is an implicit consensus at the federal and state levels that funding should, by and large, be geared towards bridging the gap between knowledge creation and application."[41] The system is characterized by dispersion of authority and funding responsibility.[42]

Federal Ministries

At the federal level, the main responsibility for science and technology policy lies with the Federal Ministry of Education and Research (BMBF). The BMBF co-funds with the Länder Germany's non-university research organizations, including Fraunhofer, and is the principal source of the institute's basic public funding. The BMBF provides funding for a broad array of research programs and projects, generally aiming at collaborative R&D between public sector research and the private sector. Many BMBF programs are managed by semi-autonomous program agencies, a mechanism that has been characterized as "industry friendly". The Federal Ministry of Economics and Technology (BMWi) promotes innovation-and-transfer projects through public-private R&D collaboration in topical areas which include energy, information technology, media, and aerospace. It operates a number of programs to support small and medium enterprises and to promote start-ups.[43]

---

[40]"In Germany, science and technology planning, policy making, and funding take place on such a wide variety of different levels, and there are so many institutional structures, that it is difficult to avoid over-simplification when one tries to draw a coherent map of the present system…not only do we have a wide variety of institutional actors, but within each institution, we also find quite diversified structures funded by a multiplicity of sources and consisting of many sub-systems." Krull, W. and F. Meyer-Kramer. 1996. Science and Technology in Germany. London: Cornell University Press.

[41]Edler, Jakob and Stefan Kuhlmann. 2008. "Coordination within Fragmentation: Governance in Knowledge Policy in the German Federal System." Science and Public Policy. pp. 267.

[42]Throughout the German research system, "no performing actor is totally dependent upon only one policy actor for its resources," so that all players can turn to more than one funding source for support, an arrangement which protects the autonomy of research actors from external direction and pressure. Winnes, Markus and Uwe Schimack. 1999. National Report: Federal Republic of Germany. Institute for the Study of Societies. TSER Project No. SOE1-CT96-1036. May.

[43]Edler and Kuhlmann. 2008. "Coordination within Fragmentation." Op Cit. p. 268.

Federal-state Arrangements

Since 1975, federal-state funding of scientific research has been governed by the "Framework Agreement on Research Promotion" (RV-Bo) pursuant to which funding ratios are assigned to the federal government for universities and non-university research organizations. Under this arrangement, the Fraunhofer receives 90 percent of its core funding from the federal government and 10 percent from the Länder. The 90:10 funding ratio has endured for decades; it is one of many "organizational arrangements" in the German innovation system which "are very rigid because they are the result of complicated power balances which nobody can disturb because of possible incalculable consequences."[44] The rigidity of the German research funding arrangements arise out of a concordat between the federal government and the Länder that was negotiated in the early 1970s.[45]

"Research Cartel"

"Fraunhofer institutes occupy a specific role in a very elaborate division of labour within the German system of research."[46]The German non-university scientific research arrangements are widely characterized as a research promotion cartel. The lines of demarcation with respect to the responsibilities of each organization have been clearly established for many years, and no significant new research actor has entered this universe for decades.[47] Competition within the institutes between internal subunits is

---

[44]Winnes and Schimank. 1999. Federal Republic of Germany. Op. Cit. pp. 14.

[45]The "Bund-Länder Commission for Educational Planning and Research" (BLK) is a standing committee comprised of representatives of the federal government and each German state with a mandate to coordinate federal-state planning in research policy, and functions as the central clearing house for the financial aspects of governmental support for research. Since 1975, when the BLK was established, major decisions concerning the organization, operation, and funding of government-supported research have required joint action by the Länder and the federal government within the complex decision making structure of the BLK. That fact has placed the public funding of university and non-university research on a stable permanent basis, shielding research organizations from short-term budget gyrations and political meddling. Moreover, because of the difficulties associated with building the necessary consensus within the BLK, selection of research themes and the allocation of funds has largely been left to the scientific community to avoid politicizing decision making. These beneficial aspects of the German federal research system are partially offset by the stasis associated with the vertical interlocking organizational structure. Changes to the status quo can be vetoed by negatively-affected Länder, forcing a constant pursuit of consensus which "often allows only decisions on the lowest common denominator." Winnes and Schimank. 1999. Federal Republic of Germany. Op. Cit. pp. 45.

[46]Written Evidence Submitted by Manchester Institute of Innovation Research, Manchester Business School, University of Manchester. (TIC 09). House of Commons Science and Technology Committee. February 17, 2011.

[47]The presidents of the research organizations meet on a regular basis among themselves and with the federal research ministry to discuss science policy, meetings which are dubbed "the holy

TABLE APP-A2-1 Germany's Key Research Organizations

Institution	Focus
Max Planck Society	Basic Research
Helmholtz Association	Basic Research ("big science")
Fraunhofer Society	Applied Research
Leibnitz-Association	Various

similarly constrained by the establishment of clear lines of demarcation between organizations and encouragement of collaboration when areas of overlap become apparent.[48]

The Max Planck organization, universities and public sector research organizations are responsible for basic research. The Helmholtz Association ensures infrastructure for large scale science projects of national importance.[49] Fraunhofer and other institutes of applied research, such as the Steinbeis Foundation, are responsible for the transfer of technology from the science base to industrial end users.[50] The Max-Planck and the Fraunhofer usually do not

alliance"—a reference to the Nineteenth Century multilateral grouping of states that emerged from the Congress of Vienna and sought to preserve existing European frontiers and the balance of power. Winnes and Schimank. 1999. Federal Republic of Germany. Op. Cit. pp. 66.

[48]The German research minister reported in 1991 that "The increasing degree of overlapping of the acquisition fields of the individual Fraunhofer establishments, acting as profit-making centers, has triggered a process of debate in the FhG. It is conducted with the goal of supporting, by means of the proper organizational measures, the increasingly necessary processes of balance and demarcation between the individual institutes and facilities." Thierbach, Dieter. 1991. Deutsche Einheit in Forschung und Technologie. Bonn: BMFT. Part. 2.6.

[49]The Helmholtz Society concentrates on long-term research which is either extremely costly or very risky and requires large research teams. Areas of focus have included high-energy physics, space technology, nuclear energy, medicine, environment technology, and biotechnology. Der Spiegel. 2008. "New Federal Funds. April 7. Wilhelm Leibniz (formerly known as the Blue List) is a designation for about 80 other large and medium-sized research organizations which engage in basic and applied research. Finally, 50 research institutes are directly organized under a federal ministry (Bundesforschungsinstitut) and similar organizations funded by the Länder.

[50]Harding. 2010. "German Technology Policy." Op. cit. pp. 228. The Steinbeis Foundation operates technology transfer centers across Germany, and performs a networking function between basic research organizations and German industry. Unlike Fraunhofer, Steinbeis does not operate large-scale research infrastructure. Within the German research community, the research demand directed at Fraunhofer is usually regarded as less scientifically interesting than the demand directed at the basic-science oriented Max Planck institutes so that in Fraunhofer, "science curiosity is clearly subordinated to societal relevance." Winnes and Shimank. 1999. Federal Republic of Germany. Op. Cit. pp. 211. At the same time, Fraunhofer scientists sometimes express a degree of disdain for basic research. One director in 1987 commented that "In the FRG we spend too much money on basic research. Currently everyone is concentrating on superconductors. That's the problem of basic research. It is unable to escape the influence of what's in vogue. Here, there is no such danger. We live on our results like a commercial firm…our job consists of demonstrating the feasibility and

compete for the same projects but will collaborate on specific projects that require the skill sets of both the basic and applied sciences.[51]

Federal High-Tech Strategy 2020

The German federal government is currently pursuing a "High-Tech Strategy 2020," an effort to coordinate existing public support for research and innovation in a manner which improves the country's international competitiveness and innovative capability. A target of this effort is to increase German R&D spending to 3 percent of GDP (it was about 2.8 percent in 2010).Reflecting the concern that German public funding of R&D is characterized as a "sprinkling can," featuring large numbers of small grants for small projects, the High Tech Strategy seeks to concentrate public spending on targeted areas of cutting edge research. The strategy also targets perceived weaknesses in the German innovation system, including an inadequate market for venture capital and direct investment capital, inadequate participation by SME's in R&D, and difficult conditions confronting technology-based start-ups and their financial backers.[52]

State Initiatives

Within the German federal system the Länder enjoy a degree of leeway for unilateral policy initiatives. They control the universities within their jurisdictions and can determine their funding level and strategic direction, and they can also provide non-university research organizations located in their territory with additional funding through research contracts and with a variety of other forms of support. Finally, they can use incentives to encourage new research institutes to choose them as sites and to attract research organizations and talented individuals from other Länder. Baden-Wurttemberg, North-Rhine-

---

economic viability of technologies of the future." Heuberger, Anton. 1988. "Applied Research the Fraunhofer Method." Industries et Techniques. February 23. JPRS-ELS-88-006.

[51]Handelsblatt. 1991. "Joseph von Fraunhofer and Max Planck Can Feel Satisfied." August 9. JPRS-EST-91-015. A recent example of Fraunhofer-Max Planck collaboration is a project involving the Fraunhofer Institute for Digital Media Technology, the Fraunhofer Institute for Integrated Circuits and the Max Planck Institute for Evolutionary Anthropology to develop a program that will recognize individual apes from photos, video, and audio footage, enabling biologists operating in the wild to track the number of individuals of a given species, to assess whether a given population is growing or declining, and to identify which factors affect the population. The Engineer. 2011. "Facial-Recognition Software Could Help to Save Great Apes." August 1. Fraunhofer offers five-year, five million Euro research contracts to Max-Planck scientists who have ideas for applications for their basic research results. Comin, Trumbull, and Young. 2012. Fraunhofer: Innovation in Germany. Op. cit. p. 10.

[52]Federal Ministry of Education and Research. 2010. Ideas. Innovation. Prosperity. High Tech Strategy 2020 for Germany.

Westphalia and Bavaria have used such methods to develop technology-intensive economies.

THE FRAUNHOFER GESELLSCHAFT TODAY

The Fraunhofer Gesellschaft is comprised of 60 research institutes active in over 250 business fields and core competencies. Its institutes are widely dispersed across Germany. Its legal form is a registered association under private law (eingetragener verein). Its annual budget is about 1.8 billion Euros. It has 20,000 employees.[53] Each Fraunhofer institute is linked with a German research university and the Directors of Fraunhofer institutes usually also serve on university faculties. The Fraunhofer has established seven institutes in the United States and research subsidiaries in Chile, Austria, Portugal, and Italy. In strictly legal terms, the Fraunhofer is under no obligation to execute research programs drawn up by government entities but in practice its research orientation is closely aligned with that of the German government and the European Union.[54]

The Fraunhofer Mission

The core purpose of the Fraunhofer is the pursuit of knowledge with which has practical utility. It manages an annual business volume of about a billion Euros, most of it involving contract research.[55] Characterizing itself as a "bridge" linking German universities and industry, the Fraunhofer is a formidable research organization in its own right, and generally owns the patents which are derived from joint university/Fraunhofer/industry collaboration. Fraunhofer institutes do not so much transfer knowledge from universities to industry as "generate relevant application-oriented knowledge themselves on demand from their clients. While this may often be strongly linked to research in universities…it nevertheless constitutes a knowledge creation sub-system of its

---

[53]The 20,000 headcount was reported in Fraunhofer Magazine in February 2012.

[54]Germany's research ministry has periodically assigned Fraunhofer clearly-defined missions in major national areas such as new materials, information technology, and the establishment of a research infrastructure in the former East Germany. Thierbach, Dieter. 1991. Deutsche Einheit in Forschung und Technologie. Bonn: BMFT. Part. 2.6. Technologie-Nachrichten Programm-Informationen. 1991. "Interim Report by the Federal Ministry of Research and Technology on the Implementation of the Unification Treaty Dated 31 August 1990 in the Area of Research and Technology. March 5; BMFT. 1986. Materialforschung JPRS-EST-86-036. November; BMFT. 1986. Forschung und Technik zum Whole Der Menschen: Jahresbericht 1984. JPRS-WST-86-015. March; "The Fraunhofer pays close attention to the research and technology policy of the European Community." Frankfurter Allgemeine Zeitung. 1985. "Expose Chips with X-Rays." JPRS-WST-86-018. October 24.

[55]Fraunhofer. 2010. Annual Report. p. 15.

own."[56] Because Fraunhofer is heavily dependent on revenue from contract research performed for industry, its research is closely correlated with the practical needs of companies. If a choice needs to be made between money from an industry project or the pursuit of cutting edge technology, the Fraunhofer institutes usually of necessity choose money.[57] Although Fraunhofer R&D projects tend to be relatively small, short-term efforts seeking incremental improvements in products and processes, its institutes have achieved numerous noteworthy innovations.[58]

Governance

The Fraunhofer's 60 research institutes in Germany are supervised by a headquarters organization based in Munich.[59] Executive direction is provided by a President who is also Chairman of the Executive Board, comprised of three Senior Vice Presidents. The Executive Board manages the institutes' business activities, develops science and research strategies, and negotiates with governmental organizations to raise funding. The Executive Board appoints the Directors of the individual Fraunhofer Institutes.[60]

---

[56]Written Evidence Submitted by Manchester Institute of Innovation Research, Manchester Business School, University of Manchester. (TIC 09). House of Commons Science and Technology Committee. February 17, 2011. Dr. Hans Kunz, Director of the Fraunhofer Institute for Materials in Bremen, explained in the 1980s, that "we start where basic research stops. In the universities, research tries to produce a few micrograms of a new material. Here we try to scale this mark up to the level of the kilogram or the ton. Our idea is to offer industries a product together with its manufacturing process." "Applied Research the Fraunhofer Method." Industries et Techniques. February 23. JPRS-ELS-88-006.

[57]Interview with Fraunhofer Institute for Process Engineering and Packaging. Freising, Germany, June 13. 2012.

[58]Spiegel Online. 2009. "Solar Energy: German Research Strives to be a World Leader." September 15; Spiegel Online. 2011. "A Stuttgart Lab's Pioneering Effort to Cultivate Human Flesh." April 15; Spiegel Online. 2007. "New Computer Program to Reassemble Shredded Stasi Files." May 10; New York Times. 2005. "MP3 Developer Yields Royalty Riches." March 10.

[59]Fraunhofer headquarters enjoys the organizational flexibility to redistribute resources among the research institutes. Prosperous institutes effectively cross-subsidize weak performers as long as the latter can be expected to prosper themselves in a reasonably foreseeable period of time. Fraunhofer headquarters is a service organization for the Fraunhofer institutes, and the institutes pay an annual service fee to headquarters.

[60]The Executive Board is supervised by a Senate comprised of roughly 30 eminent representatives of the fields of science, industry, and government. The Senate appoints the Executive Board, decides the Fraunhofer's basic research policy, and makes decisions with respect to the establishment, merger, or dissolution of individual Fraunhofer institutes. The Senate is elected by the General Assembly, an entity which meets annually and is open to members of the Executive Board and Senate as well as individuals and legal entities interested in supporting the work of the institute. Fraunhofer. 2010 Statute of the Fraunhofer Gesellschaft. The Statue is not public legislation but a set of internally generated institutional by-laws. Fraunhofer. Structure and Organization of Fraunhofer-Gesellschaft: Scientific and Technical Council. Accessed at <*http://www.fraunhofer.de/en/about-fraunhofer/structure-organization/WTR.html*>. The Executive Board is advised by the Scientific and Technical Council, an internal advisory organization. The Council is comprised of the Directors

Each Fraunhofer research institute is managed by a Director and Steering Committee. The Fraunhofer seeks Directors who combine scientific achievement and entrepreneurialism, and the Directors typically serve on boards of directors of German companies. The individual Fraunhofer institutes enjoy a high degree of autonomy and are allowed to choose their research fields and topics, their projects and the handling of project results. Individual institutes are free to determine their own internal organizational structure, set up separate profit centers, and distribute basic funding. The institutes are responsible for balancing their own budgets. If an institute experiences an operating loss in a given year, the shortfall is made good with funds from headquarters. Institutes that incur losses for a number of consecutive years risk closure.[61] Each institute is advised by an Advisory Board comprised of representatives from industry, the scientific community, and the public sector. Each institute must review its strategic plan every five years, and, at five year intervals are subject to external audits.[62]

Growth in Contract Revenue

The Fraunhofer benchmarks its performance on the demand for its services from industry, manifested in revenue from research contracts. During the period 2007-2011, its annual industrial contract revenue grew by 78 million Euros, an increase of over 23 percent, and in 2011 hit an all time high of 406 million euros. Total contract revenue (public and industrially funded research) has grown steadily and reached 1.515 million Euros in 2011.

Relationships with Universities

Each Fraunhofer institute is linked with a German university with a research capability in the sciences and/or engineering which addresses the institutes' area of focus. The Director of each institute usually is also a member of the faculty, and typically identifies the most promising university students to

---

and Senior Managers of the FhG institutes and an elected representative of the research staff from each institute. The Council advises the Executive Board with respect to research and human resources policy, the appointment of directors, and the establishment or closure of institutes.

[61]Five Fraunhofer institutes were closed or turned over to other organizations in 1983. In general such closures are rare. VDI Nachrichten. 1984. "Between Science and Economics: Fraunhofer Presents 1983 Financial Statement." November 16. JPRS-WST-85-01.

[62]Interview with Fraunhofer Institute for Interfacial Engineering and Biotechnology IGB. 2012. Munich, Germany. Jun 15 2012. Establishment of new Fraunhofer institutes begins with a search for a university with a good reputation and strong capability in a given field. A Fraunhofer "project group" approaches the university about setting up a new institute in a collaboration. At present, a Fraunhofer project group is exploring the prospects for establishing a new institute in Mannheim in collaboration with medical schools in that city. If this project moves forward, a Fraunhofer medical institute will be operating with a staff of about 50 people within 5 years. Interview with Fraunhofer Institute for Production Technology and Automation IPA. 2012. Stuttgart, Germany. June 14.

steer to positions at Fraunhofer.[63] Most Directors spend the majority of their time at the institute rather than in teaching activities.

The relationship between one of Germany's leading universities of technology, RWTH Aachen, and Fraunhofer, illustrates the close interrelationship the institute develops with universities. RWTH Aachen, Germany's oldest technology-oriented university, is co-located with four Fraunhofer institutes in the Aachen area, Fraunhofers ILT (lasers), FIT (applied information technology), IPT (production technology), and the Fraunhofer Institute for Molecular Biology and Applied Ecology IME. Adapting ideas from Stanford and Silicon Valley, RWTH Aachen allows companies to locate on campus, participate in university activities (including teaching) and enroll employees in university courses. A total of 18 chairs and institutes at RWTH Aachen and two Fraunhofer institutes (ILT and IPT) have established a "Cluster of Excellence…Integrative Production Technology for High Wage Countries," the purpose of which is "to engage process engineers and materials scientists to develop new concepts and technologies for sustainable manufacturing."[64]

Another example of the close Fraunhofer-university collaboration is the Dresden Innovation Center Energy Efficiency (DIZE^{EFF}) which involves the Technical University of Dresden and the Fraunhofer-Gesellschaft. This involves a total of 13 collaborations between 8 university institutes and four Fraunhofer institutes in the area of energy efficiency. This effort "tightly connects basic research at [the University] with Fraunhofer's competencies to transfer technologies and innovations to industry."[65]

Funding

It is commonly said that one third of Fraunhofer's revenues are contributed by the federal and Länder governments, one-third derived from

---

[63]One Fraunhofer scientist who also serves on a university faculty observes that "my marketing effort is my lecture. I convince students to work [at Fraunhofer] 15-20 hours per week. I'm positioned to find the best. They can stay until they get their PhD (5-8 years), and use our excellent network to find jobs. We don't pay a lot but the experience is good." Interview with Fraunhofer Institute for Production Technology and Automation IPA. 2012. Stuttgart Germany. June 14.

[64]Fraunhofer ILT. 2010. Annual Report. pp. 33-34. Fraunhofer ILT, which specializes in laser technology, works closely with three RWTH Aachen University Chairs: Laser Technology (LLT), Technology of Optical Systems (TOS), and Nonlinear Dynamics of Laser Processing (NLD). The Fraunhofer ILTs' "knowledge of current industrial and scientific requirements in the optical technologies flows directly into the planning of the curriculum." Undergraduates and postgraduates "can put their theoretical knowledge into practice through project work at the three chairs and at the Fraunhofer ILT." University courses are jointly dawn up by the university and the ILT institute. In addition to these collaborations, Professor Thomas Taubner, who oversees a working group at Fraunhofer ILT studying new nano-optic concepts using lasers, also holds a junior professorship at RWTH Aachen for "Nano-optics and Metamaterials," where he supervises research by students in physics involving new imaging techniques with nanometric spatial resolution.

[65]Fraunhofer Institute for Material and Beam Technology. 2011. Annual report 2011. pp. 110.

industrial contracts, and one-third from public research contracts. This Drittelsung (literally, "third solution") model has won extensive praise from foreign observers. "The Fraunhofers are quite safe in their funding, longer term, with those three separate pots of money to take."[66] While Fraunhofer spokesmen emphasize the revenue derived from contract research, the sine qua non of the Fraunhofer model is a major and sustained investment of "core" public funds at the national and state level.[67] In Germany, the "Fraunhofer, for a long period, across different governments, have maintained a very significant chunk of public money in the structure," a reflection of the relative stability of the German political system.[68]

The Fraunhofer's one-third, one-third, one-third model is not, as sometimes believed, an indicator that the private sector accounts for two-thirds or even one third of the institute's revenue. Governments substantially outspend private industry with respect to the contract research which the Fraunhofer institutes perform, accounting for over 45 percent of the total, versus 39 percent for private industry. Moreover, a substantial part of "industry" funds are comprised of government grants and other funding raised by companies and paid to Fraunhofer for contract research.[69] Table APP-A2-2 depicts the Fraunhofer's revenues by source for the year 2010, the most recent year for which such information is publicly available. The figures indicate that federal and state spending accounted for 1,080 million Euros out of total revenue of 1,727 million Euros, or 62.5 percent of the total revenues.[70] Private industry contributes about 26.3 percent of total Fraunhofer revenues. Even these figures

---

[66]Ridgway, Keith. 2010. Testimony to the House of Commons, Committee on Science and Technology. December 20.

[67]"The overriding lesson from Germany is that such centers must be funded consistently and at a significant scale over time if they are to deliver the economic impact of which they are capable." Written Evidence Submitted by Rolls Royce (TIC 82). House of Commons Science and Technology Committee. January 12, 2011.

[68]Andrew Miller, the Chairman of the Science and Technology Committee commented with respect to the Fraunhofer's funding model: "The other part of the Fraunhofer is, of course, longevity. Because the German political system doesn't fluctuate like our system does, that is perhaps one of the reasons why there has been long-term continuity. Miller, Andrew. 2011. Formal Minutes. House of Commons, Ev. 28. January 12.

[69]In 1992, Rudolf Simon, the research director of Meissuer and Wurst, a Stuttgart-based producer of water and air filtration systems, commented on a contract research project performed by the Fraunhofer for his firm. "The Fraunhofer helped us—this is a very important point—to get the funding for our research activity from the government. Without this funding, I must say we would have been very reluctant to give subcontracts to the institute." *New Scientist.* 1992. "German Innovation, British Imitation." November 21. In the 1980s, the federal research ministry provided grants to small and medium companies to participate in contract research projects at Fraunhofer institutes to develop applications for hybrid integrated circuits in sensor systems. "Technologie Nachrichten—Management Informationen. 1987. "BMFT Subsidizes Thin Film Technology Research." JPRS-ELS-87-015. March 2.

[70]A 2011 PowerPoint presentation by Anke Hellwig, a Fraunhofer Headquarters executive, the figure for 2010 for federal and Länder funding was 406 million Euros. Hellwig, Anke. 2011. "Fraunhofer: A Non-Profit Organization with Entrepreneurial Spirit."

TABLE APP-A2-2 Fraunhofer Gesellschaft Revenue by Source

		2010 (Millions of Euros)
Basic (Core) Funding		
Federal Government		490
Lander Government		63
Revenue from FhG Activities		
Federal Government		
	Project Funding	323
	Contracts	11
Lander Government		
	Project Funding	190
	Contracts	3
Revenue from the Private Sector		454
Research Funding Organizations/Other		111
Increase in work in progress		36
Other operating income		39
Interest and other income		1
Total		1,727

SOURCE: Fraunhofer Gesellschaft Annual Report 2010.

probably understate the amount and proportion of public funding because they do not break out funding from the European Union programs as a separate line item.[71]

[71]EU funding may be included in "research funding organizations and other funding sources", it may be reflected in "revenue from the private sector, or both. One Fraunhofer Institute, Fraunhofer IGB in Stuttgart, reported in 2011-12 that 4.8 percent of its revenue from contract research are derived from the EU. However, FhG IGB staff indicate that other EU funding is provided to SMEs which use it to fund contract research at FhG IGB and it is counted under revenue from "industry and trade associations." Interview with FhG IGB. 2012. Stuttgart, Germany. July 14. A 2011 Fraunhofer presentation set the EU contribution to the institutes' revenues at 65 million Euros in 2010. Hellwig, Anke. 2011. "Fraunhofer: A Non-Profit Organization with Entrepreneurial Spirit."

Basic or core funding is provided by the federal and Länder governments in the form of grants which are not tied to any particular activity or expenditure on a 90/10 federal/state ratio. The Fraunhofer headquarters allocates nearly two-thirds of its basic funding to individual FhG institutes pursuant to a formula which creates an incentive for institutes to raise additional funding from the EU and from industry—e.g. the more external funding raised by the individual institute, the more basic funding it will receive from headquarters. Another ten percent of the basic funding is awarded to institutes on the basis of competitive bids for proposals to enter new research fields. 10 percent of the basic fund is used to purchase new equipment.[72] The remaining 15 percent is retained by the FhG Executive Board for use for special projects and the opening of new institutes.[73] In recent years, the Fraunhofer has been able to establish new research units with public economic stimulus funds from the federal government and the Länder, as well as the European Regional Development Fund (ERDF).[74] The Fraunhofer also derives income from donations, which are tax-deductible like charitable contributions.[75]

Infrastructure capital outlays are reported separately from Fraunhofer's operating budget, which reports 143 million Euros in "current capital outlays." This figure reflects investments in equipment and other fixed assets to maintain the operations of existing Fraunhofer units. Total capital outlays in 2011 were 379 million Euros (infrastructure capital outlays plus current capital outlays).

---

[72]The Fraunhofer expenses its acquisitions at the time of purchase, so its books do not reflect depreciation expense.

[73]This funding formula was established by a former FhG director, Alexander Imbusch, in the 1970s. Comin, Diego, Gunnar Trumbull, and Kerry Young. 2012. Fraunhofer: Innovation in Germany. Harvard Business School Monograph 9-711-022. January 6. The basic funds are a powerful tool for attracting private money because partners benefit "from the investment capital which precedes work on any particular application." Duff, Michael A. 1998. The Fraunhofer Society: A Unique German Contract Research Organization Comes to America. Office of Technology Policy, U.S. Department of Commerce. Washington, DC: U.S. Department of Commerce. October. pp. 23.

[74]The ERDF is a fund administered by the European Union which is intended to address regional disparities in Europe. Most of the ERDF funds allocated to Fraunhofer infrastructure projects in 2010 targeted at projects in the former East Germany. The German government portion of this funding is divided 50/50 between the federal and Länder governments. Major outlays have recently been made for a new research center in Dresden developing 3-D wafer level semiconductor packaging and interconnected technologies, expansion of the Fraunhofer Institute for Applied Polymer Research IAP's Application Center Polymer Nanotechnologies in Potsdam-Golm, and expansion of the Fraunhofer Center for Silicon Photovoltaics CSP in Halle. The Fraunhofer IAP's Application Center supports innovations such as flexible displays based on organic light-emitting diodes (OLEDs). The Application Center can support about 100 employees and has a usable floor space of 2600 square meters. Fraunhofer-Gesellschaft Annual Report 2011, pp. 19-20.

[75]Interview with Fraunhofer Institute for Process Engineering and Packaging IVV. 2012. Friesing, Germany. June 13.

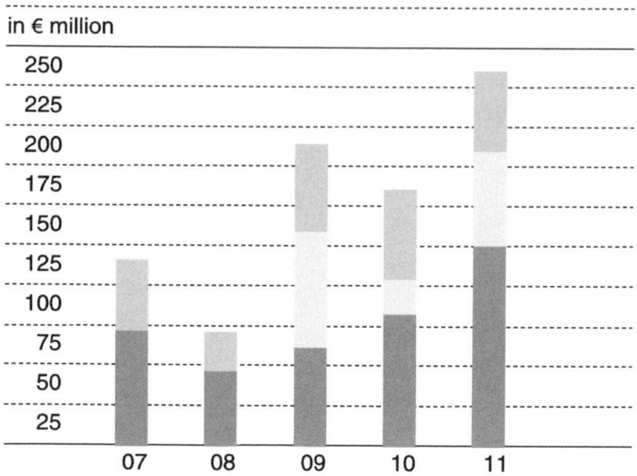

	2007	2008	2009	2010	2011	
■	72	47	62	83	126	
	–	–	73	22	59	
■	45	25	55	57	51	
=	**117**	**72**	**190**	**162**	**236**	in € million

■ Federal government and host *L nder* of the institutes
Economic stimulus programs I and II
■ ERDF (European Regional Development Fund)

FIGURE APP-A2-1 Major infrastructure capital expenditure and funding sources 2007-2011.
SOURCE: Fraunhofer Annual Report 2011, p. 20.

Contract Research for Industry

German and foreign firms contract with Fraunhofer institutes for research on specific themes. The standard form of collaboration is a one-off bilateral contract between a company and a Fraunhofer institute to achieve an agreed technological objective. The research is governed by signed contracts covering research activity and the ownership of intellectual property rights.[76]

[76]Dr. Stefan Noken, an executive at Hilti AG, a German construction company and a Fraunhofer alumnus, offers some criticism of the institute's contract procedures. "Sometimes it can take months to get a contract signed. I understand that Fraunhofer needs to stake out its claims and ensure it retains rights to the results of its work but it is also important to understand that companies want to

These projects frequently involve more than one Fraunhofer institute. Fraunhofer prices these contracts on the basis of the cost of running the project, plus a markup of about 15 percent, but the price does not include the historic costs incurred by the institutes to develop the knowledge used in the project. The price an individual institute charges for a working hour varies from institute to institute based on each organization's cost structure, including infrastructure costs. The price is independent of the success or the failure of the final project. The Fraunhofer institutes are engaged in 6-8,000 industry contract research projects in any given year. The Fraunhofer institutes sign confidentiality agreements with its industry partners to protect their proprietary information.[77] Some of these projects evolve into long-term strategic relationships between a company and a Fraunhofer institute, involving multi-stage collaborations over many years.

The Fraunhofer engages in collaborations with industry that extend beyond one-off contract research. Some technological challenges are sufficiently complex that they require multiple partners, and may engage numerous companies and Fraunhofer institutes. In some cases, Fraunhofer institutes perform contract research for industry associations.[78] The Fraunhofer invests in and supports off-site R&D centers located on companies' premises.[79] It creates pilot manufacturing lines to permit the development and testing of processes in industry-relevant settings, reducing the risks associated with scaling-up of production.[80] It forms joint ventures with companies.[81] Pursuant to "special

---

ring-fence a certain degree of exclusivity for themselves...[S]pending months honing the contractual agreements is neither in our interests, or indeed in Fraunhofer's interests, so I would like to see a more pragmatic approach in that area." "Pass Me the Hilti." Fraunhofer website. *<http://www.fraunhofer.de/en/range-of-services/reference.html>*.

[77]Comin, Diego, Gunnar Trumbull, and Kerry Young. 2012. Fraunhofer: Innovation in Germany. Harvard Business School Monograph 9-711-022. January 6. p. 10-11.

[78]In contract R&D for associations, the government contributes funding to the association, and companies that want to get access to the technology being developed must pay to join the association. Interview with Fraunhofer Institute for Process Engineering and Packaging IVV. Friesing, Germany, June 13, 2012.

[79]An example is Roth and Rau's R&D center at the company's headquarters in Hohenstein-Emsttal, Germany, an 11-million Euros project to develop products and processes to support more efficient solar cells. Fraunhofer-Gesellschaft contributed 3 million Euros to this project to equip the center with technology for research projects to be performed jointly by Roth & Rau and Fraunhofer institutes in Saxony. PVTech. 2009. "Roth & Rau Opens New R&D Centre to Boost Solar Cell Efficiencies." May 11. "Fraunhofer research scientists and engineers are often to be found at this cutting-edge research facility." Dr. Sylvia Roth, Roth and Rau Executive at *<http://www.fraunhofer.de/en/range-of-services/reference.html>*.

[80]The Fraunhofer Institute for Electron Beam and Plasma Technology FEP developed pilot lines to permit Biofilm SA to develop vacuum roll-to-roll production processes for polypropylene food packaging films. This project was co-funded by BMBF and the Saxon State Ministry for Economic Affairs, Labor, and Transport. Also involved in the project were Applied Materials (coating equipment), Vacuum Technologies Dresden (hollow cathode sources for plasma activation) and ISA GmbH (power supply system). Fraunhofer FEP, "Plasma Technology of the Fraunhofer FEP Becoming Established in the American Packaging Market." Press release, April 20, 2012.

cooperation contracts," companies can be given "guest" status at Fraunhofer institutes and operate their own laboratories and offices on the institutes' premises.[82] In 2005, the Fraunhofer entered into a public-private partnership with the U.S. semiconductor producer Advanced Micro Devices (AMD) and Germany's semiconductor maker Infineon to open the Center for Nanoelectric Technology (CNT) near Dresden, with the two companies assuming responsibility for operating the site.[83] The Fraunhofer Institute for Cell Therapy and Immunology IZI manufactures cell-based therapeutic products for pharmaceutical companies such as Northwest Biotherapeutics and Cognate BioServices Inc.[84]

Fraunhofer institutes are organized in parallel structures of thematic "research units" and "business units." Contacts with industry are made by the business units, which are the point of entry into the Fraunhofer system. The business units' expertise tends to lie in the area of potential applications and market potential of the technologies in which they specialize. When a company asks a business unit for technical help, the unit assesses the potential cost of the project, its business feasibility, and the terms under which the parities will work. The business units draw up a list of Fraunhofer research units from which assistance will be required, e.g. materials, software, packaging, equipment, and other specialties. While the business unit maintains relations with the industrial client, the research units are the point of contact for the project with university faculty, students, and laboratories.[85]

Small and Medium Enterprises

The Fraunhofer's core mission is to enable small and medium enterprises (SMEs) to gain timely access to scientific and technological innovation. Such companies typically have little or no internal research capability and must rely on external sources for such support. The Fraunhofer

---

[81]In 2012 the Fraunhofer Institute for Ceramic Technologies and Systems IKTS and FuelCell Energy Inc., a manufacturer of fuel cell power plants announced an MOU to form a joint venture to develop the European market for Direct Fuel Cell stationary power plants. FuelCell Energy has established a legal entity in Germany for the joint venture and will be the majority owner. Fraunhofer IKTS will contribute research and development resources and use local resources and relationships to develop the market. FuelCell Energy, "FuelCell Energy Announces Cooperation with Fraunhofer IKTS to Develop the European Market for Stationary FuelCell Power Plants." Press release, February 22, 2012.

[82]Fraunhofer ILT. 2010. Annual Report. pp. 20.

[83]Deutsch Press-Agentur. 2005. "Germany Opens Nanoelectrics Centre in Dresden." May 31.

[84]Fraunhofer IZI, "Fraunhofer IZI Received Manufacturing Authorization for DCVax®-L Brain." Press release, July 25, 2012; "Fraunhofer IZI Received Manufacturing Authorization for Tumor Vaccine CVacTM." Press release, October 28, 2011.

[85]Interview with Fraunhofer Institute for Process Engineering and Packaging IVV. Freising Germany. June 13, 2012.

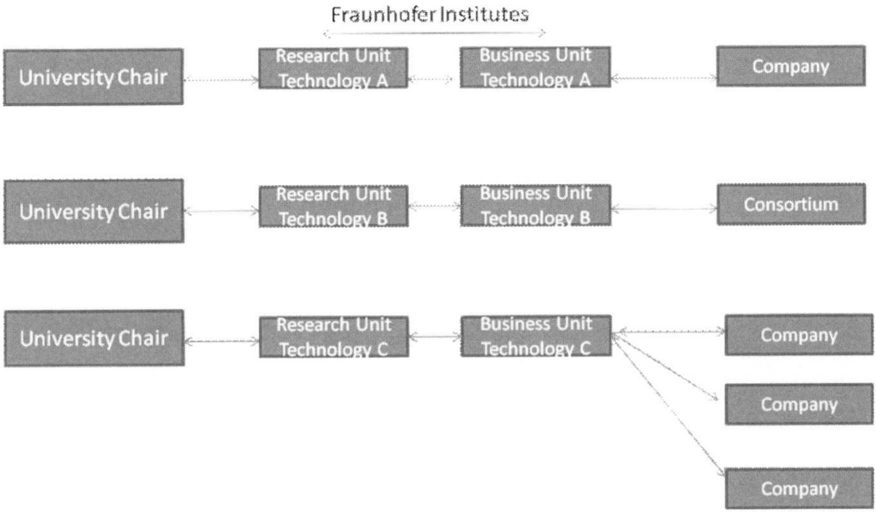

FIGURE APP-A2-2 Organization of Fraunhofer Institutes.

institutes accept contract research and develop technological solutions for problems that SMEs cannot resolve themselves. The Fraunhofer also assists SMEs by bringing them into contact with broader research and financial communities that can support their development efforts. In 2008, Fraunhofer contract R&D projects involving firms with less than 1,000 employees accounted for 43 percent of the total. Nevertheless, some studies have observed that Fraunhofer institutes "tend to pursue longer-term R&D for larger corporations and have often had a difficult time reaching and serving SMEs".[86] The Fraunhofer's figures for 2008 indicate that over two-thirds of its contract R&D for industry is conducted for firms of 250 or more employees.

The term Mittelstand is sometimes used to refer to all German SMEs, but it more accurately describes a subset of German SMEs "characterized by

[86]Shapira, Philip and Stuart Rosenfeld. 1996. An Overview of Technology Diffusion Policies and Programs to Enhance the Technological Capabilities of Small and Medium Enterprises. August. "[T]he fact that small and medium-sized companies represent only 10 percent of total Fraunhofer earnings suggests that in an era of shrinking public funding, the organization will be more inclined to go to greater lengths to satisfy the clients that represent over half their earnings—large industry and public bodies." Mitchell. 1998. Fraunhofer Society. Op. Cit. pp. 33. The establishment of a system of pre-competitive research sharing for SMEs through the Industrial Research Associations (AiF) occurred in part "to counterbalance the large company financial bias of Fraunhofer and Max-Planck Institutes." Harding, Rebecca. 2000. "Reliance in German Technology: Innovation Through Institutional Symbiotic Tension," Industry and Innovation. December. pp. 233.

TABLE APP-A2-3 Fraunhofer Contract R&D by Company Size (2008)

Number of Employees	Percent of FhG Contract R&D
1-250	32.5
251-500	6.1
501-1000	4.4
1001-10,000	18.0
10,001-100,000	24.4
1000,000+	14.6

SOURCE: Hellwig, Anke, "Fraunhofer: A Non-Profit Organization with Entrepreneurial Spirit," 2011.

certain convictions and attitudes in the context of socio-economic and political processes."[87] These firms are typically family-owned businesses which have been doing business for many years, in some cases for five or six generations.[88] Their leadership has a strong sense of social obligation and a close and paternalistic relationship with employees. They are deeply rooted in small towns and rural areas and less likely to engage in outsourcing or to move offshore.[89] They tend to plan for the very long term based on the recognition that the enterprise was founded by family members and will be passed on to future generations. They commonly specialize in very specific niche product areas and in many cases have become globally dominant in those niches through a process of constant incremental improvements in their products and production methods.[90]

---

[87]Meyer-Stamer, Jorg and Frank Waeltring. 2000. Behind the Myth of the Mittelstand Economy - The Institutional Environment Supporting Small and Medium-Sized Enterprises in Germany. Institute for Development and Peace at the Gerhard-Mercator- University Duisburg. Report 46.

[88]"About 95 percent of German businesses have a family at their helm." Deutsche Welle, 2011. "Germany's Family Businesses Reach Out to Global Markets." May 6. Mittelstand "prefer to keep a distance from the stock markets and they are not that strongly affected by international trends of the capital markets." They derive much of their capital from Germany's 50 public savings banks and 1,200 credit cooperatives, originally set up to help small firms adjust to industrialization, which have proven less susceptible to the financial crisis than the country's big banks. Christian Science Monitor. 2009. "Many Small Businesses in Germany Thrive During Downturn." April 10.

[89]While some German machine builders have moved production overseas to take advantage of lower labor costs, many others view outsourcing with skepticism. Mittelstand producer BueMi in Gevelsberg near Wuppertal, manufactures precision parts for machine tools and supplies companies like Thyssen Krupp, Siemens, and Bombardier. Its Managing Director, Frank Mittag, commented in 2010 that with outsourcing there is a "risk of important know-how seeping out." Medium-sized German companies which have established operations in China have seen their operations "quickly replicated by companies next door." In addition, cost advantages associated with lower labor costs abroad are often offset by the costs of quality assurance programs, production downtimes, and marketing. Deutsche Welle. 2010. "German Machine Builders Leery of Outsourcing." September 17.

[90]Oliver Wack and Phillip Boring quoted in, Xinhua. 2012. "What's Behind the Success Story of German Manufacturing Industry?" February 23; Meyer-Stamer, Jorg and Frank Waeltring. 2000. Myth of the Mittelstand. Op. Cit. pp. 9.

The Fraunhofer is widely viewed as an important element underlying the competitiveness of the Mittelstand.[91] The actual proportion of Fraunhofer R&D that is directed toward projects involving Mittelstand is difficult to quantify, not only because the term "Mittelstand" is not well defined but also because these firms "are private and relish their obscurity," and little information about specific Fraunhofer projects and the Mittelstand participants is made public.[92] However, anecdotal information about individual companies indicates that the Fraunhofer can play an important role in sustaining their international competitiveness. Moreover, the innovation support provided by the Fraunhofer organization helps SMEs in the capital markets, enabling financial institutions to take a longer-term view of such companies.[93]

Roth & Rau, located near Dresden, is an "archetypical" family-owned Mittelstand company. It employs about 1,100 workers and specializes in niche products and technology in photovoltaics, such as production machinery and coatings which are a requisite for solar panels. Roth & Rau's management made a strategic decision that the company would not sell coated solar wafers as an end product. "What the company sells are the machines that make the coating process, and, equally important, the know-how to make the machines work." A major market for the company is China and the management understands that the machines will be copied. "[The company] can't do anything about it. But they also know duplicating the precision-coating process isn't easy, and this has given them an estimated five-year jump on Chinese capability." The key to "keeping the gap", as a company executive explains, is innovation and continuous advances in the efficiency of production.[94] To achieve this, Roth & Rau is "working closely with various Fraunhofer institutes."[95]

---

[91]Hans Dieterle, an official at BDI, the federation of German industries, said in 1992 that "Especially for the Mittelstand, the Fraunhofers are very attractive. They find new equipment and can figure out what they never could on their own. In comparison with universities and all the other institutes, firms are much more satisfied with the Fraunhofer." *New Scientist.* 1992. "German Innovation, British Imitation." November 21. The Mittelstand are "very important in the context of the Fraunhofer model as these are the companies that are very commonly the funders, customers and economic beneficiaries of the work of each of the Fraunhofer Institutes, alongside the smaller number of industrial giants." Written evidence submitted by the University of Bristol (TIC 43), House of Commons. December 2, 2010.

[92]Venohr, Bernd and Klaus E. Meyer. 2007. The German Miracle Keeps Running: How Germany's Hidden Champions Stay Ahead in the Global Economy. Berlin: Berlin School of Economics, May. Pp. 5.

[93]Member of Parliament Gavin Barwell, citing interviews with the Federal Ministry of Education and Research in House of Commons: Science and Technology Committee. Formal Minutes: January 12, 2011.

[94]Marquand, Robert. 2011. "Germany—the new mini-superpower" Christian Science Monitor. January 30.

[95]"Using Plasma to Enhance Solar Cell Efficiency." Fraunhofer Website. The CEO of Roth & Rau, Dr. Dietmar Roth, is a member of the Board of Trustees of the Fraunhofer Institute for Solar Energy Systems ISE in Freiburg. Renewable Energy World. 2011. "Meyer Burger Acquires Roth & Rau Shares, Extends Photovoltaic Equipment Reach." April 11.

Roth & Rau employs over 100 research personnel, most of whom are based in its technology center in Hohenstein-Emstthal. "Fraunhofer's research scientists and engineers are often to be found at this cutting-edge R&D facility," reflecting the fact that the company "carries out joint projects with a variety of Fraunhofer institutes in a number of technical fields."[96] One example is a pilot line for processing and characterizing silicon wafer-based solar cells that operate in the company's technology center in a collaboration with the Fraunhofer institute for Ceramic Technologies and Systems IKTS. The purpose of this project is to develop new generations of processing equipment, to enhance crystalline standard cell technology and develop manufacturing technologies for high-efficiency cells.[97] Roth & Rau also supplies industrial machinery to Fraunhofer itself.[98]

Large Companies

Germany is home not only to the Mittelstand but to numerous globally-competitive large vertically integrated enterprises with strong ties to Germany's major banks.[99] While the Fraunhofer's core mission is the provision of applied research support for small and medium enterprises, it derives a larger proportion of its contract research revenues from large companies. A 1987 French survey of the Fraunhofer observed that "the big industrialists do not hesitate to entrust [Fraunhofer] with very high level projects. Siemens, Daimler-Benz, VW are omnipresent in their labs."[100]

---

[96] Marquand, Robert. 2011. "Germany—the New Mini-Superpower" Christian Science Monitor. January 30.

[97] "Using Plasma to Enhance Solar Cell Efficiency." Fraunhofer Website.

[98] TendersInfo. "Roth & Rau Microsystems GmBH Wins the Contract for Supplying Industrial Machinery." January 17. In 2009, Roth & Rau decided to enter the thin-film photovoltaics market, based on the recognition that thin-film PV cells are less costly to manufacture than silicon wafer-based cells. The company engaged the Fraunhofer Institute for Electron Beam and Plasma Technology FEB in a project funded by the Saxon State Ministry for Science and the Arts. The Fraunhofer FEP team worked with the company to build a complete pilot manufacturing line at the institute to produce thin film cells with the highest possible efficiency, with a target completion date of 2015. Four Roth & Rau employees work directly on the pilot line at the institute in order to facilitate continuous knowledge transfer to the company.

[99] Die Computer Zeitung. 1986. "Siemens, Fraunhofer Automate Masking for X-Ray Lithography. JPRS-WST-86-005. January 31. Spiegel Online International. 2008. "We Are Not Driving the Price Hike", July 16. "Research Product by ThyssenKrupp VDM, RWE Power and Fraunhofer Institute Significantly Reduces Development Time for "New Materials."" ThyssenKruppAG Press Release. 2008. May 29.

[100] Industries et Techniques. 1987. "Applied Research: The Fraunhofer Method." October 20. JPRS-ELS-88-006. Dr. Anton Heuberger, Director of the Fraunhofer institute of Microstructural Techniques in Berlin commented in 1987 on a massive semiconductor lithography collaboration involving Siemens, Telefunken, Philips, and Eurosil that "Fifteen researchers from Siemens are working with us on a full time basis. We're actually at risk of becoming a sort of subsidiary. But we are careful to protect our autonomy."

Box APP-A2-1

Mittelstand Niche Strategy

CleanDieselCeramics GmbH is a medium-sized German maker of particulate filters for Diesel engines based in Grossrohrsdorf in Saxony. Stricter emissions standards which came into effect in 2011 will require retrofitting of existing Diesel engines. CleanDieselCeramics and the Fraunhofer Institute of Ceramic Technologies IKTS observed that the market for automobile particulate filters was a huge mass market involving millions of vehicles but that the market for certain specialized vehicles like construction machinery might be as small as 3000 units. Fraunhofer and CleanDieselCeramics concluded that there was "a gap in the market which basically involved developing a cost-effective manufacturing process that also works for small and medium scale quantities." The company and the institute used a material patented at the Fraunhofer IKTS, a porous silicon carbide ceramic, and developed highly efficient ceramic diesel particulate filters for non-road applications. The parties also developed and refined a manufacturing process for the filters which was tested and perfected on a pilot production line at Fraunhofer IKTS. The company hired staff for its own production facility in Grossrohrsdorf who worked on the pilot line at the institute for six months, enabling them to bring their know-how to the company's commercial production line.

Publicly-funded Contract Research

The Fraunhofer receives public revenues for contract research which is linked to specific projects being undertaken for the federal government and the Länder. In some cases, these funds are augmented by EU funding. The Fraunhofer is prohibited from signing full confidentiality agreements with respect to these contracts, which may in some cases deter participation by the private sector. Publicly funded contract research undertaken by the Fraunhofer generally has commercial potential while serving broader public goals, such as public health, safety, and environmental protection.[101]

---

[101]Examples of a number of publicly funded projects were reported in the Fraunhofer Institute for Interfacial Engineering's Annual Report for 2011-12: (1) Differentiation of mesenchymal stem cells. This project, funded by the BMFB, studies mesenchymal stem cells which can differentiate into bone, fat, cartilage, and other forms of tissue, and seeks to identify the external stimuli that lead to this differentiation. (2) Detection of human fungal pathogens. This project, funded by the BMFB, seeks to develop early-detection techniques utilizing a lab-on-a-chip device for deadly yeast and mold fungi that infect humans, particularly those with compromised immune systems. (3) Rotating disk filter for anaerobic wastewater treatment. This project, funded by BMWi, seeks to develop a

Box APP-A2-2

Fraunhofer and Volkswagen

Fraunhofer has been involved with many of Germany's leading companies in multiple projects of many years' duration, which commonly lead into follow-on collaborations. The Fraunhofer has been working with Volkswagen, for example, on factory automation projects for decades.[a] Volkswagen established a subsidiary in Saxony, Volkswagen Sachsen GmbH in the early 1990s in the wake of German reunification. Fraunhofer concurrently established the Fraunhofer Institute for Machine Tools and Forming Technology IWU in Dresden and Chemnitz. Since then Volkswagen and Fraunhofer IWU have worked on a "whole series of joint research and development projects in the field of production engineering." The Fraunhofer IWU Advisory Board has always included Volkswagen executives and the Advisory Board has been chaired by Professor Hochem Heizmann of Volkswagen since 2007.[b] In 2008, Volkswagen concluded an agreement with Fraunhofer-Gesellschaft to establish a joint "Center of Excellence for Automotive Production" in Chemnitz which would develop and test flexible resource-efficient production methods for automobiles.[c] The center has become an "integral part" of the Research Factory for Resource Efficient Production, a project jointly operated by Fraunhofer IWU, Volkswagen, and Audi. In 2002, the Land government of Saxony, Fraunhofer IWU and Volkswagen launched the Mechanical Engineering and Automotive Initiative (MAINE), which established an association comprised of Volkswagen plants in Saxony and about 20 automobile parts suppliers and engineering companies.[d]

[a]The Fraunhofer IPK developed Volkswagen's robotized windshield-bonding system in the 1980s and a similar system for Audi. Industries et Techniques. 1987. "Applied Research: The Fraunhofer Method." October 20. JPRS-ELS-88-006.
[b]"Mobility Made in Germany." Fraunhofer Website.
[c]Volkswagen and the Fraunhofer Institute for Machines Tools and Forming Technology IWU reported in 2009 that they had developed a new method for punching holes in the press-hardened steel body work based on electromagnetic pulse technology (EMPT) which can perform an operation in one-seventh the time taken by a mechanical puncher with lower energy consumption. Fraunhofer Research News. 2009. "Electromagnetic Fields as Cutting Tools." December 1.
[d]MAINE was designed to develop and test new manufacturing and information technologies associated with automobile engine and body manufacturing. Fraunhofer IWU supervises and

rotating disk filter unit which will make the purification of urban and industrial wastewater cost-effective. (4) Multifunctional PEGs. This project, co-funded by the Baden-Wurttemberg Ministry of Science, Research, and the Arts and the Peter and Traudl Engelhorn Foundation, seeks to develop new forms of biocompatible materials utilizing polyethylene glycol (PEG), a substance with broad application in tissue engineering, pharmaceuticals, and cosmetics. Fraunhofer IGB Annual Report 2011-12.

coordinates this project, which has involved themes such as residue-free engine manufacturing, fully automated laser welded car body manufacturing, development of lightweight construction components for the car body and powertrain, development of new techniques for manufacturing crank shafts and B-pillars, and machine system design and process development. Presentation of Mattis Potz. 2006. Fraunhofer Gesellschaft. Piacenza. July 7. Current MAINE projects involve the development of innovative cutting and joining techniques for high strength steel, and incorporation of newly-developed techniques into mass production lines. "Mobility Made in Germany." Fraunhofer Website. Op. Cit.

EU-funded Research

The Fraunhofer Institutes derive a portion of their funding from European Union programs supporting R&D:

- Framework Program Seven (FP7): This program, the EU's principal vehicle for funding research in Europe, provides grants for technological development and demonstration projects with "European value-added."[102] Grants extended under the Seventh Framework typically involve participants from other European Countries.[103]
- Competitiveness and Innovation Framework Program (CIP): This program is designed to enhance competitiveness of small and medium enterprises through support of innovative activities.
- Eurostars: The EUREKA Eurostars Program supports R&D by small and medium enterprises to develop new products, processes, and services.

Although most EU funding for the Fraunhofer is linked to specific projects, one EU fund functions as a form of core funding which is used like federal/Land core funding to support research infrastructure. The European Regional Development Fund (ERDF is a structural fund promoting investments in areas which have experienced economic decline which has supported Fraunhofer R&D projects in the former East Germany and economically

---

[102]Peplaser, a 3-year EU Seventh Framework R&D project launched in 2009 established a consortium comprised of 11 research groups in four European countries, including Bulgaria and Poland. The project sought to develop production processes and equipment for the solid-phase synthesis of peptides. The Fraunhofer-Gesellschaft zur Foerdunug der Angewandted was a participant. 3 million euros of the projects 4 million euros was provided from EU seventh framework funds. Warsaw Voice. 2009. "Polish Firms in EU Biotechnology Project R&D." October 28.

[103]The EU is providing grant funding under the Seventh Framework to the Fraunhofer Institute for Interfacial Engineering and Biotechnology IGB for a project to develop fertilizer pellets for organic farming that also repel insect pests. Project partners include agricultural associations from Germany, Norway, Spain, Lithuania; SMEs from the United Kingdom, Germany, and Norway; Nor-Tek, a Norwegian research organization; and the University of West Hungary in Mosonmagyarovar, Hungary. Fraunhofer IGB Annual Report 2011-12. pp. 100-101.

disadvantaged areas in Western Germany.[104] In 2010, the ERDF allocated 57.2 million Euros for Fraunhofer research infrastructure investments. Fraunhofer maintains an office in Brussels and it is regarded as a formidable force in lobbying for EU funding.[105]

Internally-funded Research

Fraunhofer "blue sky" research projects are entirely funded internally. The Fraunhofer-Zukunftsstifung (Fraunhofer Future Foundation) is a foundation which the institute established to use the large operating surplus that was generated from the success of the mp3 project. The foundation invests these funds in research projects that may not bring in revenues over the short run. These funds are allocated to individual FhG institutes through a competitive bidding process.[106]

Staffing and Training

The Fraunhofer's workforce has been growing substantially, a direct reflection of the growing number of research projects and contract volumes.

---

[104]The Fraunhofer Institute for Machine Tools and Forming Technology IWU joined with ThyssenKrupp Engineering in a recent project to develop technology for advanced production in the aircraft industry. This project was funded by the European Regional Development Fund and the Free State of Saxony. One British academic commented before the House of Commons that, "the Fraunhofer have tapped into European Regional Development Funds [and] last year [2010], they drew tens of millions of Euros to establish new capital facilities." Aerospace Manufacturing. 2012. "The Refinement of Alignment." January.

[105]"[T]he lobbying power of the Fraunhofer Institutes at the European level should not be underestimated and should be a force to be reckoned with if similar institutions wish to influence or benefit from European initiatives in Research and Development." House of Commons, Science and Technology Committee. 2011. Technology and Innovation Centres. Second Report of Session 2010-2011. February 17. pp. 29. During the period 2007-2011 fully 10 percent of all EU funding in the field of photonics went to the Fraunhofer Heinrich Hertz Institute HHI in Berlin, notwithstanding intense competition. Fraunhofer HHI, "EU Funds Photonics R&D in Berlin." Press release, 2012. The EU funding process is highly complex and potentially prohibitive for small and medium enterprises. Accordingly, the Fraunhofer Institutes' business development departments work with German SMEs to explain how the system works and how to negotiate with the EU for funding. One Fraunhofer institute operating in the life sciences sector indicates that for an average program involving the institute supported by EU funding, it takes nine months to set up a project involving a novel idea or concept, about one year to decision. The approval process is reportedly gradually getting shorter. Interview with Fraunhofer Institute for Interfacial Engineering and Biotechnology IGB. Munich, Germany. June 15, 2012.

[106]As of mid-2012 roughly a dozen research projects funded by the foundation were under way. One such project, "RIBOLUTION," involves the potential use of a class of molecules known as non-(protein)-coding ribonucleic acids (ncRNAs) to identify novel diagnostic indicators for diseases such as prostate cancer and chronic obstructive pulmonary disease. This project involves five Fraunhofer institutes as well as the company GlaxoSmithKline and the Universities of Dresden, Leipzig, and Charite Berlin. Fraunhofer IGB. 2012. Fraunhofer IGB Annual Report 2011-2012. pp. 62-63. Interview with FhG IGB. 2012. Stuttgart, Germany. June 14.

Box APP-A2-3

Market-Oriented Strategic Preliminary Research

In addition, Fraunhofer Gesellschaft allocates "headquarters money" to projects that are estimated to be 2-3 years away from market application. This program, designated "Market-Oriented Strategic Preliminary Research (MAVO)", draws on 10 percent of the institutes' core public funding. The money is allocated from a "central pot" but individual institutes sometimes contribute some of their own funds. These projects, which may involve multiple institutes, are not expected to pay for themselves within a two year time frame, are seen as a way to induce FhG institutes to compete with each other in the development of new ideas with future potential as well as a mechanism for encouraging different FhG institutes to interact.[a]

[a]A current example of a MAVO project is "ProLignocell," a project to develop new sustainable processes for the utilization and material development from lignocellulose, a cellulosic substance which protects plants from decay caused by microorganisms or enzymes. The project involves collaboration between four Fraunhofer research institutes. Fraunhofer IGB. 2012. Fraunhofer IGB Annual Report 2011-2012. pp. 86-87; Interview with FhG IGB. 2012. Stuttgart, Germany. June 14.

Headcount grew by 5,978 between 2007 and 2011, an increase of over 41 percent.

Fraunhofer Institutes are each staffed with an average of 300-400 people. There are exceptions to this pattern—the Fraunhofer Institute for Solar Energy Systems (ISE), for example, reported a headcount of 1139 staff members in 2011.[107] The institute's permanent staff is a minority of the total workforce averaging 20-40 percent with most employees operating under various temporary employment contracts.[108] The majority of the institute researchers are PhD students and post-docs who commit to work half-time at a Fraunhofer institute. While a few of these individuals become part of the institute's core staff, most leave for positions in industry after 3-6 years, a turnover rate that

[107]Fraunhofer ISE. 2011. Annual Report. pp. 11.

[108]An institute does not need approval from headquarters to hire temporary employees, and some "temporary" workers have remained at institutes for many years pursuant to successively-renewed 2 and 3-year contacts. Some of these "temporary" workers rise to senior management positions. Interview with Fraunhofer Institute for Process Engineering and Packaging. Friesing, Germany, June 13, 2012.

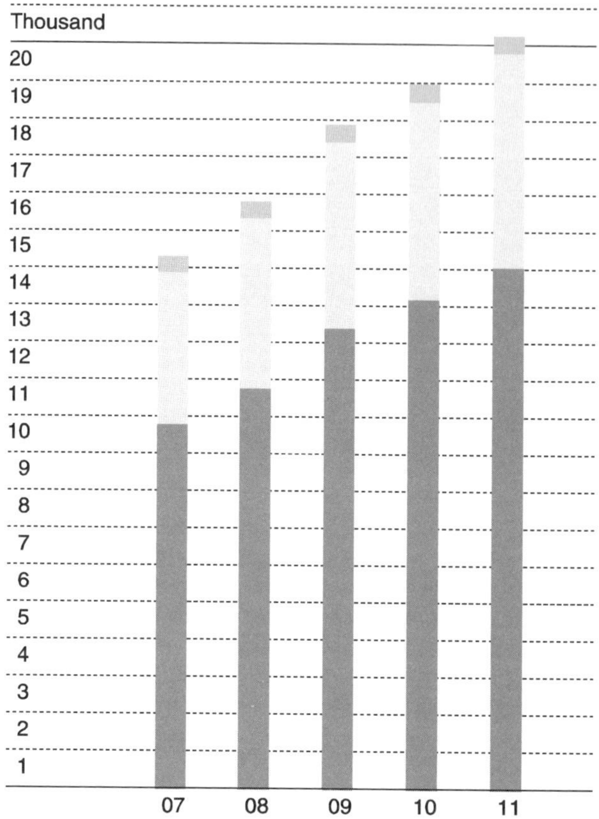

	2007	2008	2009	2010	2011
	9815	10784	12410	13202	14073
	4087	4584	5009	5313	5765
	446	455	488	487	488
=	**14,348**	**15,823**	**17,907**	**19,002**	**20,326**

Scientific, technical and administrative personnel
Graduates, students, school pupils
Trainees

FIGURE APP-A2-3 Growth in the Fraunhofer-Gesellschaft's workforce, 2007-2011.
SOURCE: Fraunhofer Annual Report 2011, p. 30.
NOTE: New basis for calculation: workforce also includes personnel on temporary employment contracts with terms of less than 18 months. Figures for previous years have been amended accordingly.

some German business partnering with the institute see as a negative.[109] But on the positive side, Fraunhofer institutes function as large scale training centers for highly skilled industrial managers and workers:

> As an employer, the Fraunhofer Gesellschaft offers a platform that enables its staff to develop the necessary professional and personal skills that will enable them to assume positions of responsibility within their institute, in industry and in other scientific domains.[110]

Networks

The Fraunhofer emphasizes the importance of networks in virtually all aspects of its operations, reflecting its view that competitive innovation increasingly requires collaborative efforts by numerous actors with specialized competencies acting in concert to achieve a common objective.[111] German

---

[109]Rudolf Smion, research director of Meissner and Wurst, a firm that contracted with the Fraunhofer for research on designing clean rooms for semiconductor production commented in 1992 that "you do not work with the institute but with the people in it. But often the researchers are with the institute for only three or four years. The people leave and then all the promises are gone." *New Scientist.* 1992. "German Innovation, British Imitation." November 21. A Fraunhofer executive conceded in 2012 that "it is difficult for us to survive the blood loss. It's usually the best people that go out." Interview with Fraunhofer Institute for Production Technology and Automation IPA. Stuttgart, Germany, June 14, 2012.

[110]Written Evidence Submitted by Advanced Manufacturing Research Centre (AMRC). University of Sheffield (TIC 35). House of Commons Science and Technology Committee. January 12, 2011. Fraunhofer is a publicly funded research organization and as such is subject to rules which limit salaries to levels that are lower than the compensation for comparable jobs in the private sector. Comin, Trumbell and Young. 2012. Fraunhofer: Innovation in Germany. Op. Cit. pp. 12. Fraunhofer permanent staff salaries are 50-70 percent of comparable jobs in the private sector. PhD candidates are paid at about half the level of full-time Fraunhofer scientists. Interview with Fraunhofer Institute for Process Engineering and Packaging. Freising, Germany. June 13 2012. "The institutes' extensive use of students bears some resemblance to the practice of medical interns in hospitals—long hours, low pay—while delivering all of the services expected of a fully-licensed and much more expensive doctor. The benefit to the Fraunhofer Institute is that they obtain access to leading-edge expertise at less than commercial rates." Mitchell. 1998. Fraunhofer Society. Op. cit. pp. 25. Students are paid one-half the amount paid to full-time workers. Interview with Fraunhofer Institute for Process Engineering and Packaging IVV. Friesing, Germany, June 13, 2012. The long-run benefits to students, however, are also substantial. By the time they leave the institute for careers in industry, they have acquired cutting-edge technological skills, business skills, and a broad network of business contacts. According to one estimate a five-year investment by students in Fraunhofer training enables them to double the income they could otherwise expect by entering industry with a PhD only.

[111]A German executive whose company works extensively with Fraunhofer comments that "[S]uccessful innovations nowadays tend to emerge from networks. Today's new products and services are developed at ever shorter intervals, which means that the window of time in which you can maintain an innovation's competitive edge is far briefer than it was in the past. But at the same time the product solutions are much more technically challenging. For example, our hammer drills are not simply mechanical products any longer, but rather they are intelligent devices with electronic, mechatronic and software technology components and functions. The only way to keep

scientific disciplines in general are highly networked, and scientists are trained to cooperate and apply jointly for funding. A Fraunhofer scientist observes that it takes a major effort to set up collaboration networks, but once they are established they are a huge competitive advantage. He comments that "our experience of the U.S. is that there is a lot less collaboration by laboratories with each other and industry."[112] Companies that engage Fraunhofer for research tap into a rich university research network.[113]

The Fraunhofer institutes also form networks with each other, with the recognition that "coordinated competencies allow quick and flexible arrangements of research work on the requirements of different fields of application to answer actual and future challenges."[114]

Fraunhofer has 23 "Fraunhofer alliances," which are combinations of Fraunhofer institutes with expertise relevant to topical themes such as cloud computing, adaptronics, embedded systems, high-performance ceramics, and optic surfaces.[115] The Fraunhofer plays a prominent role in international networks formed to address specific technological themes.[116]

---

pace with this increasingly dynamic business environment is by working within innovation networks which we can draw upon as needed. We see the Fraunhofer-Gesellschaft as a solid component of our network and we collaborate with a great number of institutes on a whole host of different projects." Noken, Stefan. "Pass me the Hilti" <*http://www.fraunhofer.de/en/range-of-services/references/reference-hilti.html*>.

[112]Interview with Fraunhofer Institute for Manufacturing Engineering. Stuttgart, Germany. June 14, 2012.

[113]Hansjorg Lerchenmuller, CEO of Concentrix solar GmbH, a maker of solar concentration modules, indicates that the Fraunhofer institutes "have great links to the University. Through Fraunhofer we can get people investigating and evaluating interesting topics in the form of dissertations and theses." <*http://www.fraunhofer.de/en/range-of-services/reference.html*>.

[114]Fraunhofer Institute for Material and Beam Technology IWS. 2011. Annual Report. pp. 120. For example, six Fraunhofer institutes form the "Fraunhofer Group Lights & Surfaces," which draws on skills in lasers, surface coating, photonics, nanotechnology, materials technology, microassembly, ultra precision engineering and carbon technology. The institutes are Fraunhofer FEP (electron beam/plasma), ILT (lasers), IOF (applied optics), IPM (physical measurement), IWS (material and beam technology), and IST (surface engineering and thin films).

[115]Within the alliances, common points of contact provide advice and coordinate the development of solutions involving one or more institutes. The Fraunhofer Battery Alliance, for example, has a designated spokesman from the Fraunhofer Institute from the Fraunhofer Institute for Chemical Technology ICT and is comprised of 19 Fraunhofer institutes.

[116]The Fraunhofer is a member of the Heterogeneous Technology Alliance (HTA), a group of leading European research organizations which has agreed to become interdependent, sharing technology and coordinating investments in the fields of micro and nano technologies. In 2011, the "European Association for Biometrics" was formed, comprised of 14 institutions from 10 countries. The EAB is chaired by Alexander Novak of the Fraunhofer Institute for Computer Graphics Research IGD. The EAB is a non-profit organization based in the Netherlands tasked with advancing the "proper and beneficial use of biometrics in Europe and abroad." Plant Biometrics. 2011. "European Association for Biometrics Founded." December 2.

Box APP-A2-4

Fraunhofer Networking—A Case Study

An academic study published in 2000 by the University of Manchester offered a case study of Fraunhofer networking in the process of developing a contract research project with a Germany company. In this case Fraunhofer ISI was working with a large engineering company to develop new, competitive industrial products over a 10-year time horizon. The first phase of discussions focused on establishing the needs of the client and its customers on the basis of interviews, a workshop, and ISI's own research. This phase produced 21 product ideas, ranging from incremental improvements in existing products to radical innovations. The second phase, another workshop was to be a forum for winnowing these ideas down to a manageable number. However, "a second purpose of the workshop was to create networks and partnerships between the client and other Fraunhofer institutes." Accordingly, representatives from four Fraunhofer institutes were present at the workshop. One institute concentrated on microelectronics themes; another addressed electrical and mechanical engineering; another emphasized its "whole systems" approach to engineering issues; while Fraunhofer ISI retained a coordinating role and a perspective on the strengths of each institute that could be deployed in the project.

What is interesting in the context of this chapter is the appearance of four, normally competitive institutes focusing their discussion on one issue. The rivalry between IEG, IPA and IPK was highly developed; IEG as an institute evolved from IPA structures and all three competed aggressively for similar client projects. Thus while the three normally came into contact through 'market-based' activities, they were here cooperating as experts on a development project...The 'hidden agenda of the day was to create an ideal collaborative partnership between the participants, which would lead to profitable product development in the third and final phase.[a]

[a]Harding, Rebecca and William E. Paterson (eds.) 2000. The Future of the German Economy: An End to the Miracle?"Manchester: Manchester University Press. pp. 98-99.

Intellectual Property

The Fraunhofer has a large patent portfolio, reflecting its policy toward intellectual property. It registered 673 new inventions in 2011 for a total of

6,131 patents and registrations as of the end of that year.[117] While the treatment of intellectual property rights is subject to negotiation when a research contract is concluded, it is the normal practice for Fraunhofer to retain patent and other intellectual property rights upon conclusion of a research project.[118] Royalties and other income from IPR go to Fraunhofer headquarters, not individual institutes. Individual scientists responsible for developing patented technology receive 30 percent of the patent revenue.[119] The industry partner or partners may receive an exclusive license, but only for the particular application that was the subject of the project. Fraunhofer remains free to license the technology to other companies for different applications. Future customers have access to Fraunhofer's large patent portfolio which enhances the value of collaborations for companies and produces licensing revenues for the institutes.[120] In recent years, Fraunhofer has placed a higher emphasis on intellectual property protection and exploitation, concentrated in areas of anticipated high revenues.[121]

Spin-offs

Fraunhofer began promoting the systematic formation of spin-off companies in 1999 as a direct way to deliver technology developed internally to the marketplace.[122] Currently, the Fraunhofer encourages the start-up of new companies from within its own institutes and continues to support these entities once they are launched.[123] The Fraunhofer transfers some of its know-how to

---

[117]Fraunhofer Magazine. 2012. "Boost in Earnings." February.

[118]In some cases, however, "if it is invented at the University, it is owned by the University." Interview with Fraunhofer Institute for Manufacturing Engineering. Stuttgart, Germany. June 14, 2012.

[119]Interview with Fraunhofer Institute for Process Engineering and Packaging IVV. Friesing, Germany. June 13, 2012.

[120]Duff, Michael A. 1998. The Fraunhofer Society: A Unique German Contract Research Organization Comes to America. Office of Technology Policy, U.S. Department of Commerce. Washington, DC: U.S. Department of Commerce. October. pp. 65. Interview with FhG IGB. 2012. Stuttgart, Germany. June 14.

[121]This assessment is based on the perception of a technology's market potential and the comparative competitive strength of proprietary clusters of patents. In 2008 over half the institutes' total revenues from licensing were derived from patents generated by two Fraunhofer Institutes, the Fraunhofer Institute for Integrated Circuits (IIS-Erlangen) and the Institute for Digital Media in Technology (Ilmenau) from research in the field of audio coding (including MP3). Miha, Andren, David Connell, and Alan Hughes. 2009. "Models of Technology Development in Intermediate Research Organizations." Centre for Business Research Working Paper 346Q. December. pp. 14.

[122]Gijsbers, Rosebaum and Vullings. Benchmarking Contract Research Organizations. Op, Cit. p. 47.

[123]Fraunhofer spinoffs are 90 percent funded by venture capital, with the remaining 10 percent split evenly between angel and bank financing. In about half of its spinoffs, Fraunhofer takes an equity stake of up to 25 percent, which becomes diluted with successive rounds of venture capital. Fraunhofer Ventures currently has shareholdings in 51 spin-off companies. Spin-offs are managed by an internal division, Fraunhofer Ventures. This entity maintains a network of contacts with certified public accountants, attorneys, business consultants, and public and private venture capital

these companies in return for a minority equity stake. It provides technical, managerial and operational support for the spin-offs through its program, "FFM—Fraunhofer Fosters Management." Over half of the spin-offs have been in information technologies, microelectronics, and photonics.

Most spinoffs are small-scale efforts involving one or two Fraunhofer employees.[124] Fraunhofer spinoffs enjoy a high degree of success to the extent that success is defined as survival over the long run. Depending on the source, the success rate is somewhere between 65 and 90 percent.[125] Whatever the actual figure, it is extraordinary in international terms. The failure rate for startups in the United States in the early 2000s was 90 percent.[126] However, the high success rate may be an indicator that only spin-offs with a high probability of success are undertaken, suggesting an aversion to high risk endeavors.[127]

International Operations

In the early 1990s the Fraunhofer-Gesellschaft began making a concerted effort to establish foreign operations, a recognition that 85 percent of the global market for the most advanced technologies was located outside of Germany. While Germany was a global leader in a number of industries, such as chemicals and machinery, it trailed other countries in emerging technological areas such as information technology and biotechnology. Establishment of

---

firms which can provide the start-up with support in its early phases. Fraunhofer Ventures emphasizes reviewing and improving business plans, securing capital, drafting partnership agreements between the spin-off and associated institutes and arranging direct support from the Fraunhofer. Astrom, Eriksson, and Eriksson. 2008. International Comparison of Five Institutes. Op Cit. pp. 122.

[124]Comin, Trumbell and Young. 2012. Fraunhofer: Innovation in Germany. Op. Cit. pp. 11.

[125]This purported success rate may be exaggerated. Fraunhofer IGB in Stuttgart reports 8 spinoffs in the past 12 years, of which 2 have survived. Interview with Fraunhofer IGB, Stuttgart, Germany, June 14, 2012.

[126]Comin, Trumbell and Young. 2012. Fraunhofer: Innovation in Germany. Op. Cit. pp. 11.

[127]Some spin-offs appear to continue essentially as extensions of the Fraunhofer itself. MEMS Foundry Itzehoe GmBH (MFI), a semiconductor foundry business spun off from the Fraunhofer Institute for Silicon Technology ISIT in 2009, specializes in the fabrication of micro-electro-mechanical systems (MEMS). Since its inception, MFI has continued to work closely with Fraunhofer ISIT in "converting the results of research projects between Fraunhofer ISIT and industrial partners into a commercial volume manufacturing environment." "X-Fab and MFI Combine MEMS Foundry Offerings." 2011. XFAB press release. February 3. In 2012 the Fraunhofer Institute for Algorithms and Scientific Computing SAI spun off SIDACT GmbH to market software programs for simulation data analysis and compression technology developed at Fraunhofer SAI. SIDACT and Fraunhofer SAI will continue to collaborate pursuant to a cooperation agreement. Fraunhofer SAI, "Fraunhofer Spin-Off Develops Data Compression Techniques." Press release, October 8, 2012.

Box APP-A2-5

Regional Innovation Clusters

Regional innovation clusters: A major element of the German federal government's high tech strategy is the creation of regional innovation clusters, and a number of Länder governments are pursuing their own cluster initiatives.[a] Pursuant to the "Pact for Research and Innovation," the Fraunhofer assumed responsibility for conceptualizing and establishing innovation clusters. It is currently pursuing nineteen cluster initiatives in various parts of Germany. These typically involve one or more Fraunhofer Institutes, a local university, and a number of large and small industrial partners.[b]

[a]In 2007, Baden-Wurttemburg's Minister of Economic Affairs, Ernst Pfister, defined "cluster" as a "geographic concentration of companies linked to each other, specialized suppliers, service providers, companies in related sectors, and supporting organizations like universities and research institutes, other scientific and educational institutions, chambers, associations, and the like." Business Baden-Wurttemburg. 2007. "The Culture of Cooperation." September.

[b]The Mechatronic Machine Systems cluster, located in Chemnitz, Saxony, is developing manufacturing tools based on mechatronics to enable companies to design processes and produce products to very high standards. The cluster combines disciplines from computer science, mechanical engineering, and electrical engineering. The research partners are the Fraunhofer Institute for Machine Tools and Forming Technology IWU and the Institute for Machine Tools and Production Processes at Chemnitz University. There are eight industrial partners including Volkswagen AG and SIEMENS VDO Automotive AG.

overseas institutes was seen as a way to bolster Germany's position in those areas and to acquire technologies that would flow back to the parent institutes for application for the benefit of German industry.[128] In 1994, Fraunhofer USA, Inc. was established in Rhode Island to manage a network of U.S. research centers, which has grown to six centers at present.[129] In 2012, Fraunhofer

[128]Mitchell. 1998. Fraunhofer Society. op. cit. pp. 39-41. The Fraunhofer Institute for Laser Technology (IGT) states in its 2010 Annual Report (p. 20) that the objective of its international collaborations is "to recognize new trends and current developments and to acquire further know-how." The Fraunhofer Institute for Material and Beam Technology IWS, based in Dresden, maintains a close working relationship with the U.S.-based Fraunhofer Center for Coatings and Laser Applications in Plymouth, Michigan. Fraunhofer IWS indicates that the advantage of this relationship is the ability to identify trends in the United States and the fact that "the research and development work performed in the United States generates additional know-how and competencies, which benefit the project acquisition in German and European markets." Fraunhofer IWS. 2011. *Annual Report.* pp. 117.

[129]Fraunhofer's U.S. operations are small in comparison to those in Germany, reporting $40 million in total revenues in 2010. The strongest-performing U.S. center is the Fraunhofer Center for Molecular Biology (CMB) in Delaware, with 2010 revenues of $20 million. CMB is engaged in the development of vaccines, protein therapeutics, industrial enzymes, and agricultural biotechnology.

unveiled plans to establish a Fraunhofer Center for Applied Photonics in strategic collaboration with the University of Strathclyde in the United Kingdom.[130]

THE FRAUNHOFER IN CONTEXT

Most foreign observers who have studied the Fraunhofer in detail emphasize that it cannot be understood outside of the context in which it operates—that is, "the culture, history, and innovation infrastructures of Germany"—and that "an indiscriminate attempt to apply the model wholesale would be inadvisable even if it were practically possible and fiscally affordable."[131] Significantly in this regard, a number of Fraunhofer institutes have been established outside of Germany but "these have not achieved the success of the institutes in Germany."[132]

In Germany the Fraunhofer-Gesellschaft exists in a dense, knowledge-intensive, highly interconnected ecosystem of universities, institutes of applied research, technology intensive companies, and sophisticated trade and industrial associations. The roots of this system can be traced back for centuries, and reflect longstanding German traditions of craftsmanship, the relentless pursuit of academic excellence in the sciences and engineering, and systematic application of scientific knowledge in the commercial realm. Germany pioneered many of the practices which underpin today's knowledge-based global economy, and until the early Twentieth Century was the acknowledged world leader in science and the translation of scientific knowledge into commercial products. Despite the massive discontinuity of the period 1914-50, the basic elements of the German scientific industrial system which emerged in the late Nineteenth and early Twentieth Century are still observable in Germany today. "Despite the

---

Fraunhofer-Gesellschaft. 2010. Annual Report. One Fraunhofer Center in the United States was not successful. The Fraunhofer Centre for Manufacturing and Advanced Materials, located in Delaware, closed its doors in 2004 and transferred its proprietary technology for metal forming to a U.S. company, Alubright of America. According to evidence submitted by Fraunhofer to the British Parliament in 2010, the Delaware center apparently did "not find its contract research market as planned." Written Evidence Submitted by Fraunhofer-Gesellschaft (TIC 71). House of Commons: Science and Technology Committee. December 3, 2010.

[130]Pharmabiz.com 2012. "Europe's Largest Contract Research Organization, Fraunhofer-Gesellschaft to Set Up UK Headquarters. June 1.

[131]Written Evidence Submitted by CBI. (TIC 34). House of Commons Science and Technology Committee. January 12, 2011. "The role and rationale [of technology integration centers] is…context dependent, which includes the presence and nature of other academic or business centres of excellence; the balance of business sectors; and the importance attached by the public and private sector to innovation within a particular nation." Written Evidence Submitted by British Private Equity and Venture Capital Association (BVCA) (TIC 54). House of Commons Science and Technology Committee. January 12, 2011.

[132]Written Evidence Submitted by the Association of Independent Research and Technology Organizations. (TIC 12). House of Commons Science and Technology Committee. January 12, 2011.

political regime breaks in 1933 and 1945 and the Second World War, the German [public research system] showed a high degree of institutional continuity."[133]

The key elements of the German innovation system are a dual system of education and training which marries the development of academic and practical competencies; layered and decentralized governmental structures which support scientific research over long time horizons; an industrial tradition featuring the systematic application of science to the development and constant improvement of products and processes; and industrial and science organizations which promote rationalization, constrain competition in favor of collaboration, and facilitate the broad diffusion of technology. These elements of the German innovation system were in place long before the Fraunhofer first opened its doors in 1949, and the place which applied research institutes like Fraunhofer were expected to occupy within the system were well understood throughout German society by the time Fraunhofer was founded. Today, many well-educated Germans understand what Fraunhofer is, what it does, and where it fits in the German economic system. Foreign assessments of the German innovation system sometimes conclude that while German successes are admirable, German methods cannot readily be transplanted because of the singular character of the German system.[134]

Scientific and Technical Education

German universities emerged as centers for "serious research and scholarship in science and technology long before their British and American counterparts."[135] In 1820, Justus von Liebig, a chemistry professor at the

---

[133]Winnes and Schimank. 1999. Federal Republic of Germany. Op. Cit. pp. 22. The current German innovation system is "embedded…within the historical development of German political economy." Harding. 2000. "German Technology Policy." Op. cit. pp. 225.

[134]The British economist David Soskice, who has studied the German innovation system, commented in a 1996 article that "attractive though the German system is, it is not the intention of this article to advocate its being copied in Anglo-Saxon economies such as the United Kingdom and the United States. This is because their institutional frameworks are fundamentally different." Soskice, David. 1996. "German Technology Policy, Innovation and National Institutional Framework." Discussion Paper ISSN Nr. 1011-9523. Berlin. September. Pp. 1. A recent Australian examination of German economic success concluded that Australia had no real counterpart to the German Mittelstand. Australia had large companies and numerous "micro-entrepreneurs" with a high failure rate. "Compared to Germany we do not seem to be able to sustain large numbers of medium-sized firms." The Conversation. 2012. "The Missing Middle: What Could Australia learn from Germany?" May 19.

[135]Chandler, Jr., Alfred. 1990. Scale and Scope: The Dynamics of Industrial Capitalism. Cambridge, MA and London: Belknap Press of Harvard University. pp. 475. Von Humboldt's reforms "marked a clear break from the prior practice of universities as mere teaching colleges as well as from the French model of specialized higher education where each school trained students only for a particular industry or profession." Siebert, Horst and Michael Stople. 2001. Technology and

University of Giessen, developed a teaching method based on a professor-supervised laboratory staffed by graduate students "which in many ways served as a model for laboratories ever since."[136] By the end of the Nineteenth Century, German universities were "providing the best technical and scientific training in the world."[137] German universities led in the development of systematic methods for acquiring and transferring scientific learning for practical applications in the medical and industrial spheres. In addition to the universities, Germany established numerous polytechnic universities, *Technische Hochschulen*, to train students for the specific purpose of employment by Germany industry.[138]

In the mid 19th, century, the states of southwestern Germany, notably Baden and Wurttemberg, concerned about the geographic dispersion of their large number of small, craft-oriented businesses in the face of industrialization, established decentralized networks of "formal, educationally-oriented, self-help institutions to disseminate information, technology and skill, thereby compensating local producers for the economic disadvantages of their regional social geography."[139] Baden was the first German state to implement a "Dual System" of vocational training with the establishment of a network of trade schools where education in utilitarian subjects such as algebra, geometry, free drawing, mechanics, and bookkeeping, taught by "instructors with practical experience in the trades they taught," was paired with workplace apprenticeship. The institutions were linked to Baden's institutions of higher learning, and in the latter Nineteenth Century, evolved into specialist schools (Fachschulen) designed to provide mid-level instruction in specific disciplines such as metalworking, engineering, and woodcutting.[140] The speed and extent to which these schools improved the quality of local production "astounded" visiting German officials and led to "general acclaim" for Black Forest wood products at

---

Economic Performance in the German Economy. Liebniz-Informationszentrum Wirtschaft. Kiel Working Papers No. 1055. April.

[136]Woodrow Wilson School. 1992. "Justus von Liebig: An Educational Paradox." Accessed at *<http://www.woodrow.org/teachers/ci/1992/Liebig.html>*.

[137]Chandler Jr. Scale and Scope. Op. Cit. p. 425.

[138]Chandler Jr. Scale and Scope. Ibid.

[139]Hansen, Hal. "Rethinking the Role of Artisans in Modern German Development. 2009. Central European History. p. 36. In Baden, local artisanal businesses included the manufacture of wooden clocks, textiles, leatherworking, and woodworking.

[140]Baden supported two of Germany's oldest universities, Freiburg and Heidelburg. It opened a polytechnic in Karlsruhe in 1825 for the training of engineers and skilled tradesmen. Ferdinand Rudtenbacher, known as the father of German mechanical engineering, joined the faculty of the polytechnic in 1841 and in subsequent decades, "he trained large numbers of machine builders and tirelessly promoted a scientific approach to the subject." The polytechnic encouraged its faculty to serve local businesses, in return for which companies made expensive equipment and workspace available for experimentation and technical research. "Instructors and employers traded information, discussed technical plans, exchanged advice, and took active roles in Karlsruhe's industrial association. By 1860, the polytechnic had established a national reputation in machine building." Hansen, Hal. 2009. "Rethinking the Role of Artisans in Modern German Development." p. 48.

the Chicago's World Fair in 1893.[141] Instead of disappearing as Germany industrialized, as was the case elsewhere, the region's small artisanal businesses adopted modern production methods, developed products of extremely high quality and as their social status rose, began referring to themselves as "middle class" (Mittelstand)."[142] By the early Twentieth Century the states of southwestern Germany had established networks of training schools that astonished visitors from abroad.[143] In 1897 the Handiwork Protection Law was enacted in the German Empire, extending the institutional practices developed in Baden and Wurttemberg to the Empire as a whole, and creating the "cornerstone" of the German vocational training system.[144]

The foundation of the German Empire in 1871 accelerated industrialization and urbanization and gave rise to increased demand for scientific knowledge. Because within the Empire the states jealously defended their control over the universities, the central government concentrated on establishing non-university research facilities and projects including a number of

---

[141]Hansen, Hal. 2009. "Rethinking the Role" Op. Cit. "Over time, Baden built educational institutions for virtually every level of society, from bureaucrats, managers, and engineers to weavers of straw, cotton and silk. In comparison with the rest of Germany, however, what most distinguished the southwest was the number and quality of the educational resources it made available to its middling and laboring classes." pp. 47-48.

[142]Reichaw, Christopher. 2011. "An Artisan Mittelstand, How German Artisans Tried to Preserve their Identity by Identifying with the Mittelstand."

[143]In 1911, a commission from Wisconsin visited Wurttemberg to develop recommendations for industrial training. The visitors found Wurttemberg a relatively small, poor, hilly state had "besides its splendid system of elementary and secondary schools, about 250 industrial schools in its towns and villages, 1 knitting school, 3 weaving schools, 2 industrial shops for actual practice in weaving, 2 technical schools for textiles and mechanical work, a large state university, a technical university, a royal building trade school (a trade school for building purposes), a great commercial college, several commercial improvement schools, a great agricultural school, many farming schools similar to our country agricultural schools here, an art trade school for industrial art, a pure art school, and many miscellaneous schools of all kinds for workmen of various grades, evening schools, continuation schools, etc. including schools in domestic economy for women. The tremendous investment by this little province is beyond anything of which we, in our prosperity, have thought." Report of the Commission Upon the Plans for the Extension of Industrial and Agricultural Training. 1911. Madison: Democrat Printing Company. Cited in Hansen, Hal. 2009. "Rethinking the Role" Op. Cit.

[144]Under this law compulsory "handicraft" (Handwerk) chambers were established and vested with extensive powers to regulate the content and quality of craft apprenticeship. The chambers closely monitored and supervised the training of apprentices in German companies. The Handiwork legislation was strongly promoted by "industrial associations" (Gewerbevereine), comprised of small manufacturing and handicraft firms that "sought to enhance their members' individual competitiveness through mutual support and the systematic promotion of technical and commercial education. The Handiwork Protection Law created a system under which chamber-supervised small business provided high-quality skills training to apprentices and journeymen which was frequently linked to mandatory school attendance, The system, which produced a flow of skilled workers into German industry from the artisanal sector, established "some of the scaffolding on which Germany's system of vocational education and training would ultimately be built." Thelen, Kathleen. 2004. How Institutions Evolve: The Political Economy of Skills in Germany, Britain, the United States, and Japan. Cambridge: Cambridge University Press. p. 46.

significant mission-oriented research organizations.[145] By 1900 Prussia and Saxony alone had over 300 technical institutes.[146]

In 1911 the Kaiser-Wilhelm-Gesellschaft (KWG) was formed to promote research in the natural sciences. Established as a private association, the KWG derived most of its funding from private sources but the Prussian government contributed building, land and salaries for the directors of the institutes.[147] Between its founding in 1911 and 1948, the KWG supported the establishment of 35 research organizations, which included some of the most important and prestigious academic institutes in the world.[148] The Institutes were a source of trained scientific and engineering talent, but also a source of scientific and technical knowledge for German industry, a linkage which "was much closer in Germany than in Britain, where it rarely existed at all, and even in the United States, where at the turn of the century, the process was just beginning."[149]

Education and Industrialization

Fredrich List, a native of Wurttemburg, was a German economist who developed and espoused a series of proposals to enable Germany to overtake England in industrial development. His 1841 book, The National System of Political Economy, is widely regarded as the first articulation of the concept of national systems of innovation.[150] He advocated protection of infant industries and a broad array of economic development policies, most of which "were concerned with learning about new technology and applying it." Due to List's advocacy as well as that of other economists, "Germany developed the best technical education and training systems in the world [which was] one of the

---

[145]In 1887, the first national government laboratory was established, the Imperial Institute for Physics and Technology, with the mission of developing precision mechanics, conducting "physical investigations and measurements," and solving scientific problems requiring specialized research facilities. Winnes and Schimank. 1999. Federal Republic of Germany. Op. Cit. pp. 20-21. Kaiser Wilhelm himself was a strong believer in the establishment of technical institutes, which had as their purpose "to secure for science footholds in the workshop." Comin, Trumbull, and Young. 2012. Innovation in Germany. Op Cit. pp. 3. These entities emphasized close collaboration between science, industry, and the state and were public-private partnerships jointly supported by governments and industry. The institutes exempted their scientists from teaching responsibilities, featured well-equipped laboratories, and allowed scientists greater freedom to pursue their research interests.

[146]Comin, Trumbull, and Young. 2012. Innovation in Germany. Op Cit. p. 3.

[147]Winnes and Schimank. 1999. Federal Republic of Germany. Op. Cit. p. 21.

[148]Seidelman, William E. 2001. "Science and Inhumanity: The Kaiser-Wilhelm/Max Planck Society," If Not Now- an e-journal.

[149]Chandler Jr, 1990. Scale and Scope. Op. Cit. p. 426.

[150]List, Freidrich. 1991. The National System of Political Economy. New Jersey: Augustus M. Kelly; Freeman, Chris 1995. "The National System of Innovation in Historical Perspective," Cambridge Journal of Economics. p. 5; Levi-Faur, David. 1997. "Freidrich List and the Political Economy of the Nation-State," Review of International Political Economy. Spring.

main factors in Germany overtaking Britain in the latter half of the Nineteenth Century and to this day is the foundation for the superior skills and higher productivity of the German labour force."[151]

In 1800, Germany was still a disaggregated array of over 300 states, principalities, free cities and other jurisdictions, many of them entirely independent. Although Germany was unified in 1871, individual states (now known as Länder) have retained considerable power under a succession of federal governments down to the present day. The concept of states' rights remains "more deeply imprinted upon German life than upon American." The legacy of political fragmentation "underlies the intricate federal-state relationship that regulates and nurtures the veritable forest of research institutions that dot the German landscape."[152]

In addition to Baden and Wurttemburg, other state governments played a key role in fostering German industrialization. Prussia, the largest German state, implemented a series of sweeping economic, administrative and social reforms in the wake of its defeat by Napoleon in 1806. One of the officials who was instrumental in Prussian reforms, G.J.C. Kunth, presided over a number of government bodies promoting manufacturing technology, including the acquisition of information about foreign machines.[153] Kunth's legacy was carried forward by his principal assistant, Peter Beuth, who became the head of Prussia's Department of Trade of the Ministry of the Interior, a post which he held from 1818 until 1845.[154] Beuth's entire career was dedicated to the modernization of Prussian industry. During Beuth's tenure, with extensive

---

[151]Freeman, Chris 1995. "The National System of Innovation in Historical Perspective," Cambridge Journal of Economics. p. 6, citing Hobsbawm, E. 1968. Industry and Empire. London: Weidenfeld and Nicolson. Burnett, Corelli. 1988. The Audit of War. Cambridge: Cambridge University Press. Landes, M. 1970. The Unbound Prometheus: Technological and Industrial Development in Western Europe from 1750 to the Present. Cambridge: Cambridge University Press.

[152]Beyerchen, Alan. 1990. "Trends in the Twentieth Century German Research Enterprise," in National Academy of Sciences, *The Academic Research Enterprise within the Industrialized Nations: Comparative Perspectives*. Washington, DC: National Academy Press. p. 79.

[153]Kunth was Director of the Prussian Department of Manufactures and Commerce, a member of the Factory and Commerce section of the General Directory (Generaldirektorem), and the leader of a committee of experts advising both of the foregoing bodies, Teknische Deputation. During the Napoleonic Wars, when Allied armies occupied Belgium and the Rhineland, he arranged for technical studies of manufacturing technologies in those regions. Kunth was also the draftsman of an 1817 memorandum which led to the abolition of internal customs duties in Prussia the following year. Henderson, W.O. 1955. "Peter Beuth and the Rise of Prussian Industry, 1810-45," Economic History Review. Vol. 8, No. 2. p. 223.

[154]Beuth reorganized and headed the advisory body, which oversaw the Prussian patent law. He established the Association for the Promotion of Industrial Knowledge in Prussia in 1821, and in the same year, persuaded the government to establish an Industrial Institute in Berlin to train engineers. Beuth extensively toured the most sophisticated factories in the world, which at that time were in Great Britain, Belgium, France, and the Rhineland, and dispatched government officials and faculty and students of the Prussian Industrial Institute to foreign countries to examine manufacturing machinery and processes. Henderson. 1955. Peter Beuth. op. cit. p. 226.

government support, Prussia acquired what one scholar characterized as the "Alpha and the Omega of modern machine-building," British machine tool technology developed in the first decades of the Nineteenth Century, which enabled the design and construction of metal-working precision machinery for all other industries." "The transfer of [machine tool] technology promoted and coordinated by the Prussian state was highly successful," and this "set Prussia (later Imperial Germany) well on the road to overtaking Britain."[155]

Germany's principal technology-based industries, chemicals, machinery automobiles and electrical engineering were all established in the late Nineteenth and early Twentieth Centuries. The modern chemical industry was created by three German firms, Hoechst, Bayer, and BASF, "one of the most impressive achievements in the annals of industrial history."[156] In the latter decades of the Nineteenth Century, these firms created the world's first large scale industrial laboratories dedicated to systematic research to develop new products and industrial processes. The industrial labs established close ties with leading research organizations. The laboratories transferred technology to the chemical firms' industrial complexes. The industrial labs recruited managerial talent including a general staff headed by doctoral level chemists:

> *In this way, they were the pioneers not only in the chemical and pharmaceutical industries but also in shaping the management organization of modern high technology enterprises.*[157]

In the 1880s, these firms revolutionized the pharmaceutical industry by developing a wide range of medicines from coal-tar feed-stocks, dyestuffs, and intermediates, including synthetic sedatives, painkillers, fever depressants, vaccines, and serums.[158] The in-house R&D laboratory "also emerged in other industries which had the same need to access the results of basic research from

---

[155]Freeman, Chris 1995. "The National System of Innovation in Historical Perspective," Cambridge Journal of Economics. p. 6-7, citing "Der Technologietransfer fur die Metallbearbeitung und die Preussische Gewerbeforderung 1820-50, in Blaich, F. (ed.). 1982. Die Rolle des Staates fur die Wirtschaftliche Entwicklung. Berlin: Blaich. British engineers pioneered in the critical innovation in machine tool technology in the early Nineteenth Century. Considerable efforts were made to keep the technology secret, and British government banned export of the machines. The Prussian government ensured that its Technical Training Institutes received British machine tools for reverse engineering and training of craftsmen who disseminated the technology through German industry. Freeman, Chris 1995. "The National System of Innovation in Historical Perspective," Cambridge Journal of Economics. p. 7.

[156]Chandler, Alfred D. 2005. Shaping the Industrial Century: The Remarkable Story of the Evolution of the Modern Chemical and Pharmaceutical Industries. p. 115.

[157]Chandler, Alfred D. 2005. Ibid.

[158]Ibid. p. 116.

universities and other research institutes and to develop their own new products."[159]

Networks

As Germany industrialized, companies forged vast networks of interfirm ties, leading to the characterization "organized capitalism." Industrial sectors were linked by professionally staffed and well-financed associations:

> *a tight weave of branch and regional associations crisscrossed the entire country, bringing together at the national level producers of similar goods and linking at the regional level producers of various goods who operated within the same geographical area. Furthermore, a plethora of special purpose business associations provided still additional links between the components of big business.*[160]

Imperial Germany had no laissez faire tradition and its commercial culture differed sharply from that of the North Atlantic world dominated by English legal and economic traditions and philosophy[161]. Contracts restraining trade were enforceable in German courts, and German industrial leaders believed that it was not only their right, but their social obligation, to combine to

---

[159]Freeman. 1995. National System of Innovation. Op. cit. pp. 8-9. Commenting on the proliferation of industrial research laboratories in the latter 1800s, a leading physicist said that the greatest invention of the Nineteenth Century was the method of the invention itself. Ibid. p. 9.

[160]Turner Jr., Henry Ashby, 1985. German Big Business and the Rise of Hitler. New York and Oxford: Oxford University Press. P. xix. In southern Germany, industrial associations of Mittelstand emerged as powerful advocates for technical and trade schooling, established clearinghouses for exchange of technical information, and engaged in a variety of technology diffusion activities, such as trade shows, technical courses, and instruction on emerging technologies and industries. The associations operated as "semi-official intermediaries between the Mittelstand and state governments with respect to the formation and operation of vocational and technical schools "since these functioned effectively only insofar as they reflected the ongoing developments in the communities of practice for which they trained. Hansen, Hal. 2009. "Rethinking the Role of Artisans" Op. Cit. pp. 59-60.

[161]German Chancellor Otto van Bismarck stated that the Germans in having once believed in Free Trade, had been the "dupes of an honest conviction worthy of the honorable capacity for dreaming in the German race," but the dream of free trade, "must not mislead a country placed between three strong states—Russia, Austria, and France—all of whom were moving, or preparing to move, away from it. The German Empire must not take its policy from Manchester." Clapham, J.H. 1928. The Economic Development of France and Germany, 1815-1914. Cambridge: Cambridge University Press. The long domination of non-capitalistic economic forms imbued the German peoples with settled habits of thought, codes of ethics, and underlying value systems which have never been brought into complete harmony with laissez-faire doctrines. Brady, Robert Alexander. 1974. The Rationalization Movement in German Industry: A Study in the Evolution of Economic Planning. New York: H. Fertig. p. 53.

restrain destructive competition.[162] By the early Twentieth Century every major German industrial sector was organized into one or more Kartells (cartels), which were integral to German industrial organization. Despite an effort by the Allied occupation to break up the web of cartels and associations after World War II, many aspects of the "organized capitalism" of Wilhelmine Germany have been carried forward into the modern federal republic.[163]

War and Occupation

It is difficult to overstate the damage which occurred to Germany's science-industrial base between the onset of World War I and the end of the Allied occupation at the end of World War II. World War I severed Germany from overseas sources of oil and saw the expropriation of German petroleum assets, effectively ending efforts to build a world class oil and petrochemical industry.[164] The Treaty of Versailles imposed restrictions on German scientific research, reflecting the role played by German science in the First World War in the development of chemical weapons and other technologies supporting the war effort. The hyperinflation of the 1920s destroyed the endowments of the KWG and similar research organizations. Economic dislocation and the burden of war reparations deprived German industry of the levels of equity capital needed to commercialize new technologies on a mass basis. The rise of Hitler to power saw a massive increase in public investment in infrastructure and industry and under the leadership of the Reich Research Council, founded in 1937, academic research was mobilized to support the war effort, and yielded dramatic new technologies, including jet aircraft, nerve gas, stealth technology, hardened armor, ballistic missiles, radar, and synthetic materials such as nylon and synthetic fuels. However, the Nazi regime's anti-Semitic policies led to an exodus of Jewish scientists, decimating the faculties of German universities and research institutes. German science was deeply involved in horrific crimes committed by the regime, tarnishing its reputation for decades afterward.[165]

---

[162]U.S. Federal Trade Commission. 1916. Report on Cooperation in American Export Trade. Part I. June 30. p. 107.

[163]Der Spiegel. 1991. "Cartel Germany: Competition in Our Ranks is Being Curbed." August 26. Wolfgang Streeck, Director of a Max Planck Research Institute in Cologne, observed in 1995 that "widespread organized cooperation among competitors and bargaining between organized groups, conducted through publicly enabled associations is probably the most distinctive feature of the German political economy. Governance is delegated either to the individual associations or to the collective negotiations between them with the state often awarding its outcome legally binding status." Streeck, Wolfgang. 1995. "German Capitalism: Does it Exist? Can it Survive? In Colin Crouch and Wolfgang Streeck, eds. Modren Capitalism or Modern Capitalisms? London: Francis Pinter.

[164]Siebert and Stople. 2001. Technology and Economic Performance. op. cit. p. 23. Chandler. 1990. Scale and Scope. op. cit. p. 519.

[165]Associated Press. 2001. "German Science Elite Apologizes for Nazi Medical Experiments." June 16. Science. 2000. "Reopening the Darkest Chapter in German Science," June.

Anglo-American bombing severely damaged German industry, although by no means destroyed it.

The Allied Occupation of Germany at the end of World War II inflicted further harm on Germany's science-industrial infrastructure. Research bans were placed on sectors such as aviation and rocket propulsion in which German science enjoyed a strong position, setting back the country's long run prospects in these fields. The U.S.-led "Project Paperclip" identified thousands of the most talented German scientists and removed them to the United States, while the Soviets forcibly relocated German scientists found in their occupation zone.[166] Berlin, the traditional center of German science, was divided between Allied occupation zones and marooned in what would become a Soviet puppet state, the German Democratic Republic. German factories in the Russian Zone were dismantled and shipped to the East, along with management and staff.[167] Between 1947 and 1950, the British pursued a similar program of industrial dismantlement in the Ruhr [168]. The American authorities mounted a "de-cartelization" effort which broke up industrial combinations and trade associations in an effort aimed at "rooting out German-style organized capitalism and replacing it with the idealized version of the American free-market system envisioned by Thomas Jefferson" [169]. German industrialists were aghast at deconcentration measures that "dismantled hundreds of viable plants, disrupted supply networks, and broke many of the links between the agricultural and industrial sectors." "Many Germans saw the Allied controls as even more oppressive than the Nazis," and those measures contributed to the subsequent German "atmosphere in which enthusiasm for state intervention was quite limited."[170]

Despite the massive discontinuities of the period 1914-50, the German innovation system reestablished itself in the immediate aftermath of World War II with a remarkable degree of continuity with respect to pre-1914 institutions

---

[166]BBC News. 2005. "Project Paperclip: Dark Side of the Moon." Quote from Major General Hugh Knerr, November 21.

[167] The Russians dismantled and relocated 92 percent of the Carl Zeiss optical works in Jena. "Carl Zeiss—A History of a Most Respected Name in Optics," Company 7. *<http://www.company7.com/zeiss/history.html>*.

[168]"[S]ome 700 million Deutschmarks worth of plant were unbolted, disassembled, cut apart with blow torches and carted away." Gillingham, John. 1991. Coal, Steel, and the Rebirth of Europe 1945-1955: The Germans and French from Ruhr Conflict to Economic Community. Cambridge: Cambridge University Press. p. 206.

[169]Gillingham, John. 1991. Coal, Steel, and the Rebirth of Europe 1945-1955: The Germans and French from Ruhr Conflict to Economic Community. Cambridge: Cambridge University Press. p. 109.

[170]Allen, Christopher S. 1989. "The Underdevelopment of Keynesianism in the Federal Republic of Germany," in Peter Hull (ed.), The Political Power of Economic Ideas: Keynesianism Across Nations. Princeton: Princeton University Press. p. 266.

and norms.[171] The KWG research institutes were reformed under the designation Max-Planck Society. The dual system of education/apprenticeship reemerged more as by habit than by central direction.[172] German training institutions, "turned out to be incredibly resilient in the face of huge exogenous shocks of the sort we might expect to disrupt previous patterns and prompt dramatic institutional innovation."[173] German industry regrouped in networks of associations, "cartel-like arrangements," and industrial combinations linked by a maze of interlocking directorates and cross-shareholdings involving banks and insurance companies.[174]

Emergence of the Fraunhofer-Gesellschaft

The creation of the Fraunhofer Gesellschaft was part of a post-war effort by the state of Bavaria to capitalize on the Allied dismemberment of Prussia and to "lure research and engineering talent from Prussia's main agglomeration of science and technology activities which was in Berlin."[175] Bavarian politics in the half century after the end of World War II were dominated by the Christian Socialist Union (CSU), a party that was deeply conservative yet committed to the promotion of science, education, and technologically-advanced industries. The Bavarian government invested heavily

---

[171]Kathleen Thelen, an academic who has studied German labor relations, vocational training and education, observed in 2004 that "one of the striking features of the system is the resiliency of core elements even in the face of enormous disruptions over the twentieth century, which of course in Germany include several regime changes, the incorporation of the working class, defeat in not one but two world wars, occupation, and transitions in and out of fascism. Although changes certainly occurred at these junctures, what is remarkable and in need of explanation are some striking continuities in key features of this system despite these disjunctures." Thelen, Kathleen. 2004. How Institutions Evolve. Op. Cit. p. 7.

[172]Plant-based apprenticeship "more or less spontaneously re-surfaced after the war recommencing most quickly and thoroughly in the craft sector as early as 1945." By the time of the founding of the federal republic in 1949, "the enterprise-based part of the dual system had been re-established on the basis of employer self-governance with distant supporting roles assigned to both the state and unions." One academic terms these developments a "re-anchoring of traditional training structures."Thelen, Kathleen. 2004. How Institutions Evolve. Op. Cit. pp. 244-251, citing Gunter Patzold (ed.) 1991. Quellen und Dokumeate zur Betrieblichen Berufsbildung. Koln: Bohlau Verlag. .

[173]Thelen, Kathleen. 2004. How Institutions Evolve. Op. Cit. pp. xiii.

[174]Der Spiegel. 1991. "Cartel Germany: Competition in our Ranks is Being Curbed." August 26. Katzenstein, Peter. 1987. Policy and Politics in West Germany: the Growth of a Semisovereign State. Philadelphia: Temple University Press. p. 88. Grant, Wyn, William Patersan and Colin Whitston. 1988. Government and the Chemical Industry: A Comparative Study of Britain and West Germany. Oxford: Clarendon Press. p. 90.

[175]Siebert and Stolpe. 2001. Technology and Economic Performance. op. cit. p. 20. The Sunday Times. 2011. "Wiped from the Map—The Compelling Story of East Prussia and its Erasure by a Brutal Post-War Past." London. June 5. The Soviet blockade of Berlin in 1948 induced a number of German companies to relocate their headquarters from Berlin to Bavaria. Milosch, Mark S. 2006. Modernizing Bavaria: The Politics of Franz Joseph Strauss and the CSU, 1949-1969. Monographs in German History. V.S.

in its universities and technical schools "but what set...Bavaria apart was its concentration of non-teaching research facilities," principally the Max-Planck Institute and the Fraunhofer Gesellschaft.[176] Bavaria methodically lured eminent scientists and technology-intensive industries to the Munich region. The Max Planck Institute moved from Berlin to Munich in 1949.[177] By the 1960s, such efforts had "created the synergy to transform Munich into a Mecca of German science."[178]

The Fraunhofer Gesellschaft was established in Munich in March 1949 as a non-profit organization tasked with raising funds from governmental bodies to support industrial research projects.[179] It was created under the leadership of the Bavarian Minister of Economic Affairs Hugo Geiger, who wanted to promote applied research in the state.[180] At a 1948 meeting convened by the Bavarian Ministry of Economic Affairs comprised of local leaders in science and research, the notion of a broad applied research organization was agreed, with a mission of identifying research topics in various subject areas, mediating between university research and industry, and securing and managing research funding. The Fraunhofer-Gesellschaft was formally launched at a March 26, 1949 meeting hosted by the Ministry of Economic Affairs which convened over 200 eminent Bavarian scientists, economists and political leaders.[181]

---

[176]The Irish Times. 2009. "Bavaria or Bust." June 8.

[177]Handelsblatt. 1991. "Joseph von Fraunhofer and Max Planck Can Feel Satisfied." August 9. JPRS-EST-91-015.

[178]Adolf Buenandt, the foremost biochemist in Germany, moved to Munich in the 1950s after Bavaria decided to offer "whatever it took" to get him, which included construction of a new building for his institute and a "private house with a garden." Warner Heisenberg, one of the foremost German scientists of the postwar era, moved his Max-Planck Institute for Physics to Munich in the 1950s when Bavaria committed to construct Germany's first atomic reactor. Milosch. 2006. Modernizing Bavaria. Op. cit. pp. 38-39.

[179]The Fraunhofer-Gesellschaft was named after Joseph Fraunhofer (1787-1826) a Bavarian optician who was a scientists, inventor, and entrepreneur, embodying the virtues which the new institute sought to promote and integrate. He developed optical instruments which enabled Bavaria to overtake England as the world center of the precision optics industry.

[180]"Die Grundig der Fraunhofer Gesellschaft," *<http://www.uni-protokolle.de>*. *<http://www.uni-protokolle.de/nachrichten/id/48304/>*. The original idea underlying the formation of Fraunhofer was to promote geological research to identify untapped mineral resources in Bavaria which could be "seeds for increased industrial development," a concept strongly supported by the American occupation authorities. Deutinger, Stephan. 2001. Vom Agarland zum High-Tech-Staat: Zur Feschichte des Forschungsstandarts Bayern 1945-1980. p. 180, 185.

[181]Egger, Christine. 2010. "Nachdenken im Auftrag: Eine Geschichte der Fraunhofer Gesellschaft." Aventinus Bavaria. October 22. The Fraunhofer's first president, Nobel Prize winner Walther Gerlach, established a pattern that still characterizes Fraunhofer institutes, retaining a university position while simultaneously serving as head of the research institute. Gerlach, a nuclear physicist, was an important player in Hitler's efforts to develop an atomic bomb. After the war, he led a movement advocating renunciation of nuclear weapons by the Federal Republic, He held a faculty position at Ludwig Maximiliens University in Munich. Vican, Jacob. 2009. "Tot, Schuldig oder Geflohen." Badische Zeitung. May 23.

The Fraunhofer struggled in its early years. Initially, Fraunhofer's main functions were administrative—it raised funds from government organizations, donors and association members for distribution to research projects relevant to industry. The German Federal Ministry of Economics formally recognized it in 1952 as the "third pillar in the German research infrastructure" alongside the Max-Planck-Gesellschaft and the Deutsche Forschungsgemeinschaft (DFG, the German Research Council), but it met with resistance and competition from established research organizations.[182] Originally largely limited to geological surveys, it was ridiculed as "the Bavarian Research Association for Mining." The organization's first president, Gerlach, was a basic scientist who disdained applied research, and as a result, was pressured into resignation by the Fraunhofer Senate in 1951.[183]

The Fraunhofer survived its shaky beginning largely as a result of public support. In 1951, it successfully applied to the German federal government for Marshall Plan funding pursuant to the European Recovery Program.[184] With the institute on the brink of liquidation in 1954, the Land government of Bavaria, which had theretofore provided "more ideological than material support," began direct funding to Fraunhofer, although the initial amount was characterized as "enough to prevent death, but not enough to sustain life."[185] In 1954, however, Fraunhofer received additional support from the Land government of Baden-Wurttemburg to establish what was to become the institute's first contract research organization, the Fraunhofer Institute for Applied Microscopy, Photography, and Cinematography (IMPIC).[186] Finally, in 1956, Fraunhofer began to receive contract research funding from the newly-created Ministry of Defense, which for historical reasons preferred to outsource

[182]Fraunhofer. 60 Year History of Fraunhofer Gesellschaft. pp. 8-9.

[183]A further shock was the 1953 establishment of a laboratory in Frankfurt by the U.S. contract research organization the Battelle Memorial Institute, which was viewed by the Fraunhofer as a direct competitive threat. Battelle is an Ohio-based nonprofit science and technology development company. It was created out of the estate of Gordon Battelle, a steel industry executive who was interested in the application of science to the metallurgical industries. Egger. 2010. "Nachdenken im Auftrag." op. cit.

[184]Walter Hirsch, an official in the Federal Ministry of Economics persuaded U.S. authorities that 41 million deutschemarks should be allocated to support applied industrial research projects of a non-military character. The Fraunhofer and the DFR applied for and were awarded these funds. The United States aid was conditioned on its ability "to stimulate the German economy." The research institutes complied by restricting their spending to German-made scientific instruments—no foreign equipment could be procured. Krige, John. 2006. American Hegemony and the Postwar Reconstruction of Science in Europe. Cambridge, MA: The MIT Press. p. 36.

[185]Deutinger, Stephan. 2001. Vom Agarland zum High-Tech-Staat: Zur Feschichte des Forschungsstandarts Bayern 1945-1980. p. 182.

[186]Baden-Wurttemburg had competed with other Länder to become the site of the Battelle laboratory. When it lost out to Frankfurt in 1953, it began looking for other ways to establish its own research facilities. Egger. 2010. "Nachdenken im Auftrag." op. cit.

its defense-related R&D.[187] For many years afterward, "military funding accounted for more than half of the Fraunhofer-Gesellschaft's total research budget."[188] The defense research work was critical in the development of Fraunhofer not only because of the revenue it generated but because it enabled the institute to develop the research infrastructure and resources to "develop promising ideas and lines of research to a level where they could be converted into innovative products and processes and offered to firms."[189]

In 1965, the German Science Council, an influential advisory body to the federal and Länder governments, recommended a broad expansion of non-academic research organizations in general and the Fraunhofer Gesellschaft in particular as the umbrella organization for applied research.[190] The recommendation was conditioned on the implementation of structural reforms and the creation of a substantial development program for the institutes. In 1967, a recommendation by the Science Council "paved the way for incorporation of the Fraunhofer Society in the institutional funding of the BMwF and a Science Council commission studying the Fraunhofer concluded that "promotion of applied research with public means" was necessary for "ensuring the technological development in the FRG."[191] The Fraunhofer grew explosively after it began receiving federal institutional funding. The number of Fraunhofer institutes doubled in the decade 1965-1975. Its budgets grew steadily, enabling it to establish new institutes and modernize existing ones.[192]

---

[187]In 1955, Germany joined NATO and the occupation was formally terminated. Germany began rearming. The Ministry of Defense recognized the need for defense-related research, but in light of the recent Nazi past, did not want to establish an internal isolated military research department. It chose instead to draw upon university and industrial research, frequently using Fraunhofer as the intermediary. Egger. 2010. "Nachdenken im Auftrag." op. cit. The Bavarian politician Franz Joseph Strauss who was Defense Minister at the time, "set-up the Defense Ministry as a patron of the Fraunhofer Foundation." Five Fraunhofer institutes were opened at the instigation of the Ministry and by 1959, the Fraunhofer had four institutes working exclusively for the ministry and six others working part time. Milosch, Mark S. 2006. Modernizing Bavaria: The Politics of Franz Joseph Strauss and the CSU, 1949-1969. New York and Oxford: Berhahn Books. pp. 114.

[188]Fraunhofer, 60 Year History. op. cit. p. 8-9. Fraunhofer Vice President Albert Maucher warned in 1960 that if the military work kept growing, at some point, "we are no longer FhG but the Army Ordinance Department. Egger. 2010. "Nachdenken im Auftrag." op. cit.

[189]Winnes and Schimank. 1999. Federal Republic of Germany. Op. Cit.

[190]The German Science Council (Wissenschaftsrat) makes recommendations on the development of higher education institutions, science and research, and the establishment of new universities.

[191]Winnes and Schimank. 1999. Federal Republic of Germany. Op. Cit. pp. 37.

[192]One key to the institutes' growth was a formula introduced into its funding model by the federal research ministry, "variable success-dependent institutional funding" (erfolgsabhangige Grundfinanzierung). Under the arrangement, Fraunhofer received for each Mark acquired through industry research contracts an equal amount from the federal and state governments, creating a powerful incentive to orient research toward industry needs and to pursue industry funding. There was no ceiling on government matching funds, and from Fraunhofer's perspective the arrangement worked so well that in the 1980s the federal and Land governments curtailed the funding model to limit the growth in government grants to the institute. Winnes and Schimank. 1999. Federal Republic of Germany. Op. Cit. pp. 69.

As the German "economic miracle" (Wirtschaftswunder) began to accelerate, Fraunhofer succeeded in attracting significant funding from industry for contract research.[193] Hermann von Siemens, CEO of Siemens AG, assumed the presidency of the Fraunhofer in 1955, a post which he held for nearly a decade. Thereafter Fraunhofer added 1-2 new institutes per year, gradually extending a network of research organizations across West Germany. In 1973, Fraunhofer received the status of a federal research organization and began receiving funding from the BMFT, now the BMBF.[194] Beginning in the early 1970s the Fraunhfer made a phased shift in emphasis away from military R&D toward civilian applications.[195]

The Fraunhofer has grown not only by opening greenfield institutes but by absorbing institutes affiliated with other research organizations. In 1970 it took over an organization originally formed in 1930 as the Prussian Institute for Wood Research, which was rebranded as the Fraunhofer Institute for Wood Research WKI.[196] In 1971, it acquired the Max-Planck Institute for Silicate Research in Wurtzburg, which became the Fraunhofer Institute for Silicate Research ISC.[197] In 1991, in the wake of German reunification, numerous research organizations based in former East Germany were absorbed into Fraunhofer.[198] In 2001, Fraunhofer acquired 8 research institutes of the German National Research Center for Information Technology (GMD) a move that was backed by the German research ministry but bitterly resisted by the GMD's management and staff, who believed (correctly) that integration with Fraunhofer would orient GMD away from basic research and towards applied research.[199]

---

[193]In 1955, the Institute received 600 million marks from German industry for contract research. Fraunhofer, 60 Year History. op. cit. pp. 8-9.

[194]National Academy of Engineering, *Technology Transfer Systems in the United States and Germany*, Washington, DC: National Academy Press, 1997, p. 321.

[195]VDI Nachrichten. 1991. "Research Institutes Seek Civilian Customers." JPRS-EST-91-003. February 2.

[196]Fraunhofer WKI, "History of Our Institute." *<http://www.wki.fraunhofer.de/en/wki-profile/wki-history.html>*.

[197]Fraunhofer Institue for Silicate Research. ISC Website.

[198]A booklet written by then-Minister of Research and Technology Dieter Thierbach in 1991 stated that "Nineteen new establishments, nine institutes and 10 branch facilities will be operated by the Fraunhofer Society in the new Länder. Eight independent Fraunhofer facilities are planned—and an institute section of the Duisburg Fraunhofer Institute for Microelectronic Circuits and Systems (IMS)—with a total of 700 employees and 10 branch facilities of existing Fraunhofer institutes with 250 employees." Thierbach, Dieter. 1991. Deutsche Einheit in Forschung und Technologie. Bonn: BMFT. Part. 2.6.

[199]Spiegel Online 2000."Faule Fusion. October 12. Spiegel Online 2000. "Stillstand be; Verhandlongen." April 19. Spiegel Online 2011. "Gelahmte Fusion." February 2.

Box APP-A2-6

Promoting Factory Automation

Beginning in the 1980s, the German research ministry implemented a succession of large-scale, long range programs to promote factory automation.[a] By the early 1990s, BMFT was spending nearly 300 DM per year on projects to promote innovation in manufacturing.[b] Ten Fraunhofer institutes with over 1,000 employees were supporting this effort, exploring themes such as robotics, micromechanics, sensors, computer-aided design and manufacturing (CAD/CAM), machine tools, software for factory controls, and factory systems.[c] In 1986, the Technical University of Berlin and the Fraunhofer Institute for Production Facilities and Mechanical Construction Technology IPK entered into a “marriage” to develop “the structures of the factory of the future,” which the Director of the Institute, Dr. Günter Spur said “will be a mechanized organism that, coalesced from individual production cells, is capable of using programmed intelligence to automatically produce goods.” Numerous German companies donated equipment to this project and the Fritz Werner mechanical engineering company shifted part of its own production operation to the 3,200 square meter joint project facility, setting up two complete processing centers. Reflecting such initiatives, in the 1980s “West Germany adopted computer integrated manufacturing more rapidly than most other countries.”[d]

[a]Technologie Nachrichten—Programm-Informationen. 1988. “Report on 1988-1992 BMFT Research Program on Manufacturing and Engineering.” JPRS-EST-88-04. July. In 1984 BMFT made available DM 441 million for investments in CAD/CAM, robots and handling systems. “There was a flood of applications” from German industry and 1,285 applications for CAD/CAM grants were approved. Produktion. 1986. “Firms, Funds up to Fall 1985 in FRG Process Technology.” JPRS-WST-86-006. February 12. Suddeutsche Zeitung. 1983. “FRG 1984-87 Plan for Computer Aided Design, Manufacturing.” November 18.

[b]Technologie-Nachrichten Programm-Informationen. 1994. “Strategies for Production in the 21st Century. September 16. JPRS-EST-94-023.

[c]Utlands Rupport. 1985. JPRS-WST-85-002. January 17; Technische Rundschau. 1985. “Funding Sources, Organization of CAD/CAM Research in FRG.” JPRS-WST-85-012. April 10; VDI Nachrichten. 1984. “Micromechanics Yields Innovative Sensors.” JPRS-WST-84-051; Bild der Wissenschaft. 1987. “Research for Factory of the Future.” JPRS-ELS-87-028.

[d]Siebert and Stolpe. 2001. Technology and Economic Performance. Op. Cit. pp. 9. Bild der Wissenschaft. 1987. “Research for the Factory of the Future.” JPRS-ELS-87-025. May 6.

CONCLUDING PERSPECTIVE

In the Nineteenth Century, German companies generated spectacular innovations that established many of the technological foundations of the

Box APP-A2-7

Bavaria—From Rural Backwater to Germany's Silicon Valley

Today Bavaria's heavy and sustained investments in science, education, and industrial promotion—prominent among which was support for the Fraunhofer—appear prescient. By the first decade of the Twenty-First Century, Bavaria had become one of the pre-eminent high technology regions in the world, and was widely characterized as "Germany's Silicon Valley."[a] The German company with the most patents, Siemens, was headquartered in Munich and virtually all of Germany's top 20 patentees were either based in Munich (Infineon, BMW), operated research institutes in the city, or were involved in research collaborations with one of Munich's research institutes. The greater Munich region was the site of 10 Fraunhofer institutes, 11 Max-Planck institutes, 11 universities, and 17 polytechnics. Bavaria's Fraunhofer institutes include one of the largest, the Fraunhofer Institute for Integrated Circuits, with 450 employees operating at sites in Nuremburg, Furth, and Erlangen.[b] High tech manufacturing accounted for over half of the employment and nearly two-thirds of the revenues generated by the Bavarian manufacturing sector. In 2005, Bavaria had the highest number of start-ups in Germany, accounting for one-fifth of the total in 2005.[c] "Even though the Czech border is just 300 km from Munich, companies stay in the high-cost Bavarian capital because of their dependence on the highly-trained staff and world class facilities."[d]

[a]The Guardian. 2011. "How Bavaria Became European Silicon Valley." March 14. See also Technologie-Nachrichten Programm-Informationen. 1991, "Report by the Federal Ministry of Research and Technology on the Implementation of the Unification Treaty Dated August 31, 1990 in the Area of Research and Technology." June 20. JPRS-EST-91-012.
[b]The Bavarian Land government invested more than 140 million Euros, or their equivalent in Deutschmarks between 1984 and 2005, to establish and expand the Erlangen facility Pressbot. 2005. "Beakstein: Fraunhofer Institute Leisten Wichtigen Betrag zum Ausbau des Technlogie-Standortes Bayern." June 30.
[c]Van Winden, William, Leo Van Den, Luis Carvalho, and Erwin Van Tuijil. 2011. Manufacturing in the New Urban Economy. Abington and New York: Routledge. Pp. 93, 95-96.
[d]The Irish Times. 2009. "Bavaria or Bust." June 8.

modern world. These include the designs and thinking of Werner von Siemens in electric power generation (which included the very idea that electricity could be a power source); the Diesel and Otto cycle internal combustion engines; and the discovery of aspirin and other pharmaceutical substances in the world's first large-scale industrial laboratories[200]. Although the modern day Federal Republic can cite many economic and technological achievements, it has generated no obvious innovations comparable to the revolutionary achievements of Germany's industrial and scientific heyday. Instead the German innovation system has built on the strong position it achieved over a century ago with continuous incremental improvement of products and process in the automobile, engineering, machinery, electric power, chemical, and pharmaceutical industries. It remains highly competitive in all of those sectors today, in contrast to a number of other industrialized countries.

The German innovation model in general, and the Fraunhofer's approach, in particular, do not offer the prospect of radical, paradigm-shattering technologies. They do, however, demonstrate that a high cost, high wage country can compete effectively in global markets in established industries through the systematic and continuous application of knowledge—without necessarily moving production processes offshore. The Fraunhofer represents a vast knowledge system which a company of any size can commission to perform commercially-relevant research at below-market cost, with a high degree of assurance in the quality of the research and the associated business counseling. Fraunhofer projects appear to involve less bureaucracy, faster time-to-market, and greater probability of success than pursuit of government research grants by individual companies. Participation in Fraunhofer research projects exposes companies to highly qualified and motivated students and employees who can be recruited. The Fraunhofer makes its technology and business networks available to its industrial partners, a benefit small companies, in particular, usually cannot replicate on their own.

The Fraunhofer spins off new companies and assists other start-ups, but it is unclear whether on a net basis it fosters or inhibits the kind of start-ups that grow into world class companies in the United States. A prominent German academic commented in 2012 that Fraunhofer "kills start-ups. It offers services that should be provided by private companies. It thwarts individual activities."[201] While the Fraunhofer itself would undoubtedly dispute the assertion, it performs tasks in Germany that might be undertaken by entrepreneurs in some other economies. In most collaborations the Fraunhofer—a large parapublic institution—owns the intellectual property, rather than a small business or university, arguably acting as an inhibition on innovation, particularly that of a paradigm-breaking variety.

The Fraunhofer is an important member in the group of parapublic non-university research organizations that has gradually been displacing German universities as sites for scientific research. German universities are suffering

---

[200]Deutsche Welle. 2011. "The Automobile at 125: From Humble Birth to Global Dominance." January 25.

[201]Interview with Professor at Ludwig Maximilians University. Munich. June 13, 2012.

from numerous problems including shortages of funds, overcrowding, and a reputation for mediocrity—and as The Economist noted in 2009, the universities "would do better if research at non-university institutes like the Max-Planck Society were brought into academia, adding teaching to research."[202] Organizations representing the interests of German universities have urged restraint in the establishment of non-university research organizations for many years based on the "concern that universities will fall back in the resource competition with non-university institutions."[203] Non-university research organizations like Fraunhofer exert a gravitational pull on individuals who might otherwise become or remain members of university faculties because they offer relief from teaching obligations and the need to pursue grant funding independently, superior equipment and research infrastructure, and the chance to make money on innovations which they originate. Recognizing the risk of a wholesale "emigration of research" from universities to non-university research organizations, the German Science Council has sought to encourage close collaborations between universities and non-university research institutions—a phenomenon which has been developed to a very high level in the Fraunhofer.

Finally, the Fraunhofer model might not work in the United States because the industrial and scientific milieu is substantially different than Germany. Systematic application of new technology in factories may prove more difficult in situations where the work force has not undergone a long apprenticeship process, has a more fragile allegiance to a particular company, and may be demoralized as a result of falling wages and benefits, downsizing, offshoring, and labor-management strife. While the United States has many innovative small and medium businesses, the proportion that possesses the peculiar structure, traditions, discipline, and values of the Mittelstand is probably relatively small. Competition at every level of the U.S. innovation system is more intense and destabilizing than is the case in Germany, and complicates efforts at network-building and collaborations. Perhaps most importantly, the ability and willingness of federal and state governments to provide a stable source of long-term public funding to non-university institutes of applied research is highly questionable.

Whatever the obstacles to a literal implementation of the Fraunhofer model in the United States, the United States has adopted and adapted other German institutional innovations ranging from Kindergarten to the modern university research laboratory system, and it would appear that limited aspects of the German system could be adopted in this country. While the United States will not replicate Germany's Dual System of vocational education, the enlightened expression of that system at the high end of the educational ladder in the Fraunhofer institutes, featuring concurrent academic study and applied

---

[202]*The Economist.* 2009. "Germany's Mediocre Universities: On Shaky Foundations."

[203]Winnes and Schmank.1999. Federal Republic of Germany. Op. Cit. p. 37.

industrial research, could be more broadly incorporated in university curricula in collaboration with local industry. The Fraunhofer funding model, which locks in public funding on a virtually permanent basis is an obvious factor underlying the institute's achievements, could be put in place for a U.S. system of institutes of applied public research, although this may seem improbable in the current impasse over government spending. The Fraunhofer has refined technology networking to what it regards as a higher level than U.S. counterparts, and the adoption of the institutes' techniques should be studied by U.S. research organizations.

From an American perspective, the German experience is also of interest because like the United States, Germany is a federal system. Some aspects of German federalism would never be considered attractive in the United States such as the interlocking of state-federal responsibilities in a manner which limits the ability of each to act independently of the other. However, in Germany, the Länder have often functioned as laboratories in which institutional innovations have been implemented which, when demonstrated to be successful, have been adopted at the national level. The states of southwestern Germany pioneered the Dual System of education and practical training which was embraced nationally with the enactment of the Handicraft Protection Law of 1897. The Prussian initiative which established the first Kaiser Wilhelm institute eventually led to the creation of a nationwide network of parapublic research organizations. A similar initiative by Bavaria and Baden-Wurttemburg in the 1950s culminated in the creation of the nationwide network of Fraunhofer institutes. The successful postwar efforts by Bavaria and Baden-Wurttemburg to promote education and research and attract technology-intensive industries finds a current expression at the federal level in the form of initiatives to create innovation clusters on a nationwide basis. There is no readily apparent reason why the U.S. federal system could not be used to explore institutional innovations in applied research at the state level with potential for national policy.

From the broadest U.S. perspective, the Fraunhofer system represents a major public investment by Germany in applied research with short-run commercial relevance. The positive effects of that investment over the long run, as manifested in the international competitive performance of German industry, are undeniable. In the U.S. government investments in commercially-oriented research are commonly controversial in the short run and politically unsustainable over the long run. But against the background of an economic environment which has seen the erosion and offshoring of traditional industries in the face of global competition, some elements of the German innovation system merit consideration by U.S. policymakers.

Appendix A3

Taiwan's Industrial Technology Research Institute: A Cradle of Future Industries

Taiwan's Industrial Technology Research Institute (ITRI) is a not-for-profit research organization established in 1973 to provide applied industrial research for Taiwanese industry. ITRI draws upon research conducted all over the world in companies, research organizations and universities and uses the knowledge to develop product prototypes and the processes, equipment and materials necessary to manufacture those prototypes. It fosters not only the creation of companies that make new products, but of entire industry chains supporting the manufacturing process, including design, materials, equipment, testing, packaging, quality control and applications. ITRI "has played an integral role in transforming Taiwan's economy from a low-tech, labor-intensive model to a high-tech, knowledge-based industrial core."[1] ITRI's formation was the most important aspect of a broad national effort to develop Taiwan's semiconductor industry, a spectacular success which has subsequently driven the growth of Taiwan's capabilities in computers, lighting, displays, telecommunications, photovoltaics, and machinery.

ITRI is located in one of the most successful technology "clusters" in the world, grouped in and around Hsinchu Science Park (HSP), a research and high technology industrial zone established by the government in 1980.[2] In addition to ITRI, two world class research universities, National Tsing Hua University (NTHU) and National Chiao Tung University (NCTU) adjoin the park, which is the site of research and manufacturing operations of over 400 high technology companies. Most points within this complex are within

[1]Allen Hsu, "ITRI Pushes Technology Sector to New Frontier of Innovation," *Taiwan Journal* (October 19, 2007).

[2]Michael Porter defines "clusters" as "geographic concentrations of interconnected companies, specialized suppliers, service providers, firms, in related industries and associated institutions … in particular fields that compete but also cooperate. Michael Porter, *On Competition* (Boston: Harvard Business School Printing, 1998) pp. 197-198.

walking distance of each other, a proximity which fosters personal interchange and cross-pollination of ideas. NCTU, NTHU and ITRI train large numbers of workers for the industries in HSP; company executives and ITRI officials teach in the two universities; and university professors turn to ITRI for assistance in developing practical applications of new ideas and sit on advisory boards of local companies. ITRI "has been praised as the incubator of Taiwan's chief executive officers of publicly held companies and talents for industries," and the same could be said of NCTU and NTHU.[3] The companies located in HSP account for about 15 percent of Taiwan's GDP, making the park one of the most productive pieces of real estate on earth.[4]

The creation of ITRI, perhaps the most important milestone in the entire course of Taiwan's industrialization, was the brainchild of an elite group of highly competent bureaucrats and business leaders, most of them holding degrees in engineering.[5] They frequently had extensive experience working for multinational high technology companies and were in a position to apply their practical experience to the development of indigenous companies and industries.[6] They were relatively unhindered by political pressure—the Kuomintang Party (KPT), which held a monopoly on political power until 1990, had a tradition of relying on "scientific" government planning when it arrived on Taiwan in 1949, and technocrats "had already won a large measure of independence from party and military control."[7] This pattern was maintained

---

[3]As of late 2006 ITRI had cultivated over 60 CEOs and 18,000 specialists for Taiwan's high tech industries. "ITRI Transforms Into a Value Creator from a Tech Follower," *Taiwan Economic News* (October 24, 2006).

[4]Interview with Han-Ping David Shieh, National Chiao Tung University, Hsinchu, Taiwan, February 16, 2012. In 2007, the Geneva-based World Economic Forum ranked Taiwan in first place worldwide in industrial clustering competitiveness, a distinction "attributed mainly to the effect of the world-renowned Hsinchu Science Park." "Taiwan Ranks 1st Place in Industrial-Clustering Competitiveness Worldwide: WEF," *Taiwan Economic News* (December 26, 2007).

[5]Eleven of the first fourteen individuals to serve as Minister for Economic Affairs in Taiwan held degrees in engineering or science. K.Y. Lin, Taiwan's chief economic planner in the 1950s and early 1960s, had a degree in electrical engineering, and of his two assistants, one was a physicist and the other a civil engineer. MOEA's Industrial Development Bureau, which created ITRI, was dominated by engineers at the time. Robert Wade, *Governing the Market: Economic Theory and the Role of Government in East Asian Industrialization* (Princeton and Oxford: Princeton University Press, 1990), p. 98.

[6]Morris Chang, with 25 years of experience at Texas Instruments, is a former head of ITRI and when he moved to set up TSMC, he founded TSMC and "loaded TSMC's ranks with American-trained managers such as Britt Brooks, Doug Chance (the successive general managers of TSMC) and other international professional manager. Among the managers, most of them are Chang's former colleagues at TI." Chang's "excellent education and work experience established his professional knowledge in the semiconductor industry and contributed to the creation of the focused business model of the pure play foundry." Similar observations could be made with respect to many members of the generation of leaders which oversaw Taiwan's economic development. T. H. Liu, S. C Hung, S. Y. Wu, and Y. Y. Chu, "Technology Entrepreneurial Styles: A Comparison of UMC and TSMC," *International Journal of Technology Management* Vo. 29 ½ (2005) p. 681. SEMI Oral History Interview, *Morris Chang* (August 24, 2007, Taipei, Taiwan).

[7]Chiang Kai-shek's government, shaken by Japan's seizure of Manchuria in 1931, came to believe that its survival depended on governing through a highly educated and professional bureaucracy.

during the decades of KMT rule but is eroding with the advent of democracy and challenges to KMT policies by the Democratic People's Party (DPP).[8]

ITRI has deep American roots. Its founder, Y.S. Sun, formulated a developmental strategy for high tech industry in Taiwan based on discussions in the early 1970s with a colleague, Dr. Pan Wen-Yuan, an electrical engineer then employed in RCA's David Sarnoff Laboratories in New Jersey. The two men determined that Taiwan should develop a semiconductor industry, that acquisition of U.S. technology would be required, and that Chinese engineers working in U.S. technology companies would be an important asset for such an effort.[9] RCA trained the first cadre of ITRI engineers in semiconductor manufacturing technology, a number of whom were U.S. residents holding Ph.Ds. from U.S. universities. The Taiwanese development effort was advised by a "Science and Technology Advisory Group" (STAG), established in 1979, led by U.S. semiconductor executives, including Pat Haggerty, former CEO of Texas Instruments and former member of the National Academy of Sciences, and B.O. Evans, former VP for development at IBM.[10] Hsinchu Science Park represented a deliberate effort by Kuo-Ting Li, a Taiwanese leader known to posterity as the "architect of Taiwan's economic miracle," to replicate the best features of California's Silicon Valley in Taiwan.[11] HSP's first director was an American, Dr. Choh Li, formerly a research director at Honeywell in

---

Chiang promoted senior officials who believed in keeping economic policymaking insulated from political interference. Chiang's philosophy was paraphrased by a KMT official as "governments and political forms are transitory, the problems facing a nation are not." Chiang's son, Chiang Ching-kuo, who succeeded him in 1972, believed in recruiting "the central economic decision-makers according to demonstrated technical abilities." Wade, *Governing the Market* (1990) op. cit. pp. 247-248

[8]In 2006, Taiwan Premier Su Tseng-chang, a member of the KMT's rival party, the Democratic People's Party (DPP), visited ITRI, praised the organization for its achievements, and commented that he envied ITRI "because the researchers can commit themselves to professional studies without political interference." "ITRI Has Contributed Greatly to Taiwan: Premier," *Asia Pulse* (May 23, 2006). In fact the DPP tends to view ITRI as a KMT stronghold supporting business interests aligned with that party. Voices within the DPP reportedly suggest that if the government plans to spend money on industries, the funds would be better directed toward fisheries and agriculture.

[9]John A. Mathews and Dong-Sung Cho, *Tiger Technology: The Creation of a Semiconductor Industry in East Asia* (Cambridge: Cambridge University Press, 2000) p. 158.

[10]STAG initially consisted of two Technical Review Boards (TRBs), one for semiconductors and one for electronics, dominated by overseas Chinese selected by Evans or his staff. Morris Chang recalls that "Pat Haggerty was really the major figure at TI. It was he that made TI a big successful company. TI was a very small unknown and not all that successful company before him. He was the one that made the key decision of entering the semiconductor business." "Oral History of Morris Chang," recorded August 24, 2007 (Computer History Museum, 2007) p. 7.

[11]Li, a former Finance Minister, consulted with Frederick Terman on how Taiwan could replicate Silicon Valley. Terman served as Dean of Stanford's School of Engineering and spearheaded the establishment of what is now Stanford Research Park. Together with William Shockley he is widely regarded as the father of Silicon Valley. Similarly, Li, the founder of Hsinchu Science Park, is credited with transforming Taiwan from an agrarian country to a high technology center. "Fred Terman, the Father of Silicon Valley," *Net Valley* (October 21, 2010).

Minneapolis.[12] Morris Chang, perhaps the most famous person to head ITRI and the founder of Taiwan Semiconductor Manufacturing Corporation (TSMC), holds two degrees from MIT, a doctorate from Stanford and served for over 20 years with Texas Instruments, including time as the company's CEO. At present, four decades after its creation, a large proportion of ITRI's total work force still holds bachelors and advanced degrees from leading U.S. universities.

THE TECHNOLOGY INTERMEDIARY

An industry analyst quoted in *The Economist* observed in 2010 that Taiwan was "the best place in the world to turn ideas into physical form."[13] That fact is substantially attributable to ITRI, which functions as the bridge between ideas and form, or as its managers express it, as a "technology intermediary" serving Taiwanese industry. Its primary function is not research but adaptation and transfer of technology from domestic and overseas laboratories to domestic companies. ITRI is "arguably the most capable institution of its kind in the world in scanning the global technological horizon for developments of interest in Taiwanese industry, and executing the steps required to import the technology—either under license or joint development—and then absorbing and adopting the technology for Taiwanese firms to use …"[14] Technology is transferred to Taiwanese industry through a variety of channels, including licensing, patent auctions, co-development arrangements, spinoffs, migration of ITRI personnel to companies, and research consortia and alliances.

Strategic Direction

Although Taiwan's National Science Council (NSC) is formally responsible for formulating the country's science and technology policy, and ITRI is subordinated to the Ministry of Economic Affairs, strategic direction has determined through a consultation process involving foreign experts and ethnic Chinese with relevant experience in multinational companies:

> *"a distinctive feature of Taiwan's technology policy making is the extensive participation of overseas technologists, mostly of ethnic Chinese origins, as advisers. Enjoying a varying level of access to the top policy circle, they help policymakers identify the industrial sectors with technological promise,*

---

[12]Constance Squires Meaney, "Taiwan's Semiconductor Industry,": in Joel D. Aberbach, David Dollar and Kenneth Sokoloff, (eds.) *The Role of the State in Taiwan's Economic Development* (Armonk, NY: M.E. Sharpe, Inc., 1994) p. 178.
[13]"Hybrid Vigour," *The Economist* (May 27, 2010).
[14]John A. Mathews and Dong-Sung Cho, *Tiger Technology: The Creation of a Semiconductor Industry in East Asia* (Cambridge: Cambridge University Press, 2000).

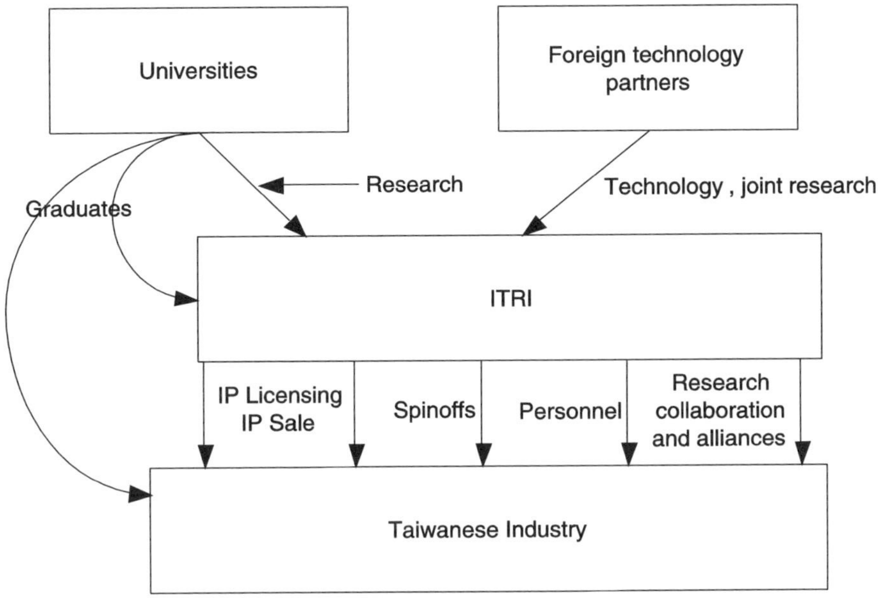

FIGURE APP-A3-1 ITRI as a technology intermediary.

> *suggest the strategies of technology transfer and adaptation, and provide technical expertise for the daily operation of research projects."*[15]

MOEA determines ITRI's strategic direction in consultation with such experts and ITRI itself. The Technical Advisory Committee (TAC), originally an informal group of expatriate Taiwanese engineers working in the United States, evolved into a standing organization of Taiwanese with international education and work experience advising ITRI on relationships with multinational corporations.[16] The Science and Technology Advisory Group (STAG), which advises the Executive Yuan (cabinet) on science and technology policy, includes international experts (occasionally including Nobel Prize winners) as well as eminent Taiwanese academics and holds an annual Industrial Science and Technology Strategy Conference, which addresses themes in areas such as electronics, telecommunications, information technology and

---

[15]Chen-Dong Tso, "State-Technologist Nexus in Taiwan's High Tech Policymaking: Semiconductor and Wireless Communications Industries," *Journal of East Asian Studies* (May 2004).

[16]Dr. Chintay Shih, former president of ITRI, commented in 2009 that "The TAC were our tutors and lighthouse. They have contributed the rest of their lives without any payback." Thirty years after ITRI's formation some of the original TAC members were still serving as consultants to ITRI. Cristina Chen, Jason Kao, Frans Nauta, Jan-Frens Van Giessel, Clement Goossens and Pipijn Veling, *Excellent Government on a Far-East Silicon Island* (December 2009).

biotechnology.[17] STAG prepares developmental technology blueprints for specific industry sectors regarded as important.[18] The Technical Review Board (TRB) of the National Science Council consists of foreign and domestic experts who concentrate on targeting particular technologies and adapting them at the operational and project level. A number of key Taiwanese experts have served in more than one capacity in such organizations.[19] This multifaceted and cosmopolitan advisory system has been widely praised, but has fostered failures as well as successes.[20]

Ding-Yuan Yang, a former ITRI official who founded Winbond Electronics Corporation in 1987, explained in a 2011 oral history interview that the founder of ITRI, MOEA Minister Y.S. Sun, would use the conferences with experts to absorb opinions on the development of technology. MOEA would then draft formal minutes of the conferences, "and then the plans would start." MOEA would give "suggestions" to ITRI for execution. ITRI would respond by proposing contracts with the Ministry, sign the contracts, and begin to execute them. "These kinds of national projects were all four-year contracts, which allowed it to have a certain continuity."[21]

Organization

ITRI was formed through the combination of three existing research centers subordinated to MOEA which were relocated to a new site in Hsinchu in 1973.[22] At its inception ITRI consisted of 400 employees and was funded by a

---

[17]STAG was at one time headed by Frederick Seitz, former president of the U.S. National Academy of Sciences. "STAG Provides Policy Suggestions," *Taiwan Today* (June 15, 2001)."The Science and Technology Advisory Group Must Take its Job Seriously," *United Daily News* (Taipei, November 13, 2009). Arthur Carty, a Canadian nanotechnology who was invited to join STAG in 2008, recalled in 2010 that in a recent week-long STAG meeting, "government officials, industry professionals, research and development experts and academics reviewed and discussed a number of investment proposals before throwing out the bad ones. We don't have anything like that in Canada," Carty said. "Taiwan's Technology Success Underappreciated: Canadian Scientist," *Focus Taiwan* (July 24, 2010).

[18]"M-Taiwan Program and Objectives of WiMAX in Taiwan," *WiMAX360°* (June 24, 2009).

[19]Dr. Hwa-Nien Yu, one of the world's foremost experts on semiconductor device technology and design, has chaired the TAC for ITRI since 1993. He also served on the TRB and has been advising ITRI in various capacities since the 1970s. He is an emeritus member of IBM Research with B.S., M.S. and Ph.D. degrees in electrical engineering from the University of Illinois.

[20]Based on advice from STAG, in 1982 the Taiwanese government decided to place a priority on the development of a biotechnology industry. Major financial and institutional commitments were subsequently undertaken. Today, thirty years after this effort began, it is not at all clear that the modest results achieved to date represent an adequate return on Taiwan's major investment. See generally Yu-Shan Wu, Academia Sinica, "Taiwan's Developmental State: After the Economic and Political Turmoid," Paper prepared for delivery at the Conference on A Decade After the Asian Financial Crisis, Thammasat University, Bangkok, February 23-24, 2007, p. 22.

[21]Interview with Ding-Yuan Yang, "Taiwanese IT Pioneers: D.Y. (Ding-Yuan) Yang," recorded February 23, 2011 (Computer History Museum) p. 12.

[22]The research centers were the Uni-Chemical Research Centre, the Metal Research Centre, and the Mineral Industrial Research Centre, all of which became internal ITRI research laboratories.

government budget of $213 million. In 1974, the Electronics Industry Research & Development Centre was established within ITRI to create a domestic semiconductor industry, subsequently being renamed as the Electronics Research & Service Organization (ERSO). In 1990, new laboratories were established for computer and communications research and biomedical engineering.[23]

Currently, ITRI's R&D activities are centered on six core laboratories pursuing "deeper and new" ideas and eight technology centers which focus on particular themes emphasizing a multidisciplinary approach and drawing on the specialized competencies of the core laboratories. ITRI's Business Development Unit is responsible for commercializing research results, technology transfer, and relations with foreign research partners.

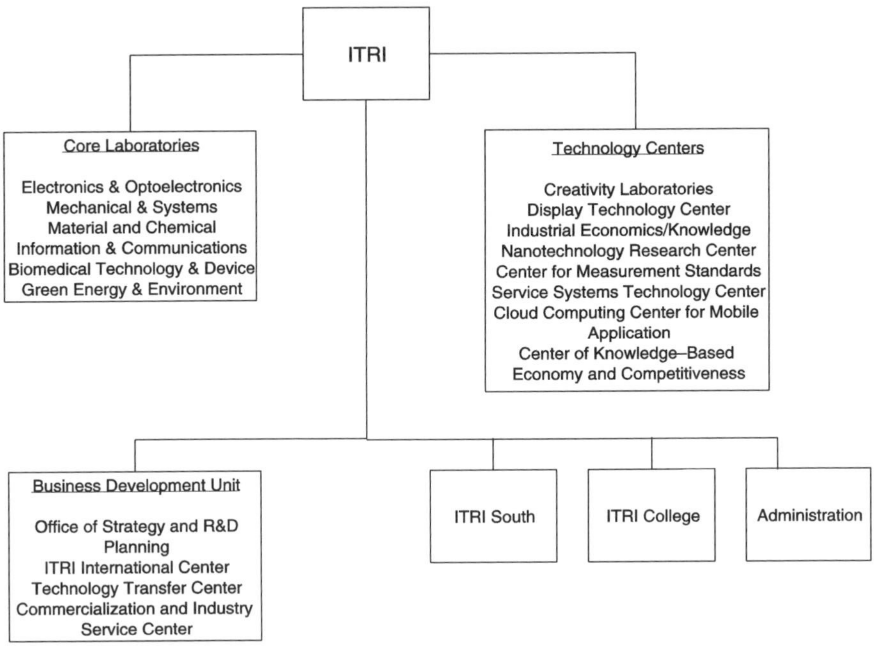

FIGURE APP-A3-2 ITRI's organization.

[23]Min-ping Huang, "The Cradle of Technology: the Industrial Technology Research Institute," in Terence Tsai and Bor-Shiuan Cheng (eds.). *The Silicon Dragon: High Tech Industry in Taiwan.* (Cheltenham, UK and Northampton, MA: Edward Elgar, 2006) pp. 27-28. ITRI undertook a fundamental restructuring of its organization in 2006 with an eye toward "silo-breaking" among its research laboratories, which were seen as competing with each other for resources and not collaborating sufficiently. ITRI's eight core laboratories were reduced to six through the merger of its semiconductor and optoelectronics labs and its materials and chemical labs. "Technology centers" were established and tasked with integrating the work of multiple ITRI core laboratories.

ITRI's budget is currently about $600 million per year, half of which is provided by the government and half by the private sector.[24]

ITRI employs 5,728 personnel, of which 1,163 hold Ph.Ds. and 3,152 masters' degrees. ITRI's staffing of research projects is "very bare bones," even in technology areas regarded has having breakthrough potential.[25]

Technology Acquisition

In ITRI's early years it relied almost entirely on the acquisition of technology and know-how from foreign high technology companies, which it disseminated to Taiwanese industry. In recent decades the research capabilities of Taiwan's universities, including the Hsinchu-based National Tsing Hua and National Chiao Tung Universities, have emerged as increasingly important sources of technology. As ITRI's reputation as an R&D powerhouse has grown, it has become possible for the institute to enter into joint R&D projects with first-tier foreign companies and research organizations.[26] Technology obtained via such collaborations remains a vital aspect of ITRI's operations.

---

[24]Private sector income is derived through activities such as sale of intellectual property to industry, provision of knowledge-based services to industry, other forms of technology transfer, and royalty payments. ITRI's budget has not grown for many years despite pleas from industry to MOEA to increase ITRI's funding. Interview with John Chen, Director, ITRI Display Technology Center, Hsinchu, Taiwan, February 14, 2012. Interview with Taiwanese semiconductor executive, Hsinchu, Taiwan, February 15, 2012. ITRI's budget is subject to periodic public pressure based on the perception that it is channeling public resources to companies and industries that are already mature and do not need public assistance. In 1994, ITRI's budget was cut in half by the Legislative Yuan based on such criticism. Douglas B. Fuller, *Globalization for Nation Building: Industrial Policy for High Technology Products in Taiwan* (MIT Working Paper MIT-IPC-02-002, January 2002), p. 12.

[25]Interview with Taiwanese semiconductor executive Hsinchu, Taiwan, February 15, 2012. ITRI recruits employees by offering them a good work environment, a large working team, and excellent career development prospects. Taiwan's system of compulsory military service has served as a major source of educated personnel for ITRI. The Ministry of National Defense assigns 400 soldiers to ITRI annually, of which 70 percent are graduates of the leading universities in Taiwan. Only 10 percent of Taiwan's soldiers are qualified for assignment to ITRI. They serve at ITRI for at least four years as an alternative to active military duty. Terence Tsai and Borshivan Cheng, *The Silicon Dragon: High-Tech Industry in Taiwan* (Edward Elger, 2006) p. 32. Taiwan has had a system of compulsory military service since 1949. Alternatives to active duty military service include "national defense service," which is available to draftees with advanced degrees, particularly in engineering and the sciences. Qualified individuals who choose this option receive three months of officer training and a commission in the reserves, followed by four years of work in a government or academic research institution such as ITRI or Academia Sinica. Annual turnover is extremely high, averaging about 10 percent but in some divisions rising to 15-20 percent, posing a continual challenge to management. Interview with Mao-Jian Wang, National Tsing Hua University, Hsinchu, Taiwan, February 16, 2012.

[26]A number if ITRI's foreign collaborations involve establishment of a research center on ITRI's premises by the foreign partner. "Festo Inaugurates Automation Parts Engineering Center at ITRI," *Taiwan Economic News* (February 9, 2004); Corning operates a research center within ITRI on glass technology including bendable glass. "Corning Inaugurates Research Center in Taiwan," *Asia Pulse* (March 13, 2006). In a recent and significant example of technology acquisition from abroad, ITRI's entry into the field of flexible electronics (bendable electronic devices and displays) was made possible by friendly technology transfer from Eastman Kodak Company to ITRI. Dr. John Chen,

TABLE APP-A3-1 ITRI Foreign Technology Partnerships

Foreign Partner	Year Initiated	Themes
Intel Corporation	2011	Memory chips
Lawrence Berkeley National Laboratory	2011	Renewable energy
Corning	2006	Optical glass, materials
Microsoft	2005	Forward versatile disc (FVD)
Alvarion Ltd.	2010	WiMAX, wireless
Spirit Aerosystems	2009	Carbon fiber
Hewlett Packard	2004	Radio frequency identification (RFID)
IBM	2009	Cloud computing
Novartis	2008	Pharmaceuticals
Applied Materials, SUSS Microtec	2009	Semiconductor manufacturing equipment, 3D ICs

SOURCES: "ITRI, Intel Announce Research Project," *Taipei Times Online* (December 7, 2011); "ITRI Signs MoU with LBNL on Renewable Energy Technology Development," *Taiwan Economic News* (August 11, 2011); "Corning Inaugurates Research Center in Taiwan," *Asia Pulse* (March 13, 2006); "Microsoft Launches Windows Engineering Center in Taiwan," *China Post* (September 10, 2005); "ITRI, Israeli Firm to Build WiMAX Testing Lab," *Central News Agency* (February 9, 2010); "Taiwan Institute to Cooperate with U.S. Firm in Carbon Fiber Research," *Central News Agency* (July 17, 2009); "HP Sets Up First RFID Center on the Island," *China Post* (April 14, 2004); "ITRI to Spearhead Green Energy and Biotech Research," *Taiwan Economic News* (September 7, 2009); "Novartis Signs Five Year Cooperation MoU with MOEA," *Taiwan Economic News* (January 11, 2008); "ITRI and Applied Materials Team Up on 3DIC Technology," *ITRI Today* (4th quarter 2009).

Technology Development

ITRI's six core laboratories do not perform basic research, but are tasked with undertaking "exploratory and pioneering" research with respect to technologies with potential commercial applications.[27] Projects can run five

currently the Director of ITRI's Display Technology Center, held numerous R&D managerial positions at Kodak between 1982 and 2006. Kodak, which had developed technology for large area roll-to-roll (R2R) fabrication of flexible displays, gave up on commercialization and sought to sell the technology "to someone who was competent," eventually selecting ITRI. Chen led an ITRI team to Kodak's labs in Rochester where a Kodak delegation worked with them to facilitate the technology transfer, which included transfer of equipment. ITRI's subsequent development of R2R manufacturing processes for flexible displays was "all based on Kodak technology" but now exceeds the highest technological levels reached by Kodak. "This was the beginning of flexible displays [in Taiwan]." Interview with Dr. John Chen, Director, ITRI Display Technology Center, Hsinchu, Taiwan, February 14, 2012.

[27]ITRI President Johnsee Lee, "ITRI Pushes Technology Sector to New Frontiers of Innovation," *Taiwan Journal* (October 19, 2007).

years or more.[28] Their competencies are concentrated in hardware-related fields, although ITRI is working to refocus on software, systems and services.[29] The core laboratories develop and test prototype products, production equipment and materials, and experiment with new applications.

- ***Electronics and Optoelectronics Research Laboratories***: ITRI's Electronics and optoelectronics research laboratories focus on manufacturing technology in the fields of nano-electronics, micro-optical-mechanical electronics, optoelectronic 3D integrated circuit packaging, 3D image processing, flexible electronics, and advanced display technology. This laboratory is currently heavily involved in research involving flexible electronics.
- ***Information and Communications Research Laboratory:*** ITRI's Information and Communications Technologies Laboratory conducts research on core technologies for the development of IT-enabled services (ITES). This laboratory coordinates with other national projects involving system-on-chip, e-learning and telecommunications.
- ***Green Energy and Environmental Research Laboratory:*** ITRI's Green Energy and Environmental Research Laboratory conducts research in the areas of energy efficiency, clean environment, renewable and alternative energy, natural resources, and energy management and policy.[30]
- ***Material, Chemical and Nanotechnology Research Laboratory:*** ITRI's Material, Chemical and Nanotechnology Laboratory collaborates with Taiwanese companies to develop materials and components for application in the fields of electronics, green energy, optoelectronics, and panel display. In addition, it develops high-tech

---

[28]Interview with John Chen, op. cit.

[29]"An Interview With ITRI President Shyu Jyuo-min," *Taiwan Today* (December 17, 2010).

[30]In 2011, a research team from the laboratory won an Angel Business Communications; (ABC) Solar Industry Award, the first Asian laboratory to achieve this honor, for development of a "green energy antenna." This device "integrates antenna transmission and solar panel power storage technology, simultaneously improving the conversion efficiency of solar energy optoelectronics and the efficiency of antenna reception and emission." ITRI has reportedly secured international patent rights for the technology and is seeking "international cooperation partners for technology transfer." According to estimates, application of the new technology at a 3-G base station with power consumption of 500W could product about 30 percent supplementary power mitigating peak hour power consumption and relieving the load on the power grid. Lillian Lin, "ITRI's Green Energy Atenna Technology Wins Solar Industry Award" *Central News Agency* (September 7, 2011). This lab has developed carbon capture and storage (CCS) technology jointly with Taiwan Cement Corp. which reportedly cuts the cost of carbon capture from the current international level of over $45 per metric ton to under $26 per metric ton. "Taiwan Unveils Microalgal Biofuel Technology," *Central News Agency* (October 2, 2010). It has also developed processes to transform microalgae into biodiesel fuel. "ITRI Wins Prominent Display Technology Award from Industry Group," *Central News Agency* (May 19, 2011).

fibers and specialty chemicals intended to transform the textile and chemicals industries into high value-added sectors.[31]

- ***Mechanical and Systems Laboratory:*** ITRI's Mechanical and Systems Technologies Laboratory develops technologies to assist Taiwanese companies in the areas of precision manufacturing, green energy and intelligent automation.
- ***Medical Device and Biomedical Technologies Laboratories:*** ITRI's Medical Device and Biomedical Technologies laboratories are performing research to create and expand a biomedical industry in Taiwan that will develop products according to the "biomedical 3Ps (preventive, predictive, personalized).

ITRI research projects are aimed at the development of technologies that can be commercialized, and involve the creation of specific product prototypes. ITRI's main site in Hsinchu, Taiwan displays scores of recent product prototypes that have emerged from its laboratories, including a number that have won international R&D awards. In most cases, the commercial potential of these prototypes is readily apparent.

Technology Integration

ITRI operates a number of technology integration centers. ITRI "hopes the centers will become spin-off companies."[32] The centers are more mission-oriented and dynamic than the core laboratories—new ones are frequently

Upstream	Midstream	Downstream	Application
Materials Components	Design simulation assembly	Manufacture process systems test run mass production	Product testing certification

FIGURE APP-A3-3 Coordinating and integrating industrial technologies at ITRI.
SOURCE: ITRI.

[31]In 2011, this lab won the silver award of Display Component of the Year by the world's leading display industry group, the Society for Information Display. The award was given for development of the first and only technology permitting mass production of flexible and transparent displays of all sizes. The ITRI team dedicated ten years to developing this technology. Oscar Wu and Mai Huang, "ITR Wins Prominent Display Technology Award from Industry Group," *Central News Agency* (May 19, 2011).

[32]Interview with Dr. Yi-Jen Chan, Director, ITRI Electronics & Optoelectronics Research Laboratories, Hsinchu, Taiwan, February 14, 2012.

formed while others spin off to become start-up ventures. Currently, ITRI technology centers are addressing topics in displays, nanotechnology, cloud computing, and measurement standards. The centers seek to establish capabilities for the integrated production of advanced technologies by fostering the creation of an entire industry supply chain.

The technology centers coordinate and integrate research and technology obtained from various ITRI core laboratories, foreign partners, and Taiwanese industry. They have discretionary funds with which they can commission R&D in the core laboratories. At present, ITRI's Display Technology Center (DTC) is performing work on flexible displays which integrates the research efforts of five separate ITRI labs as well as technology licensed from foreign companies. DTC Director John Chen observes that "we are the program office or integrator … we have the fab here so we can concentrate on process integration strategy. We don't work on materials, we rely on our colleagues [in ITRI's Material, Chemical and Nanotechnology Research Lab] to develop the materials we need."[33] DTC's main facility is a 3,124 sq.m second generation laboratory pilot line (glass substrate size 370x470mm2) which has been used to produce flexible 20-inch thin film transistor liquid crystal displays (TFT-LCDs).

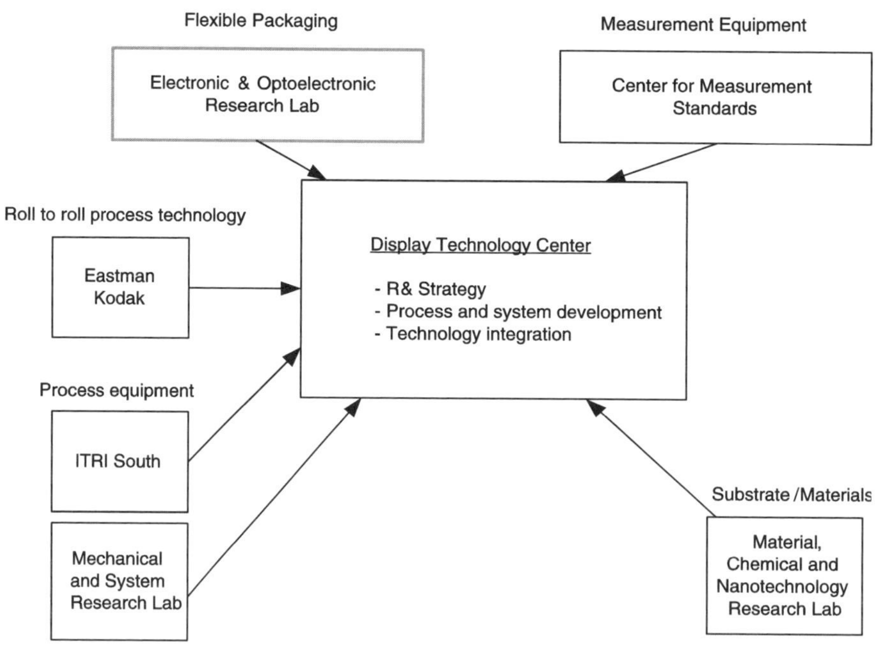

FIGURE APP-A3-4 Flexible display program integration.

[33]Interview with Dr. John Chen, Director, DTC, Hsinchu, Taiwan, February 14, 2012.

TABLE APP-A3-2 Developing ITRI Industrial Chain for Flexible Electronics

Topic	Companies
Materials	Chang-Chun
Equipment	Shuz-Tung
	Gallant Precision Machining
Panels	AUO
	CMO
	Chi Mei Optoelectronics
System	Elan Microelectronics Corp.

DTC commonly engages in contract services, joint R&D, technology transfer, cross-licensing, and evaluation and verification of customers' flexible displays materials, equipment and systems. It is currently partnering with local Taiwanese companies to establish the foundation for a complete industrial chain for the manufacture of flexible electronics products.[34]

DTC Director Chen observes that the multidisciplinary approach to R&D practiced in ITRI's technology centers is the key to the launch of new industries:

> *The biggest strength of ITRI is the multidisciplinary cooperation. We create a complete manufacturing supply chain in its early stages. That is the secret. Then you can scale it up, then you have a complete supply chain for the industry. So DTC does not just work with display companies, but also materials suppliers and equipment makers.*[35]

Decades of application of this holistic approach has transformed Taiwan into a major manufacturing center for electronic information products. Barry Lam, former head of Kinpo Electronics and founder of Quanta Computer, commented about the Taiwanese IT industry chain in a 2011 oral history interview:

> *The supply chain is very complete in Taiwan. We have semiconductor foundries here in Taiwan. We have good design houses. We have many good assembly houses. We also have many components, such as CD-ROM drive. We gave up on the hard drive business at the time, so we didn't catch the momentum when the industry was blooming. Assembly was done mostly in Thailand or Southeast Asia. We pretty much know how to make most components in mass production. So from components ... that's why even when we make resistors and capacitors, our price is still very cheap. For components, from PCB boards to chips, we can make it*

[34]ITRI Display Technology Center presentation (February 14, 2012).
[35]Interview with DTC Director John Chen, Hsinchu, Taiwan, February 14, 2012.

all in Taiwan. So, we can even complete the design in Taipei. And, why? Because all the vendors are concentrated in Taipei. Taipei is not big, so it's easy to deal with everything in Taipei and everything can be done here. This is good, isn't it?[36]

Technology Diffusion

ITRI's technology is transferred to Taiwanese industry through numerous channels. ITRI research organizations participate in a large number of topical R&D alliances designed to ensure that the participating companies follow, monitor, and in some cases participate in ITRI research projects in a manner which enhances the prospect they can successfully absorb and apply new technologies.[37] ITRI licenses technology to domestic companies, frequently on more favorable terms than could be obtained from foreign sources.[38] It performs contract R&D for companies.[39] Since 2005 it has auctioned off blocks of intellectual property on an exclusive basis.[40] ITRI's OpenLab is an incubator for new companies which provides facilities, utilities, business services and research support to start-up companies—in 2005 OpenLab was honored as the "best incubator in Asia" by the Asian Association of Business Incubation.[41] Among other services ITRI offers pilot production opportunities to companies prior to the commercialization of products, which enables companies to use ITRI's pilot lines for process verification, product development and semi-commercial production. The newly-formed ITRI College provides customized training programs for Taiwanese companies, as well as government officials and researchers from developing countries in

---

[36]Interview with Barry Lam, "Taiwanese IT Pioneers: Barry (Pak-Lee) Lam," recorded March 2, 2011 (Computer History Museum, 2011) p. 29.

[37]An important aspect of ITRI's industry-coalition building is organizing member companies to specialize in specific developmental areas so they do not duplicate effort. In 2003, for example, ITRI formed a "SIP alliance" with twelve of Taiwan's leading information technology firms pursuant to which they would share their silicon intellectual property, an initiative that arose out of the recognition that many local IC design houses were doing the same work. "From the government's point of view, this is a waste of resources," an MOEA spokesperson commented. "12 Top IT Firms Form SIP Alliance," *China Post* (April 9, 2003).

[38]In 2006, ITRI's Electronics and Optoelectronics Research Laboratories licensed indium tin oxide transparent electrode technology to nine Taiwanese producers of light-emitting diodes (LEDs) on terms that "would help the island's LED makers pare their spending on intellectual property licenses from foreign sources." "ITRI Lab Licenses Advanced Electrodes to LED Makers," *Taiwan Economic News* (July 7, 2006).

[39]ITRI performs contract research for both domestic and foreign companies to develop new products, improve manufacturing processes and to comply with environmental and safety requirements. In addition, it offers professional measurement, inspection and certification services to help companies address technical obstacles in world markets.

[40]"ITRI Puts Up 111 Patents for Sale," *Taiwan Economic News* (May 18, 2006).

[41]"ITRI OpenLab to be Honored as Best Incubator in Asia," *Taiwan Ajourd'hui* (June 10, 2005).

innovation competencies and six technological areas.[42] Finally, former ITRI employees taking positions in Taiwanese high tech companies bring extensive know-how and in some cases, technology which they are expected by their former employer to commercialize.

Spin-offs

ITRI's most dramatic method for technology diffusion is the spin-off of parts of itself to form new companies or to join existing companies (the latter being referred to as "spin-in").[43] ITRI spinoffs entail transfer of employees, technology and in some cases equipment to a new company, coupled with the provision of a wide range of incubation services (legal and business counseling, provision of office space and utilities at modest cost, and introductions to sources of financing and to potential customers). Since ITRI's founding in 1973 it has formally spun off a total of 162 companies, and informally contributed to the launch of several hundred others.[44] As of early 2012, roughly two dozen spin-offs were under consideration with five in the incubation phase.

Picking Winners

ITRI's venture capital subsidiary, the Industrial Technology Investment Corporation (ITIC) plays a key role in facilitating ITRI spin-offs.[45] When ITRI laboratories seek to spin off a promising technology ITIC screens the proposal, assessing whether the technology has commercial potential, and sizing up the business prospects for the new venture and the competition it is likely to encounter. ITIC works closely with ITRI's Commercialization and Industry Service Center (CIS) in assessing whether a given proposal has commercial

---

[42]ITRI College does not issue degrees because its training is for the purpose of hands-on application by companies in an industrial context. Technological training is offered in IT and telecommunications, optoelectronics, electronics, biomedicine, device research, energy, the environment, materials and chemical engineering and mechanical and systems R&D.

[43]A recent example of a spin-in arrangement is the 2008 formation of Hi-Tech Energy Co to produce lithium batteries. Hi-Tech Energy is a joint venture between ITRI and Taiwan's Welldone Co., a producer of batteries. The joint venture was "led by a team of ITRI specialists." R&D at the joint venture was to be led by Yang Mo-hua, "an ITRI battery export." "Taiwan Spearheads Lithium-Battery Module Effort," *Taipei Times* (May 15, 2008).

[44]"ITRI Encourages Formation of Spin-in Technology Venture Firms," *Taiwan Economic News* (January 31, 2012).

[45]ITIC was established in 1979 as a wholly-owned subsidiary of ITRI to facilitate the development of high technology industries and the upgrading of traditional industries. ITIC was established because at the time Taiwan lacked a venture capital industry. ITIC provided capital to a number of Taiwanese companies that were spun off by ITRI, including United Microelectronics Corporation (UMC), Acer Display Technologies (Now AU Optronics), and Taiwan Mask Corporation. ITIC currently provides funds to over 50 companies and operates an incubation center with 30 start-ups representing various high tech sectors. As of early 2012, ITIC had four funds supplying venture capital to start-ups. ITIC has a staff of 23 professionals, half of whom have ITRI backgrounds and half previously held positions in industry. Twelve individuals comprise ITIC's venture capital team. Interview with Ching-Jiunn Chang, Vice President, ITIC, February 13, 2012.

TABLE APP-A3-3 Examples of ITRI Spin-off and Spin-in Ventures

Year	Company	Products
1999	Taiflex Scientific	Electronic materials
1997	Prolific Technology	ASIC designs
2005	Neo Precision Technology Co.	High-speed spindles for machine tools
2006	Amos Technologies Inc.	Radio-frequency identification (RFID)
2003	Phalanx Biotech Group	Gene chips
2008	Hi Tech Energy Co.	Lithium battery modules
2001	Phison Electronics Corp.	Flash memory
2004	DelPoint	Solar cells

SOURCE: "Taiwan to Roll Out High-Speed Spindles for High End Machines Tools," *Taiwan Economic News* (November 21, 2005); "Mosel Vitelec, ITRI Set Up Venture Firm to Tap into RFID Market," *Taipei Times* (November 8, 2006); "Gene Chip Venture Phalanx Inaugurated," *China Post* (January 24, 2003).

prospects. If the answer is affirmative, ITRI will provide funding and incubation services, and in some cases will recruit one or more outside experts with business experience to join the new venture.[46]

ITIC provides early stage financing to some ITRI spin-offs, usually in amounts under $US 1 million. ITIC takes an equity stake in new ventures which it normally expects to sell at some point in the future. ITIC's decision on whether to provide financial/support to a given start-up is taken independently of ITRI and ITIC is not obligated to invest in any ITRI spin-off. As of early 2012, ITRI had 24 spin-offs under consideration, of which ITIC would typically approve funding for one third of the group and turn the others down. The success rate for ITRI spin-offs funded by ITIC is currently about 3 out of 10.

Incubation

The ITRI OpenLab Incubation Center, near Hsinchu Science Park, provides temporary support for ITRI spin-off and startup companies. The Incubation Center features nearly 140,000 square feet of leasable space which can be used for offices, research space, or product assembly. Companies in the incubator receive management advice, technical consulting services, legal help, and access to ITRI's research facilities and pilot production lines.

[46]Interview with Ching-Jiunn Chong, Vice President, ITIC, Taipei, Taiwan, February 13, 2012.

K.S. Pua, an ITRI alumnus who with four other ITRI staffers co-founded the successful start-up Phison Electronics Corp., a maker of flash memory systems, recalled the importance of ITRI's support in the early days of his company between 2000 and 2003.[47] The new company was allotted space at ITRI's Incubation Center, which "not only provided office space, but also facility management," so that "we didn't have to worry about the trash, water, power, or other sundry things." The rent was "comparatively cheap." ITRI loaned Phison laboratory equipment that the start-up could not have otherwise afforded. Its legal adviser assisted Phison in defending a lawsuit. The new company assembled its products "right in the office," achieving a turnover of $36.7 million in 2002. The Incubation Center Director, C.J. Chang, advised Phison to apply for a grant from MOEA's Small Business Innovation Research (SBIR) program and assisted the company in drawing up the application. ITRI helped Phison to recruit qualified employees, connected it with investors, and vouched for it in its pitches for funding. Phison had 12 employees when it entered the Incubation Center in August 2000 and 60 when it left in January 2003.[48]

A key milestone in Phison's history was the formation of a close relationship with Toshiba, which invested in Phison, provided a source of demand for its USB flash drives and assisted the company in securing patents and fending off legal challenges.[49] Initially Toshiba had no idea who or what Phison consisted of, so ITRI made a presentation to Toshiba introducing the company. On the basis of the presentation and ITRI's established high technology bona fides, Toshiba decided to invest in the company, becoming its largest shareholder.[50]

THE HSINCHU TECHNOLOGY CLUSTER

ITRI's contribution to Taiwan's industrial development has been substantially enhanced by its physical presence in a technology cluster in the Hsinchu area. Innovation or technology clusters are concentrations of knowledge-based companies pursuing research, development and production of advanced technology products in proximity to each other. The most successful clusters draw upon nearby universities for talent and research, and benefit from

---

[47]In 1998, Pua's team at ITRI designed a controlled which was sold to Kodak. Interview with K.S. Pua, *Hsinchu*, Taiwan, February 14, 2012.

[48]K.S. Pua, *Driven to Success: Tech Star Phison's US $1 Billion Journey* (Taipei: Common Wealth Magazine, 2012) pp. 46-49.

[49]Characterizing Toshiba as an "umbrella" and a "shield," Pua states that "Toshiba was the 'door deity' that protected us from the outside world," Pua (2012), *Driven to Success,* (2012) op. cit. p. 40. Toshiba has been investing in Taiwan since the 1950s, when it acquired a five percent equity stake in Tatung Co., which was then Taiwan's only integrated electronics company. Toshiba licensed technology to Tantung, enabling the company to develop competencies in high-end compressors, CRT picture tubes, and LCDs. Today, Tantung is a diversified multinational corporation producing consumer electronics products.

[50]Interview with K.S. Pua, Phison Electronics Corp., Hsinchu, Taiwan, February 14, 2012.

an array of supportive government policies, particularly the provision of research infrastructure. The most well-known innovation clusters in the United States are Silicon Valley, Research Triangle in North Carolina and Boston's Route 128. Technology clusters are particularly important in "bringing the upstream, midstream and downstream aspects into one complete series of connections, which possess high speed communication exchange, technical support and other competitive advantages," and in sectors with comparatively long industrial chains, such as semiconductors and photovoltaics, the cluster effect is particularly powerful.[51]

Taiwan's Hsinchu technology cluster has evolved into one of the most productive in the world. According to a recent estimate firms located inside HSP are 66 percent more productive than firms located outside.[52] In 2010, 139,416 people were employed in HSP itself.[53] At the end of 2010, HSP contained operations of 449 companies which generated over $40 billion in that year. The semiconductor industry was dominant in terms of percent of total revenue.[54]

In addition, many small technology companies are located just outside the HSP itself and contribute to the cluster effect. At present HSP receives around 1,000 visiting missions per year seeking to understand how such a technological-industrial complex can be established and managed.[55]

The Hsinchu technology cluster has been extensively studied by academics and by other countries and regions seeking to replicate its success. John A. Mathews, an Australian professor of competitive dynamics, has written

TABLE APP-A3-4 Hinschu Park Revenues by Sector

Sector	Percent of HSP Revenue
Semiconductors	68.44
Optoelectronics	19.08
Computers	5.99
Telecommunications	3.28
Precision machinery	2.22
Biotechnology	0.42

[51]Hwa Meei Liou, "Overview of the Photovoltaic Technology Status and Perspective in Taiwan," *Renewable and Sustainable Technology Reviews* (2010).
[52]John A. Mathews, "The Hsinchu Model: Collective Efficiency, Increasing Returns and Higher-Order Capabilities in the Hsinchu Science-Based Industry Park, Taiwan," Keynote Address, Chinese Society for Management of Technology 20th Anniversary Conference, Hsinchu, Taiwan, December 10, 2010.
[53]Interview with Mao-Jiun Wang, National Tsing Hua University, Hsinchu, Taiwan, February 16, 2012).
[54]Presentation by HSP administration, Hsinchu, Taiwan, February 15, 2012.
[55]Japan International Cooperation Agency, *Report on Taiwan Mission* (GRIPS Development Forum, April 7, 2011).

extensively about the Hsinchu phenomenon, concluding that "firms generate mutual advantages by clustering together," and collectively form "powerful engines of wealth generation." In Hsinchu, these advantages have been enhanced by the presence of ITRI and the two science-oriented universities—"companies attracted to the Hsinchu park have been able to count on the universities for supplies of skilled professional staff, while they have been stimulated by exposure to the technological innovations emanating from ITRI."[56]

The basic elements of a successful cluster are sometimes referred to today as "Marshall's trinity," based on the pioneering nineteenth century work of the British economist Alfred Marshall in his study of the Sheffield industrial district of Britain.[57] According to Marshall the trinity consists of three basic elements:

- ***Supplier linkages:*** Cluster-based firms foster economies with respect to transportation, communications, supply chain, infrastructure and logistics.
- ***Labor:*** Specialized activities within the cluster foster growth of a pool of skilled labor, permitting increased specialization by enterprises located in the cluster.
- ***Knowledge spillovers:*** As Marshall expressed it, in an industrial district the "secrets of trade are in the air," and presence in the cluster enables firms to access market intelligence, new designs and new applications.[58]

All of these factors are present in abundance in the Hsinchu technology cluster and are critical elements in its success.

Supplier Linkages

Hsinchu Science Park has greatly facilitated the development of Marshall's "supplier linkages" by encouraging technology-intensive firms to locate in proximity with each other. Such arrangements reduce transportation costs with respect to materials and parts.[59]

---

[56]Mathews, *Hsinchu Model* (2010) op. cit. p. 10.
[57]Paul Krugman, a U.S. economist and winner of the Sveriges Riksbank Prize in Economic Sciences (e.g. the Nobel Prize in Economics) reiterated the importance of Marshall's trinity in *Geography and Trade* (Cambridge, MA: MIT Press, 1991).
[58]The synopsis of Marshall's work is drawn from Mathews, *Hsinchu Model* (2010) op. cit. p. 13
[59]Agglomeration of firms from a given sector's supply chain in a cluster enables them to specialize, achieving increased economies of scale, higher capacity utilization with respect to the specialized machinery they employ, and improved skill sets. "If the minimum efficient scale of production varies across a range of products through different stages of the production process, manufacturers can choose an optimal combination of operations by working closely with many specialized suppliers." Eric Y. Cho and Hideki Yamawaki, *Clusters, Productivity and Exports in Taiwanese Manufacturing Industries* (Presented at Gerald R. Ford School of Public Policy, University of

Although companies in the supply chain cooperate, they also compete—any company that falls short of expectations can quickly be replaced by another. Competition continually drives down costs and forces companies to improve yields. Reflecting these dynamics, the HSP Administration has assembled data which reportedly demonstrates that firms within the park generate an average value-added (revenues minus costs as a percent of total revenues) of 50 percent, versus 30 percent for Taiwanese manufacturing enterprises located outside of the park—or a productivity edge of 60 percent.[60]

HSP contains complete or nearly complete industry supply chains in a number of sectors, most notably semiconductors. The presence of materials, design, equipment, fabrication, and testing companies in the park is cited as a major locational and competitive advantage by semiconductor industry executives.[61]

Skilled Labor

The second element of Marshall's trinity, pool of skilled labor, is available to companies operating in HSP thanks in substantial part to the presence of two research universities adjacent to the park.[62] National Chiao

TABLE APP-A3-5 Hinschu Park Industry Supply Chain for Semiconductors

Sector	Number of Companies in HSP
Wafer fabrication	17
Fabless design	129
Equipment	17
Lead frame	3
Wafers	6
Test	3
Mask	5
EDA design tools	5

Michigan, October 16-17, 2009) pp. 6-7. Ding-Yuan Yang, founder of Winbond, explained the dynamic as follows, "Taiwanese companies may not coordinate well enough, but each company clearly defines its own focus. And [they] break down the PC industry into parts. Each company does what it does best. Some do the keyboards, some do the monitors, some do the motherboards and some do the casing. This is what I call the ability to innovate." Interview with Ding-Yuan Yang, "Taiwanese IT Pioneers: D.Y. (Ding-Yuan) Yang, recorded February 23, 2011 (Computer History Museum, 2011).

[60]Mathews, Hsinchu Model (2010) op. cit., p. 16.

[61]Presentation by HSP administration, Hsinchu, Taiwan, February 15, 2012.

[62]In 2010, 130,000 people were employed in HSP. Of these, 65 percent had junior college or higher degrees (compared with 18 percent in Taiwan's manufacturing sector as a whole). 17,000 HSP employees held master's degrees and 1,200 had doctorate degrees. "Hsinchu Science Park, a Bastion for Growth, Innovation and Cluster-Based Industries," *China Post* (December 15, 2010).

Tung University (NCTU) was established in Hsinchu in 1958 and adjoins Hsinchu Science Park. NCTU supplied much of ITRI's original R&D staff as well as much of the engineering talent for HSP when the park was created.[63] NCTU was the first university in Taiwan to operate a semiconductor research laboratory, and one of its professors, Simon Sze, is the author of *The Physics of Semiconductor Devices,* otherwise known globally as "the semiconductor Bible."[64] At present 65 percent of the CEOs, Presidents and General Managers of companies located in HSP are NCTU graduates.[65] It compares its relationship with HSP to that of Stanford with Silicon Valley.[66] The University has colleges of engineering, biological science, computer science, photonics, and electrical and computer engineering, and research centers specializing in nanotechnology, photonics, bioinformatics, biomedical electronics, intelligent information/communication, brain research and interdisciplinary science.[67] Its students commonly work as interns in HSP and join companies in the park upon graduation. Many top ITRI managers also serve as NCTU faculty members.

National Tsing Hua University (NTHU), adjacent to NCTU, has produced two Nobel laureates in physics and one in chemistry. NTHU is regarded by some as the premier research university in Taiwan, holding the highest citation rate for papers published by faculty in either Taiwan or China. Nearly 50 percent of the university's research budget is attributable to grants won by NTHU faculty.[68] NTHU includes colleges of life sciences, electrical engineering and computer science, technology management, engineering, and nuclear science. Researchers at NTHU have reported recent breakthroughs in nanotechnology and semiconductors.[69]

---

[63]John A. Mathews and Mei-Chin Hu, "Enhancing the Role of Universities in Building National Innovative Capacity in Asia: The Case of Taiwan," *World Development* (Vol. 35 No. 6, 2007), p. 1012.

[64]Pao-Long Chang, Chintay Shih and Chiung-Wen Hsu, "The Formation Process of Taiwan's IC Industry—Method of Technology Transfer," *Technovation* 14(3) (1994) p. 163.

[65]NCTU alumni who have played a prominent role in Taiwan's high technology industrial development include Dr. Stan Shih, founder of Acer Group; Dr. Robert Tsao, founder and former chairman of United Microelectronics Corporation; Dr. Ken Kao, founder of WiFi leader D-Line Corporation; and Dr. F.C. Tseng, Vice Chairman of Taiwan Semiconductor Manufacturing Corporation. Source: NCTU presentation, February 16, 2012, Hsinchu, Taiwan.

[66]Interview with Ming-Jan Yao, Professor and Associate Dean, School of Management, National Chiao Tung University, Hsinchu, Taiwan, February 16, 2012.

[67]NCTU is the site of a number of national laboratories, including the National Chip Implementation Center, The National Nano Device Laboratories, the National Measurement Laboratory, the National Space Organization, the National Center for High Performance Computing, and the National Synchrotron Radiation Research Center.

[68]Mathews and Hu, "Enhancing the Role of Universities" (2007) op. cit. p. 1013.

[69]"Taiwanese Researchers Make Nanotechnology Breakthrough," *Central News Agency* (August 18, 2011). The Hsinchu technology cluster is also supported by Taipei-based National Taiwan University (NTU), Taiwan's leading university for teaching and research, which is about one hour's drive from Hsinchu. In 2011, NTU established a research center jointly with Intel on machine-to-machine (M2M) communication, a cutting-edge field which aims to integrate billions of machines to exchange information and make correct decisions independently. The Intel-NTU Connected Catext Computing Center, the project vehicle, will study smart sensing, "green sensing" and context

ITRI itself is a magnet drawing experienced and talented overseas Taiwanese back to the Hsinchu technology cluster. [70]

Alumni networks from the two science universities and NTU have played a central role in the development of the semiconductor and PC industries in the Hsinchu technology cluster. [71]

The work discipline, efficiency, adaptability and élan of Taiwan's high technology workforce are frequently cited by the founders' generation of the country's information technologies industries as a key element in the country's success.[72]

Knowledge Spillovers

The third element in Marshall's trinity, "knowledge spillovers," occur in HSP in part due simply to the proximity of research institutions and companies to each other, but the presence of ITRI and the universities amplifies

---

analysis. NTU researchers are collaborating with ITRI in a number of key technology areas, including biomedicine, systems-on-a-chip, and wireless communications. Interview with Dr. Si-Chen Lee, President, NTU; Dr. Ching-Hua Lo, Vice President for Administrative Affairs, NTU; and Dr. Ji-Wang Chern, Dean for Research and Development, Taipei, Taiwan, February 17, 2012.

[70]Ding-Yuan Yang, who served as head of planning and marketing for ITRI, observed in a 2011 oral history interview that "In the past 10 to 20 years, many friends or classmates from abroad came back to Taiwan because of fast development in Taiwan. We suggested that some come back to ITRI first for a few years and then to go into industry. The projects that ITRI supported provided a beacon, or a rallying cry; it was if they raised a flag for people to rally to. So many people kept coming back to Taiwan. For they had a specific target to go for, when they come back from overseas, lot of people who studies abroad come back to Taiwan for the same reason." Interview with Ding-Yuan Yang, "Taiwanese IT Pioneers: D.Y. (Ding-Yum) Yang," recorded February 23, 2011 (Computer History Museum, 2011) p. 28.

[71]A TSMC executive has observed that "engineers in the Hsinchu Science Park not only work very hard, but they share the same backgrounds. Most of them graduated from the same university, took the same classes, were taught by the same professors, and had similar work experience at ITRI." Anna Lee Saxenian, "Taiwan's Hsinchu Region: Imitator and Partner for Silicon Valley" (Stanford Institute for Economic Policy Research Discussion Paper No. 00-44, June 16, 2001) p. 27. Saxenian, Dean of the University of California at Berkeley's School of Information, who has extensively studied HSP, observes that "it would not be unusual for a Taiwanese engineer to call a former classmate in the middle of the night to get help solving a technical problem, and many started companies with their classmates. Acer managers are far more likely to contact former teachers, classmates or colleagues when they encounter a business problem … than turn to a private consulting firm or institution." Ibid.

[72]Barry Lam, informally known as the "laptop king, served as president of Kinpo Electronics, which he built into the world's largest contract calculator maker, and later founded Quanta Computer, a maker of notebooks. Lam later recalled that National Taiwan University's electrical engineering program provided a stream of "good engineers from my school [Lam was an NTU alumni] who were eager to learn new skills. We had low employee turnover and we had a lot of talented engineers, who were focused and honest. So the team spirit was very high. This was the fundamental competitive edge of Taiwan, the foundation was strong. How should I put it? Employees are extremely loyal to companies they work for, due to the profound Confucian influence on Taiwanese people. We had tremendous motivation. We could accomplish anything we set our hearts on." Interview with Barry Lam, "Taiwanese IT Pioneers: Barry (Pak-Lee) Lam," recorded March 2, 2011 (Computer History Museum). p. 9.

spillover effects. ITRI scans the entire world for promising new technologies and devotes substantial resources and effort to spreading information about them to local companies and educational institutions. Its numerous R&D alliances are structured to draw upon knowledge gleaned from international markets and research efforts and to deliver that knowledge to domestic firms. Knowledge spillovers from ITRI also occur as its employees take positions in Taiwanese companies—between its founding in 1973 and 2008, an estimated 13,000 ITRI staff moved into industry, many of them remaining with the Hsinchu cluster.[73]

Research spillovers from the universities were given a powerful boost by the enactment in 1999 of legislation patterned on the U.S. Bayh-Dole Act of 1980, which allows universities to own intellectual property developed out of government-funded R&D.[74]

Fostering New Clusters

Demand for sites in HSP long ago exceeded the Park's capacity, and Taiwan has opened Southern Taiwan Science Park (1996) and Central Taiwan Science Park (2003). HSP has opened science parks in five satellite locations.[75]

Taiwan's *Statute for Industrial Innovation,* enacted in 2010, emphasizes the formation and upgrading of knowledge-based industrial parks "to facilitate

TABLE APP-A3-6 New Clusters and Their Focus Areas

Park	Industries
Jhunan Science Park	Biotechnology, optoelectronics
Tongluo Science Park	IC design, SiP, digital life, avionics, biomedical
Longlan Science Park	Optoelectronics
Hsinchu Biomedical Science Park	Biomedical
Yilan Science Park	Knowledge-based series

[73]Mark Dodgson, John Mathews, Tim Kastelle and Mei-Chih Hu, "The Evoloving Nature of Taiwan's National Innovation System: The Case of Biotechnology Networks," *Research Policy* 37 (2008) p. 435. Anna Lee Saxenian observed in her 2001 study of Hsinchu and Silicon Valley that "In the Hsinchu region, as in Silicon Valley, engineers move frequently between firms and between sectors; and there is a community of senior engineers who move not only between firms, but also between the public and private sector, between universities and the private sector, between the manufacturing sector and the venture capital industry, and between Taiwan and Silicon Valley. They meet regularly at alumni gatherings and class reunions, professional association meetings, industry conferences and trade shows, and a variety of related special events … The intense communications within this community fosters imitation, joint problem solving and transfer of information and know-how about management, technology, the job markets, and new firms and products." Saxenian, "Taiwan's Hsinchu Region" (2001) op. cit., pp. 26-27.

[74]*The Science and Technology Basic Law* and ancillary implementing regulations provide that IPR developed whole or in part with government funds belongs to the research organization. The new rules have made it possible for academic researchers to reap substantial rewards through commercialization of research results. Mathews and Hu, "Enhancing the Role of Universities" (2007) op. cit. pp. 1008-12

[75]Hsinchu Science Park *Annual Report* (2010) pp. 12-13

clustering and innovation." The government is encouraging the formation of parks dedicated entirely to specific themes, such as biomedicine, logistics, and media. The new law authorizes local governments to establish industrial parks without central government approval. The minister of Economic Affairs indicates that such policies are being implemented to foster "an overall economic environment that is conducive to innovation, the creation of value, and the free flow and application of knowledge."[76]

CHALLENGES

Taiwan appears to the outside world as an extraordinarily successful high technology powerhouse. However, its leaders recognize that vulnerabilities exist which could substantially erode the gains which the country has made since the 1970s. The overwhelming majority of Taiwan's businesses are small enterprises which lack the resources and scale for global competition. The country is arguably not well prepared for an era of global patent warfare, in which infringement litigation can suppress or destroy technology-based start-up companies. A longstanding talent shortage has been exacerbated by an exodus of skilled workers to mainland China. And the American connection, long an important aspect of Taiwan's high tech development, may fade as the number of Taiwanese students pursuing advanced degrees in the U.S. progressively declines.

ITRI Funding

ITRI's budget has not grown for a decade. Its leadership has put a good face on the situation by stating that lean budgets force the institute to rely on other research organizations, which is a key element of its mission. But ITRI is not growing apace with Taiwanese industries that it is tasked with supporting, and the detrimental effects of its limited budgets are evident in a number of areas.

Some observers believe that ITRI is trying to focus on too many technologies with two small a budget, diluting the impact of its efforts. Its low compensation levels for staff have contributed to a manpower shortage (see below). ITRI reportedly would like to attract more foreign expertise to Taiwan but the available funding is inadequate to create attractive compensation packages. Underfunding may explain what observers characterize as the "underpatenting" of ITRI's technologies, which exposes licensees to litigation.

---

[76]"Statute for Industrial Innovation Changes Thinking on Upgrading Industries," Council for Economic Planning and Development website, <*http://www.cepd.gov.tow/encontent/m1.aspx?No+0014260*>.

The Talent Shortage

Notwithstanding the fact that Taiwan enjoys an excellent science and engineering educational infrastructure, the country has long faced the prospect that it will not have enough skilled people to staff its laboratories, research centers, and high tech manufacturing businesses. Additional talent is available in Southeast Asia and mainland China, but Taiwanese investment in these regions has resulted in a major outflow of engineers and skilled managers from Taiwan itself to other parts of Asia, particularly China, where a substantial part of Taiwan's high tech production based is now located.[77] Although nearby Korea and Japan have large pools of skilled labor, attracting talent from these countries has proven difficult.

An executive at AU Optronics, a major producer of flat panel displays, said in a 2005 interview that his industry needed about 5,500 additional skilled workers per year, but that available talent equaled only about 70 percent of that total. Top universities like National Taiwan University, National Tsing Hua University and National Chiao Tung University between them were turning out 3-4,000 graduates per year and "cannot meet market demand."[78] A 2005 ITRI study found that the country's semiconductor industry would need 37,500 new skilled workers over the next three years but that the island's educational system could supply only 21,800 during that time frame.[79]

ITRI itself has been affected by the talent drain. Companies based in China and Singapore reportedly are prepared to pay five times ITRI levels to lure its researchers, one factor underlying ITRI's annual average manpower turnover of 10 percent.[80] In April 2011, ITRI Chairman Tsai Ching-yen warned that ITRI had been hurt by a "talent drain" crisis. He said that 38 ITRI-type organizations had been established in China in recent years and that "all these mainland institutions have set their sights on luring our researchers and engineers away." He noted that a 68-year old retired ITRI department chief

---

[77]"Taiwan's Wafer Fabs Need More Manpower," *Central News Agency* (March 26, 2011); "Industries Facing Shortage of Skilled Workers, Poll Finds," *Taipei Times* (April 8, 2005); "Taiwan Legislator Warns of High-Tech Exodus," *South China Morning Post* (July 5, 2000); "Taipei's Talent Exodus," *Time Asia* (May 21, 2000); "High-Tech Talent Flows to China," *Taipei Times* (March 25, 2002). In the fall of 2010 Taiwan's President Ma Ying-jeou reportedly ordered a study by the country's National Security Council (NSC) of the talent shortage. The study found that roughly 700,000 Taiwanese citizens resided in the Shanghai area of China alone. National Taiwan University Professor Chen Tain-jy, who participated in the NSC study, commented that "Taiwan is gradually losing its trained professionals … Singapore is using high pay to recruit Taiwan's medical and health care personnel, Hong Kong is luring away Taiwan's medical and health care personnel, and South Korea is targeting Taiwan's technical talent … How to keep talented and experienced people at home has become a serious issue that Taiwan cannot afford to ignore." "Taiwan Faces Serious Brain Drain Crisis," *Central New Agency* (April 18, 2011).

[78]"Industry Facing Shortage of Skilled Workers," *Taipei Times* (April 8, 2005).

[79]"Talent Shortage Escalates in Taiwan's High-Tech Industries." *Taiwan Economic News* (April 11, 2005).

[80]"Talk of the Day—Anti-Fat Cat Clause Hinders Manpower Recruitment," Central News Agency (September 12, 2011).

"was recently hired by China with a lucrative compensation package that included the promise of a villa in two years. The new Chinese employer "mainly hopes the former ITRI official can use his connections in Taiwan's high tech industry to help recruit more Taiwanese talent to work in China."[81]

Enterprise Scale

Taiwan's economy is dominated by small businesses, which account for over 97 percent of the country's enterprises and employ over 77 percent of the work force.[82] These companies typically have a small capital base and only a few employees, and have proven particularly vulnerable to the global economic downturn which began in 2008.[83] While many small and medium enterprises are innovative, they frequently lack the resources and skills needed to bring new products to the global market, to build brand recognition, and to respond to legal and regulatory challenges outside of Taiwan. One of the most important considerations underlying the creation of ITRI was the recognition by the government that Taiwan's small companies could not afford the equipment, training and other costs associated with advanced R&D.[84] While ITRI has partially offset this intrinsic disadvantage, turning ITRI research "into a successful product by SMEs has been a struggle."[85] Even small Taiwanese startups that have achieved technological breakthroughs have proven vulnerable to competitive challenges from large multinationals because they "lack critical mass."[86]

The government of Taiwan commonly seeks to offset the fact that most companies are too small to undertake expensive research by forming them into

---

[81]"Taiwan Faces Serious Brain Drain," *Central News Agency* (April 18, 2011). ITRI faced a manpower crisis beginning in March 2011 when Taiwan's Executive Yuan released an "anti-fat cat" decree to the effect that the monthly salary of high-caliber researchers at ITRI should not exceed the level earned by deputy-ministerial level officials, or about $5,500. The new rule was hampering recruiting efforts. ITRI researchers reportedly earn about half what they would be paid for comparable work in the United States. ITRI Chairman Tsai Ching-yen commented that "such rigid restrictions will only become more difficult for us to retain our researchers in some cutting edge fields." "Talk of the Day—Anti-Fat Cat Clause Hinders Manpower Recruitment," Central News Agency (September 12, 2011).

[82]"SME Competitiveness Seen Growing With High Management Efficiency," *Central News Agency* (October 3, 2008).

[83]"Taiwan's Suffering SMEs Cry for Help," *Taiwan Economic News* (December 3, 2008).

[84]Tommy Shih, "Scrutinizing on Economic Development Model—The Taiwanese Semiconductor Industry Revisited," Industrial Marketing and Purchasing Group (IMP Group) (2009), p. 10.

[85]Interview with Taiwanese semiconductor executive, Hsinchu, Taiwan, February 14, 2012.

[86]Interview with Taiwanese semiconductor executive, Hsinchu, Taiwan, February 14, 2012. In 1984, Dr. Bo-bo Wang, established Microtek, a Taiwanese enterprise which developed the first computer-affiliated scanner in the world. Roughly 20 similar Taiwanese companies entered the field, and for a time Taiwan was the world's leading producer of scanners. However, when major image-processing firms entered the field (Cannon, HP), Taiwanese scanner firms were driven from the market. T-J. Chen, "The Emergence of Hsinchu Science Park as an IT Cluster," in S. Yusuf, K. Nabeshihua and S. Yamashita (eds.), *Growing Industrial Clusters in Asia: Serendipity and Science* (Washington, D.C., The World Bank, 2008) p. 70

technology alliances. ITRI bases the alliances on its laboratories, which it uses to teach small firms to catch up with leading edge technologies sufficiently to enable them to perform contract work for the industrial chains of larger Taiwanese and foreign high tech enterprises. As a result, larger Taiwanese firms like Tantung, Acer and Mitac can "rely on hundreds of loosely affiliated domestic suppliers to which they can pass on an endless variety of low-margin, yet quite demanding manufacturing and design tasks."[87] ITRI commonly encourages SMEs in such alliances to divide up research tasks, to specialize and to avoid duplication of effort.[88] Organizing successful technology alliances is challenging and a number of them have failed due to divisions among the participating companies.[89]

The government has also established a panoply of programs to support SMEs. Beginning in 1996, MOEA's Small and Medium Enterprise Administration has utilized an SME Development Fund to promote the establishment of small business incubators in Taiwan.[90] An SME Credit Guarantee Fund guarantees 70-90 percent of commercial bank loans to SME's for purchase of machinery, plant construction, and modernization of equipment.[91] In 2011, Taiwan's Financial Supervisory Regulatory Commission set a target for domestic banks to offer domestic SMEs a total of $7.33 billion in

---

[87]Dieter Ernst, "What Permits David to Defeat Goliath? The Taiwanese Model in the Computer Industry," (Alborg University, International Business Economics Research Paper Series, 1997) p. 9.

[88]"We bring them together in a formal alliance, even though they are competitors. So we find a way to divide the work. Company A makes equipment A. B makes B. Each company must see a role for itself. It's a huge task, but doable." Interview with Dr. John Chen, Director, ITRI Display Technology Center, Hsinchu, Taiwan, February 14, 2012. In 2011, a group which included ITRI and 19 other organizations formed a wide bandgap electronic devices industry alliance, which would address the manufacture of silicon carbide (SiC) and gallium nitride (GaN) compound semiconductors for end uses such as LEDs and electric cars. Alliance members were "divided into groups to develop substrate materials, epitaxy wafers, devices, modules and inspection technologies according to their professions. They [planned] to complete packaging and test verification on 600-volt devices by the end of the year." "Wide Bandgap Alliance Formed in Taiwan," *Taiwan Economic News* (March 9, 2011). Jonney Shih, Chairman of ASUSTEK, a maker of computer hardware, observed in a 2011 interview that the standardization of the personal computer industry constituted a "good opportunity" for Taiwan because it permitted the rationalization of small businesses into aligned groups of specialty companies. "Because of the standardization, you can divide [the products into sections]. You can do just the motherboard. You can do the power supply. You can do the graphics card. I think that gave the Taiwan industry a very good opportunity." Interview with Jonney Shih, "Taiwanese IT Pioneers: Jonney (Chang-tang) Shih," recorded February 15, 2011 (Computer History Museum, 2011) p. 16.

[89]In 1993, ITRI promoted formation of a technology alliance to develop thin film transistor-liquid crystal display (TFT-LCD) technology. The alliance sought to raise $500 million to build a TFT-LCD production facility. However, the members disagreed over which technology to use, and the alliance was disbanded. Greg Linden, Jeffrey Hurt and Stefanie Lenway, "Advanced Displays in Taiwan and Korea" (Working Paper 109, BRIE, December 1997).

[90]Chia-Chi Sun, Grace T.R. Lin, and Gwo-Hsiung Tzeng, "The Evaluation of Cluster Policy by Fuzzy MCDM; Empirical Evidence from Hsinchu Science Park," *Expert Systems with Applications* (2009), p. 11903.

[91]"SME Credit Guarantee Fund Raises Ceiling on Guarantee," *Taiwan Economic News* (November 7, 2008). This program reportedly benefits 150,000 SMEs per year.

2012; banks with "good lending performance" would be rewarded "by prioritizing approval of their applications to set up branches in China or overseas."[92] SMEs also can qualify for management and technical consulting support.

The Small Business Innovation Research Program (SBIR) was established by MOEA in 1999 to provide matching fund support to local start-up companies pursuing innovative technological research with respect to improving existing technologies or developing radical new ideas. The program is funded at a level of about $30 million. The funds can be used to perform research and to collaborate with research organizations like ITRI in joint research projects. The SBIR organization is comprised of a working group which picks promising research proposals to be reviewed by an examination committee consisting mainly of university professors. Individual SBIR grants can be up to amounts of $340,000.[93] Taiwan's SBIR program has come under considerable domestic criticism for being overly bureaucratic.[94]

Intellectual Property Protection

A major challenge confronting Taiwan's high tech industries is the increasingly widespread phenomena of patent litigation used as a business strategy to block new industry entrants.[95] The *Economist* reporting on an infringement action by Apple against HTC, a Taiwanese maker of smartphones, commented in December 2011 that—

---

[92]"SME Loans to Taiwan Hist 5 Year High of NT $4.05 Trillion as of October 2011," *Taiwan Economic News* (December 29, 2011).

[93]Christina Chen, et. al., *Excellent Government on a Far-East Silicon Island* (December 20009) p. 21.

[94]Paul Kuo, founder of a software startup, tried to apply for SBIR support but "the massive amount of paperwork was overwhelming" for his small and understaffed firm. Even Chung, CEO of Xinosys Co., a winner of the 2011 Asia Red Herring Award honoring innovative high tech companies, was an applicant to the SBIR program but said that the "panel of judges seemed out of touch with reality." Jamie C. Lin, co-founder of appWorks Ventures, a company fostering Internet startups noted in a discussion of SBIR that "start-up companies need to consistently change direction, to adjust to the needs of their users. The government's set-a-goal-then-execute-it style of funding programs contradicts the logic of entrepreneurship. The result is a startup that successfully executes a bad plan." A government spokesperson responded to this criticism that "the SBIR program's rigorous evaluation process ensures taxpayer's money is well spent." "Taiwan—A Growing Market for Startup Companies," *Central News Agency* (November 27, 2011).

[95]Roger Borovoy, a lawyer with Fish & Richardson, P.C., a major U.S. patent litigation firm commented in a presentation at ITRI College in 2006 that intellectual property "can not only be sued as a defensive tool but also an offensive tool by businesses barring competitors from using similar patents." "Taiwan's Patent Commercialization Rate Far Below Global Average," *Taiwan Economic News* (August 2, 2006). Morris Chang, CEO of TSMC, observed in 2008 that the light-emitting diode (LED) industry was "an industry rife with pitfalls and the biggest one involves patents," a reason his company was unlikely to invest in the sector. "Epistar, Formosa, Epitaxy Settle Patent Dispute out of Court," *Taiwan Economic News* (August 8, 2008).

> *Apple's victory is only the latest episode in a fierce war in which just about everyone you can think of seems to be suing everyone else for patent infringement ... "This really is the first global patent war," says Joshua Walker, chairman of Lex Machina, which compiles and analyzes data on intellectual property litigation.*[96]

The plaintiffs are typically "big players in their fields" who can afford the legal fees associated with securing patents and patent litigation whereas the defendant companies are often small Taiwanese companies lacking comparable resources.[97] Morris Chang, the founder of TSMC, recalled in a 2007 interview that the decision to establish a company devoted purely to manufacturing itself was driven in part by the recognition that Taiwan had "no strength in IP" and that manufacturing was "the part that was least vulnerable to IP attacks from other companies. Most IP disputes were about circuit designs."[98]

The increasing complexity of the technologies being commercialized by Taiwanese companies is amplifying their exposure to patent claims. Hand-held devices are now comprised of huge bundles of software and hardware—according to one recent estimate there are now 250,000 patents relevant to a single smartphone.[99] Payment of royalties pursuant to licensing agreements is an alternative to litigation, but also entails increased costs. As of 2008, Taiwanese companies were paying over $2.3 billion annually in royalties to foreign patent owners.[100]

Dr. Da Hsuan Feng, Senior Vice President of National Tsing Hua University observes that "one of the weak points of the entire Taiwan island is a shortage of IP legal competency." He indicates that a need exists for the universities and ITRI to combine and integrate their IP and to market it as a whole—"you can't commercialize with one patent."[101] A number of professors from NCTU and NTHU make the point that Taiwan's patent pools, to the extent they exist, are insufficiently broad and should be comprised of larger numbers of patents.

The best defense for a company against patent lawsuits is ownership of comparable patents, preferably bundled into packages of multiple patents

---

[96]"World Patent War 1.0," *The Economist* (December 19, 2011).

[97]ITRI notes on its website that "Taiwan businesses have been confronting international conglomerates tens or hundreds of times their size in patent disputes that are entangled in an endless cycle of business competition, patent infringement litigation, settlement negotiations, and cross-licensing." ITRI, "Pioneer for Taiwan's Brokered Patent Transactions—ITRI," *<http://www.itri.org.tw/eng/econtent/references/references02_05.asp>*.

[98]SEMI Oral History Interview, *Morris Chang* (August 24, 2007, Taipei, Taiwan).

[99]"Estimate by RPX, San Francisco-based patent aggregator and licensor, in "Mobile Phone Makers Wage Was to Protect Their Patents," *BBC News* (October 23, 2011).

[100]"Alliance Formed to Help ICT Firms with Patent Issues," *Taiwan Economic News* (August 19, 2009).

[101]Interview with Dr. Hsuan Feng, National Tsing Hua University, Hsinchu, Taiwan, February 16, 2012.

covering all aspects of a new technology. In 2005, ITRI adopted a practice long under way in the United States and Europe of auctioning off bundles of patents on an exclusive license basis.[102]

ITRI organizes Taiwanese industries, often comprised of small and medium enterprises, into "alliances" to exert collective strength "in talks with international players over patent licensing," securing better terms and lower royalty rates.[103] ITRI defends its own patents in periodic litigation and is forming an infrastructure to support Taiwanese companies' efforts to manage and defend their intellectual property.[104] In 2011, ITRI announced that it would form an IP management company, "IP Bank," to provide assistance to Taiwanese firms engaged in legal battles over intellectual property.[105] A U.S. patent expert familiar with the IP Bank plan commented that the organization's primary role would be "defensive" in character, identifying and preventing litigation and averting it by helping companies acquire patent portfolios that protect them from lawsuits.[106]

---

[102]ITRI has a large portfolio of intellectual property—as of early 2012 it held 14,571 patents. In 2009 alone, it was awarded 397 U.S. patents, the largest number for any research organization in the world. Jackson Chang and X.L. Kao, "ITRI Ranked as Leading Patent Recipient Among Research Institutes," *Central News Agency* (February 12, 2010).

[103]In 2008, ITRI organized Taiwanese manufacturers of compact discs, DVD players and chips into the "Blu-ray DVD alliance," which represented the companies in negotiations with foreign firms over licensing and collaboration deals. "Taiwan Organizes Blue Ray Alliance to Strengthen License Bargaining Chip," *Taiwan Economic News* (March 25, 2008). In 2009, ITRI and the Taipei Computer Association announced the formation of an "information industry mutual-benefit patent alliance" to create "a large depot of patents for domestic information and communication technology, to help companies to plan patent strategies and to secure the "assistance of experts in patent-infringement law." "Alliance Formed to Help ICT Firms With Patent Issues," *Taiwan Economic News* (August 19, 2009).

[104]In 2010, ITRI filed four patent infringement actions in a U.S. District Court in Texas against Korea's LG Electronics Inc., alleging infringement of 22 ITRI patents on mobile phones, LCD televisions, Blu-ray disc players and air conditioners. "Taiwan's ITRI Takes LG to Court Claiming Patent Infringement," *Taipei Times Online* (December 2, 2010).

[105]"Seeking Government Aid is Futile," *Taipei Times Online* (September 5, 2011); Jeffrey Wu, "Good Idea to St Up IP Bank in Taiwan: Expert," *Central News Agency* (September 4, 2011).

[106]"Good Idea to Set Up IP Bank in Taiwan: Expert," *Central New Agency* (September 4, 2011). ITRI officials said that the new entity would seek to strengthen local high-tech businesses' defenses against litigation by obtaining international patent rights, filing annulment suits against competitors' patents, and shifting the battleground from the United States to China. ITRI officials note that about 30 percent of patents challenged in patent infringement suits in the United States were found to be invalid, and that it could be less expensive to file annulment suits than to buy patents. With respect to shifting the battleground to China, they commented that legal action had become a business strategy "used by corporate giants to expand market share and stifle the advances of competitors." Taiwanese companies have heavily invested in China and patents in that country cost about one-tenth the cost in the United States. "Patents obtained by Taiwanese businesses in China would then make it possible to block the sale of competitors' products that violate those patents helping improve the market share of the Taiwanese companies in China and reinforce their global status." "Taiwan Drafts Strategy for IT Patent War," *Central New Agency* (August 28, 2011).

SECTORAL CASE STUDIES

The government of Taiwan has promoted industrial development since 1953 through multi-year development plans targeting what were viewed as strategic sectors. The Third Four Year Plan (1961-64) called for the development of heavy industry, chemicals and electronics sectors, and major industrial growth subsequently occurred in steel, electronics, petrochemicals and shipbuilding.[107] The onset of the 1973-74 oil crisis and high inflation brought economic growth nearly to a halt, and led the government to seek an industrial strategy placing more emphasis on non-energy intensive, high technology industries. Arguably, the formation of ITRI in 1973 as part of this broader effort represents the key event in Taiwan's modern industrial history.[108] ITRI's first major sectoral development effort, in semiconductors, established the foundation for promotion of related sectors in which the country's growing competency in microelectronics could be leveraged, including displays, lighting, computers, telecommunications and photovoltaics.[109, 110]

Semiconductors

In the early 1970s the island's microelectronics sector consisted solely of offshore assembly and test operations of foreign manufacturers and one small local firm making bipolar transistors. ITRI founder Y.S. Sun worked with a network of Taiwanese and Chinese-American engineers to formulate a developmental strategy for semiconductors. Sun created ITRI in 1973 as an arm of MOEA charged with conducting applied industrial research and providing technical support to domestic companies. Within ITRI an organization was set up in 1974 to develop semiconductor technology, the Electronics Research and Service Organization (ERSO).

---

[107]The institutional framework for Taiwan's economic development has been established through a succession of industrial policy statutes providing for incentives to designated sectors. The first of these, the *Statute for the Encouragement of Investment* (1960-1990) authorized tax, fiscal and duty drawback incentives for categories of industries that were specified by the Executive Yuan (cabinet) through periodic decrees. This measure has been superseded by the *Statute for Upgrading Industries* (1991-2010) and the current *Statute for Industrial Innovation.* The latest promotional statute is distinguishable from its predecessors in its emphasis or innovation, its goal of shifting Taiwan from hardware manufacturing toward "soft power" (biotechnology, industrial services and high value-added services), and the promotion of national brands.

[108]See generally, Robert Wade, *Governing the Market: Economic Theory and the Role of Government in East Asian Industrialization* (Princeton and Oxford: Princeton University Press, 1990) pp. 75-112.

[109]Chun-yen Chang, who founded Taiwan's first semiconductor research center at National Chiao Tung University, observed in a 2011 interview that "[Y]ou can see that all the Taiwan high tech industry was originally from … the success of semiconductor industries in Taiwan. We spun off [from the semiconductors] to LCD displays and then to the computer business." Interview with Chun-yen Chang, "Taiwanese IT Pioneers: Chun-yen Chang," recorded February 16, 2011 (Computer History Museum, 2011) p. 18.

[110]"I'm Willing to Start from Scratch," *Commonwealth* (June 18, 2009).

Technology from the United States

ITRI charged ERSO with developing complementary metal oxide semiconductor (CMOS) technology, which it saw as key to the production of components for application in consumer electronic products. Controversial at the time, the choice of CMOS technology as opposed to then-prevalent NMOS method proved to be an important strategic move because CMOS' lower cost and power consumption made it attractive for use in personal computers, workstations, and other product sectors which would grow rapidly in the 1980s.[111]

ITRI established a partnership with RCA of the United States in 1975. RCA's semiconductor manufacturing technology was obsolete, but RCA agreed to supply Taiwan with complete production technology, including complementary metal oxide semiconductor (CMOS) process design, product specification and testing technology, and manufacturing management and cost accounting know-how as well as a 6-12 month training program for ITRI engineers. The fact that RCA's technology was outdated had advantages as well as drawbacks—most importantly, the mature process technology was already thoroughly tested and proven.[112] RCA supplied CMOS technology and trained a group of 37 Taiwanese students in the fundamentals of semiconductor manufacturing, the "RCA 37." "In certain respects the training of related ITRI staff at RCA was the decisive moment in the [Taiwanese] semiconductor industry's history."[113] Out of the RCA 37, "virtually the entire senior echelons of the subsequent semiconductor industry in Taiwan [were] formed."[114]

---

[111]CMOS is a form of semiconductor technology that performs electronic functions more slowly than several alternatives, but involves lower power consumption and waste heat production and lower fabrication costs. At the time Taiwan obtained the technology CMOS devices accounted for about 10 percent of the semiconductor market. By the end of that decade, according to one estimate, CMOS devices accounted for roughly half of the semiconductor market. Korea, which had initially pursued an alternative technology in promoting its semiconductor industry, later converted to CMOS. Chiang, "Management of Technology Programs" (1990) op. cit. p. 11. Ding-Yuan Yang, who established ERSO, commented later "Thank the Heavens for having blessed Taiwan that seized the chance to develop the technology of CMOS technology from such an early stage." Interview with Ding-Yuan Yang, "Taiwanese IT Pioneers: D.Y. (Ding-Yuan) Yang," recorded February 23, 2011 (Computer History Museum, 2011) p. 7.

[112]Shih, "Taiwanese Semiconductor Industry Revisited" (2009) op. cit. p.13. RCA supplied 7.0 micron complementary metal oxide (CMOS) technology and the product specifications and design and testing technology, as well as assistance in constructing a pilot integrated circuit manufacturing plant and suggestions for equipment specifications. Chang, Shih and Hsu, *Technovation* (1994) op. cit. p. 165.

[113]Interview with K.S. Pua, Phison Electronics Corp., Hsinchu, Taiwan, February 14, 2012. Morris Chang, the founder of TSMC, pointed out in 2007 that although the RCA deal was important, at the time Taiwan acquired RCA semiconductor technology. "RCA was not a first tier semiconductor company," and that Taiwan subsequently fell further behind industry leaders like Intel and Texas Instruments, who "Set the pace, and ITRI couldn't keep up the pace," because it lacked a real commercial base and volume sales. SEMI Oral History Interview *Morris Chang* (August 24, 2007).

[114]Mathews, "Hsinchu Model" (2010) op. cit. p. 18.

Ding-Yuan Yang was one of the founders of ERSO and led the original team of trainees to RCA. Wang had studied physics and chemistry at Princeton University, where he was struck by the nature of the education process. He recalled later that—

> *Princeton attached importance to experiments. I've talked about this experience many times. I took 50 hours non-credit in a machine shop making things. Some of my classmates who majored in chemistry were required to learn glassblowing. This is when I knew that when you get to do truly advanced and sophisticated research, you have to be able to design and make all the equipment yourself, instead of purchasing existing equipment, machines and tools.*[115]

The ITRI team, which included a number of other Princeton graduates, took this approach toward the RCA technology transfer, seeking to master every aspect of the manufacturing process and applying and proving skills in a factory environment. Robert Tsao, then an ITRI engineer who participated in the project, said later that "the purpose of this project was not only to conduct pure research but also to transfer over an industrial technology. Here, the phrase 'industrial technology' indicated not only a transfer of the technology itself but also the ability to carry out economically efficient mass production of goods."[116] Ding-Yuan Yang recalled that the four project leaders learned not only—

> *engineering and manufacturing, but the four of us also learned about designing, testing, quality control, procurement accounting and database management. It was a holistic and complete packaging of training. If we only transferred these techniques in a lab, we wouldn't have been able to attain industrial production. [The idea was] to make a demonstration plant, in the transferring of the technology. You have to get into a factory, bring in some products and transfer some of them out to verify the effectiveness of the transfer of manufacturing technology. Only when you reach a certain yield rate can the transfer be considered valid.*[117]

---

[115]Interview with Ding-Yuan Yang, "Taiwanese IT Pioneers: D.Y. (Ding-Yuan) Yang, recorded February 23, 2011 (Computer History Museum, 2011).

[116]Interview with Robert Tsao, "Taiwanese IT Pioneers: Robert H.C. Tsao," recorded February 17, 2011 (Computer History Museum, 2011) p. 3.

[117]Interview with Ding-Yuan Yang (2011) op. cit. p. 8. The ITRI approach placed broader demands on the research team than a simple transfer of manufacturing techniques. Ding-Hua Hu, one of the ERSO team members, recalled that "There were a lot of problems involved. For example, the supply of gases. It wasn't something you could just buy off the street and all different kinds of chemical raw materials. So we had to digest the transferred RCA technology and we did that very well." Interview with Ding-Hua Hu, "Taiwanese IT Pioneers: Ding-Hua Hu," recorded February 10, 2011 (Computer History Museum, 2011) p. 16. The RCA deal did not include some elements in the

When ITRI's trainees returned to Taiwan, a pilot facility for manufacturing integrated circuits was established in ERSO to stimulate production and promote R&D, the Integrated Circuit Model Factory (1977). The pilot line produced its first integrated circuits in 1977. The operation of the pilot line served as practical confirmation for the trainees who had served at RCA and the "hidden knowledge" they had gained at RCA was passed on to other trainees who had not been picked for the original RCA mission. The pilot line enabled the RCA trainees to experiment with the actual manufacturing of semiconductors and pass the know-how to Taiwanese manufacturers. Ding-Yuan Yang commented that—

> *Now looking back, setting up an IC demonstration plant in a research institution was something truly unique in the world. Normally a private plant is set up, and they transfer technologies by themselves. Using the power of the government to establish a demonstration plant in a research institution, to plant it as a seed to spread it out, was a very unique method.*

ERSO's pilot line ramped up its production, and volume and quality soon surpassed that of RCA's devices. In the technology transfer agreement, RCA had guaranteed that that the yield rate would be at least 17 percent, but the yield rate of the ERSO demonstration plant was 70 to 80 percent—a surprise to RCA. Robert Tsao attributed this success to use of new equipment and the high quality of the production team:

> *Those who were hired by this project were all very outstanding, such as Ding-Yuan Yang, Chintay Shih and Jian-Jun Chang, each of whom had doctoral degrees from Princeton University. Even in the semiconductor industry, it would be almost a waste of their great talents to use them on manufacturing alone. Therefore, it was natural that the yield rate was high.*[118]

---

industrial chain and these were the subject of subsequent ITRI initiatives. Hu recollects that "They [RCA] didn't teach us photo mask technology. So we worked with an overseas Chinese, named Stephen Lin. We signed a contract with him and introduced his company, IMR, to acquire that technology and that was how Taiwan got a number of listed photo mask companies. In addition, RCA was not expert in quality assurance, so we got help through TAC to get in touch with other overseas Chinese, then we sent L.H. Chiu along with another person over to HP to learn about quality and reliability … and we found a Chinese guy, named William Mao in the testing equipment … Then Chin-Chu Chang went over there with a small group." Interview with Ding-Hua Hu (2011) op. cit. p. 16.

[118]Interview with Robert Tsao (2011) op. cit. p. 5. Ding-Hua Hu made similar observations, attributing the high yield rate to the work force operating the demonstration plant. "The factory couldn't be completely automatic, so some of the facilities required people's judgment to make

ITRI continued to purchase new semiconductor manufacturing equipment, dispatched engineers abroad for training, and invited foreign specialists to participate in its development efforts.[119]

Creation of UMC

Despite investment incentives and technical support from ITRI, Taiwanese companies remained reluctant to invest in the semiconductor business because of the perceived risk. In response the government created Taiwan's first major semiconductor company in 1980, United Microelectronics Corp. (UMC). UMC in effect spun itself off from ITRI, taking with it a substantial proportion of ERSO's equipment and 31 of its personnel.[120] They included Robert Tsao, a vice-director of ERSO and one of the RCA 37 who eventually became UMC's chairman. The government took a 44 percent equity stake. ITRI provided UMC with semiconductor mask design services, relieving the company of the need to send its IC designs to foreign mask making firms.[121] Dr. Chintay Shih, who served at ITRI between 1976 and 2003, recalls that "we helped them [UMC] to design and build their first site in the Science Park."[122] ITRI lined up UMC's first customer, a Taiwanese watchmaking company.[123] UMC began operations in 1982 and grew rapidly to become one of the world's leading semiconductor producers.[124]

---

minor detail adjustments. Every piece of equipment had their own characteristics so in other words, people were the greatest invention, and God is the greatest creator—the truth was that you could count them as an integral part of the system in the place, at that moment, and at that time. Because you saw them as a link. Otherwise if the machine acted up, then there would be a discrepancy in the control. In that kind of situation, people were the most important asset. And when it came to people, discipline was imperative." Interview with Ding-Hua Hu (2011) op. cit. p. 17.

[119]Between 1979 and 1983, ITRI's process technology was upgraded from 7.0 to 3.0 microns design rules, computer-aided design programs were introduced, and an independent semiconductor mask design capability was acquired with assistance from California-based International Materials Research (IMR). Between 1983 and 1988, process technology was further upgraded to 1.0 micron design rules, the pilot line was converted to production of very large scale integration (VLS1) devices, and a Common Design Center was established to develop and disseminate application-specific integrated circuit (ASIC) technology. Chang, Shih and Hsu, *Technovation* (1994), op. cit. p. 166. VLSI refers to devices with over 100,000 transistors on a single chip.

[120]Among other personnel, ITRI transferred to UMC a manufacturing supervisor, a testing manager, a sales manager and a sales supervisor, a quality control manager and a circuit design manager. ITRI also trained UMC employees in technologies which included process and equipment engineering, quality assurance, testing, industrial engineering and production and materials control. Chang, Shih and Hsu, *Technovation* (1994) op. cit. p. 167.

[121]Chang, Shih and Hsu, *Technovation* (1994) op. cit. p. 167.

[122]Interview with Chintay Shi, "Taiwanese IT Pioneers: Chintay Shih," recorded February 21, 2011 (Computer History Museum, 2011) p. 13.

[123]Shih, "Taiwanese Semiconductor Industry revisited" (2009) op. cit. p. 14.

[124]By 2000 UMC had achieved such proficiency in semiconductor manufacturing that it entered into a manufacturing joint venture with Hitachi of Japan in which UMC actually transferred manufacturing know-how to Hitachi, a dramatic reversal from prior decades when Japanese firms shunned relationships with Taiwanese firms, which they regarded as technologically inferior.

Promoting Design Capability

In 1985, ERSO established the Common Design Center (CDC), a computer-aided semiconductor design facility intended to encourage start-up design firms. ERSO also provided computer-aided design tools and logic cell libraries to nine Taiwanese universities to encourage the establishment of IC design programs. ERSO published IC design manuals and sponsored training classes on computer-aided IC design. Between 1983 and 1990 it spun off more than a dozen firms with IC design capability. It transferred IC design specifications to Syntek Semiconductor Corp., UMC, and Silicon Integrated System Company.[125] In 1997, ITRI spun off Prolific Technology, Inc., an ASIC design house which in 2004 was named "most competitive fabless company in Taiwan" by Merrill Lynch.[126]

ITRI's most effective single initiative in promoting the IC design industry, however, was the creation of TSMC, the world's first semiconductor foundry in 1987.

Creation of TSMC

Around 1984, Taiwan's Premier Yu Kuo-hwa offered Morris Chang—who had moved from Texas Instruments to a position as President of General Instruments—a job as head of ITRI, telling Chang "that he particularly wanted to use my ability to transfer technology from just research results to economic benefits for Taiwan industry."[127] Taiwan was facing difficulty in growing its integrated circuit design sector further because of the lack of local manufacturing capability. The idea of a "pure play" semiconductor foundry had been discussed in industry circles for years—that is, an enterprise which provides contract manufacturing services to IC designers but does not sell its own IC products—but the idea was generally regarded as commercially unfeasible.[128] With substantial government support, Chang led the establishment of the world's first semiconductor foundry, Taiwan

---

"Hitachi, UMC Join for Rapid Production of 300 mm Wafers," *Kagaku Kogyo Nippon* (December 28, 1990); "Hitachi-UMC Tie-up to Generate Test Cases that Will Forecast the Future of LSI Technology and Business," *Nikkei Microdevices* (February 2000).

[125] Huang, "Cradle of Technology" (2006) op. cit. p. 41.

[126] "Taiwan Design House Innovates to Succeed," *EE Times Asia* (January 1, 2006).

[127] "Oral History of Morris Chang," recorded August 24, 2007 (Computer History Museum, 2011) p. 9.

[128] Chang was one of a number of Taiwanese leaders who had read the works of Carver Mead, an influential U.S. computer scientist who argued that semiconductor design could be separated from the manufacturing process. Mead used the analogy of the printing business and postulated that there could be a ":silicon press" to fabricate semiconductors for a service fee for design companies. Coincidentally in the early 1980s, ITRI President Hsian-Chi Fang had a daughter who was one of Mead's students, and who suggested that Mead be invited to Taiwan. Mead traveled to Taiwan and his ideas had a major impact at ITRI. Former ITRI head Chintay Shih recalls that "I was thrilled first time I heard about Mead's concept." Interview with Chintay Shih (2011) op. cit. pp. 14-15.

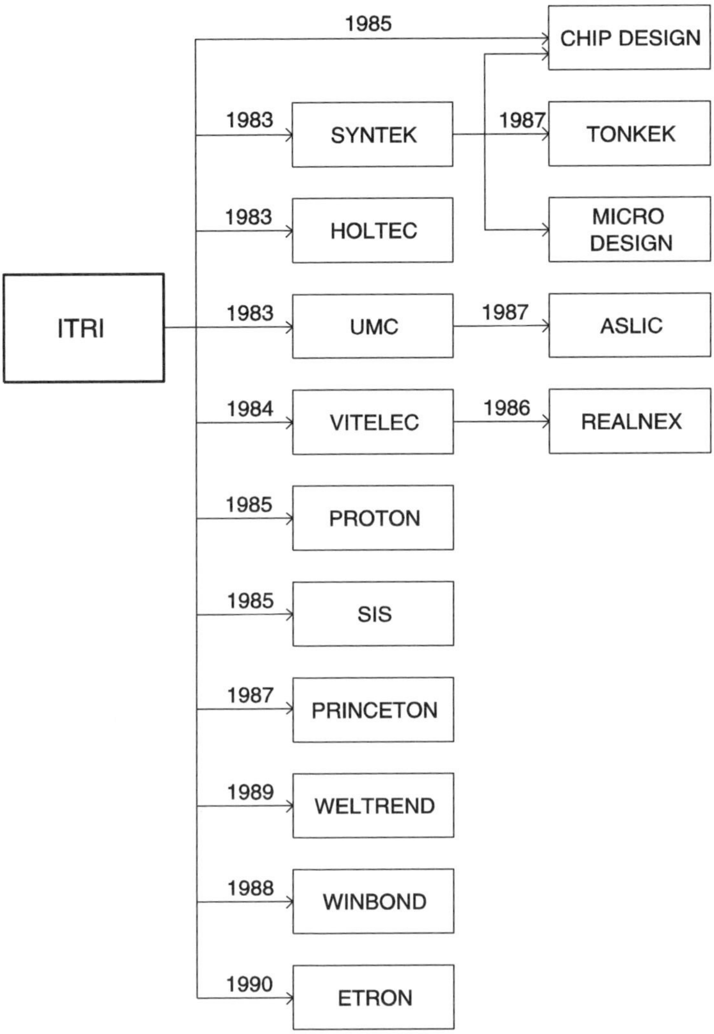

FIGURE APP-A3-5 ITRI's role in promoting IC design.
SOURCE: Chung-Yuan Liu, ITRI, "Government's Role in Developing a High Tech Industry: The Case of Taiwan's Semiconductor Industry," *Technovation*, 13(5), 1983.

Semiconductor Manufacturing Corporation (TSMC), becoming its first CEO in 1987. "The idea came from everybody's brainstorming."[129]

[129]Interview with Chintay Shih (2011) op. cit. p. 18.

Like UMC, TSMC was spun-off by ERSO, which in 1986 set up a six inch VLSI preparatory manufacturing plant and in 1987 transferred it to TSMC. Robert Tsao, the founder of UMC, recalls that ITRI provided TSMC with what was effectively a "free foundry" to support its start-up. In a 2011 oral history interview, he commented:

> *So after [ITRI] spending five years and 100 million U.S. dollars on construction, the second demonstration plant was built, and about 500 to 600 trainees were all sublet to TSMC. Therefore, TSMC had foundry at first and they only needed to spend two million U.S. dollars on subletting the foundry. Apart from this ITRI even gave TSMC 7 million U.S. dollars to subsidize the cost of subletting the foundry ... In other words the first 3 and a half years of running TSMC was for free and it went very well under the protection of Ministry of Economic Affairs.*[130]

Morris Chang, brought with him 130 ERSO professionals and process technology from ERSO for manufacturing six inch semiconductor wafers. TSMC's initial shareholders were the government (40 percent), Philips of the Netherlands (26.6 percent) and employees and local private investors (19 percent). The government's share of TSMC's capital was funded by the Development Fund of the Executive Yuan, a discretionary development fund at the disposal of Taiwan's Cabinet administered by Finance Ministry.[131]

Philips' participation in TSMC as an equity investor and technology partner was instrumental in the company's success. Philips licensed key manufacturing process technology to TSMC.[132] Philips was a captive customer for TSMC and provided an outlet for its production at a time when the stability of demand was perceived as a major risk factor for any company pursuing a pure play foundry business model.[133] The fact that a world-class company like Philips, one of the largest electronics firms on the globe, was sourcing semiconductors from TSMC drew the attention of companies like Intel and Texas Instruments, which subsequently became TSMC customers. Philips was a multinational company with practical global business experience and experience in intellectual property management. Philips' IP umbrella, based on a web of

---

[130]Interview with Robert Tsao, "Taiwanese IT Pioneers; Robert H.C. Taso," recorded February 17, 2011 (Computer History Museum) p. 9.

[131]The Development Fund also invested in semiconductor firms Powerchip and Vanguard. ITRI's venture capital subsidiary, ITIC, did not invest in TSMC because "we had no money" at the time. Interview with Ching-Jiunn Chang, Vice President, ITIC, Taipei, Taiwan, February 13, 2012.

[132]Shih, "Taiwanese Semiconductor Industry Revisited (2009) op. cit., p. 14.

[133]Shih, "Taiwan's Semiconductor Industry Revisited" (2009) op. cit. p. 14.

cross-licensing agreements transferred to TSMC, protected the company from patent lawsuits during its initial years of operation.[134]

The nature of TSMC's business served to enhance its technological capabilities. Initially semiconductor firms such as VLSI Technologies passed on technologies to TSMC because without them TSMC could not have fabricated their devices. Customers also provided feedback which was used to refine TSMC's fabrication methods. The fact that TSMC needed constantly to shift its production to different kinds of devices contributed to the company's cumulative manufacturing learning.[135] TMC grew rapidly, largely by absorbing other ITRI spin-off companies, and became one of the most efficient and versatile semiconductor manufacturers in the world.

The formation of TSMC as a pure-play foundry, as well as UMC's subsequent conversion into a foundry, changed the entire business model of the global semiconductor industry in a manner that worked to Taiwan's advantage. Before the advent of foundries the semiconductor industry was vertically integrated, with manufacturing capital costs constituting a huge barrier to entry. The emergence of foundries enabled "fabless" design firms with no factories of their own—or associated capital costs—to challenge industry leaders with innovative designs involving modest investment costs. Numerous Taiwanese IC design firms sprouted up, utilized TSMC and UMC foundry services, and reaped enormous profits. A large number of expatriate Taiwanese "engineers-turned-entrepreneurs returned from Silicon Valley to drive this process forward.[136]

Winbond and Vanguard

Winbond Electronics Corporation was formed in 1987 to produce semiconductors memory devices. Most of its original personnel came from ERSO, which also supplied the company with licensed technology.[137] Winbond has subsequently become Taiwan's largest maker of branded integrated circuits.[138] Subsequently "a new company started almost every year, and many

---

[134]Interview with Chintay Shih, "Taiwanese IT Pioneers: Chintay Shih," recorded February 21, 2011 (Computer History Museum, 2011), p. 18. Interview with TSMC executive, Hsinchu, Taiwan, February 14, 2012; Shih, "Taiwan's Semiconductor Industry Revisited" (2009) op. cit. p. 14. In 2007, TSMC disclosed that Philips would sell its stake in the company in increments through 2010 as part of its strategic decision to exit the semiconductor business. "Philips to Offload US $8.5 bn Holdings in BMC by 2010," *Taipei Times* (March 10, 2007).

[135]Douglas B. Fuller, *Globalization for Nation Building: Industrial Policy for High-Technology Products in Taiwan,* (MIT Working Paper MIT-IPC-02-002, January 2002), p. 8.

[136]By 2001, an estimated 4,292 Taiwanese engineers returned from overseas positions were working in Hsinchu Science Park. Chen, "Emergence of Hsinchu Science Park" (2008) op. cit. p. 78.

[137]Winbond was headed by Dr. Ding-Yuan Yang, who had served in various positions in ERSO including analyst of ERSO's business ventures. According to Dr. Yang about 200 personnel were transferred from ERSO to Winbond, mainly from the IC demonstration plant, and Winbond "paid a lot of money to ERSO" to license technologies. Interview with Ding-Yuan Yang (2011) op. cit. p. 30.

[138]Mathews, "Hsinchu Model" (2010) op. cit., p. 20.

were spinoffs from ERSO."[139] Vanguard International Semiconductor Corporation (VIS) was spun off of ITRI in 1994 with TSMC and the Development Fund of the Executive Yuan as major investors. Three hundred and ten employees left positions at ITRI to join the new company.[140] VIS was originally established to manufacture dynamic random access memories (DRAMs) but in the early 2000s transformed itself into a pure-play foundry on the TSMC model.

Taiwan Mask Corporation

By the late 1980s Taiwan had a significant number of IC design houses producing application-specific integrated circuits (ASICs), devices which are tailored to specific end-use applications. ERSO was supplying some of these firms with semiconductor masks from its internal operation. In 1988 ERSO spun off the Taiwan Mask Corporation (TMC), transferring its own personnel, equipment and technology to the new company. ITRI concluded 10-year technology transfer and development agreement with TMC.[141]

Taisel

In 1994, a joint venture was formed by the government-owned China Steel Corporation and the U.S. company MMC, Taiwan International Standard Electronics, Ltd. (Taisel), to produce polished and epitaxial wafers for integrated circuit fabrication. "The government's effort to create a vertically integrated industry was evident."[142]

Asia Pacific Microsystems

Asia Pacific Microsystems Co (APM) was founded in 2001 by a team led by ITRI Executive Vice President Lin Min-shyong. APM took over a five inch wafer foundry for making CMOS chips from Winbond located in HSP and converted it into a six inch foundry for micro-electric mechanical systems (MEMS) devices in 2002. MEMS devices include sensors, microstructures and actuators which are used in consumer, medical, automotive, industrial, and telecommunications applications.[143]

---

[139]Genda J. Hu, Taiwan Semiconductor Manufacturing Company, "Government-Industry Partnerships in Taiwan," in Charles W. Wessner (ed.), *Securing the Future: Regional and National Programs to Support the Semiconductor Industry* (Washington, D.C.: National Academies Press, 2003) p. 154.

[140]Chen-Dong Tso, "State-Technologist Nexus in Taiwan's High-Tech Policymaking Semiconductor and Wireless Communications Industries," *Journal of East Asian Studies* (May 2004).

[141]Chang, Shih and Hsu, *Technovation* (1994) op. cit. p. 169.

[142]Chang, "Emergence of Hsinchu Science Park" (2008) op. cit. p. 76.

[143]"APM Opens Taiwan's First MEMS Wafer Foundry," *Taiwan Economic News* (November 21, 2002).

Technology Transfer

In addition to the creation of spin-off companies, ITRI contributed substantially to the development of the semiconductor industry by transferring technology to companies and undertaking joint R&D with several firms. During the 1980s ITRI transferred semiconductor process technology to Hualon Microelectronics and Advanced Device Technology Inc. (ADT).[144] In addition to ERSO, a number of other ITRI labs were engaged in R&D involving semiconductor equipment and materials, with an eye toward creating the foundation of a complete industry supply chain.

Local Content Initiatives

In 2007, Minister of Economic Affairs told Taiwanese semiconductor manufacturers that they should "adopt locally-made equipment as much as possible in order to help boost Taiwan's chip-making equipment industry,"

TABLE APP-A3-7 Creating the Supply Chain—Participating Companies

Theme	ITRI Laboratories	Participating Companies
Process technology	ERSO	Winbond, Mosel, Vitelec, etc.
Design	ERSO Communications/Computer lab	Via, AcerLab, SunPlus, Eltran, etc.
Materials	Materials Research lab Union Chemical Laboratories	Taisel, others
Equipment	Mechanical Industrial Research Lab	GPM, Union, Tong-Tai, etc.
Testing	ERSO Center for Measurement Standards	Siliconware, First, etc.
Packaging	ERSO Communications/Computer lab Materials Research lab Energy & Resources Lab	AES, Siliconware, Precision, Oriat, etc.

SOURCE: Huang, "Cradle of Technology" (2006) op. cit. p. 42.

[144]In 2003, ERSO announced the development of wafer level chip scale package technology (WLCSP), a high performance form of chip packaging that was less costly than conventional silicon packaging. "ERSO Unveils New Chip Packaging Technology," *Taiwan Economic News* (May 30, 2003). In 2002, TSMC and ERSO announced a plan for joint R&D to develop magnetoresistive random access memory (MRAM) chips, which have applications in portable medical electronic equipment, aerospace and internet-related products, with the objective of developing product prototypes. "BMC Cooperates With ERSO to Develop MRAM Products," *Taiwan Economic News* (March 21, 2002). Three years after the start of this effort, ITRI and TSMC disclosed that they had succeeded in integrating MRAM and CMOS technology, and unveiled a low-power, high density 1-kilobit MRAM chip. "BMC, ITRI Team Up on MRAM," *Taiwan Economic News* (January 10, 2005).

ITRI's Electronics and Optoelectronics Research Laboratories (formerly ERSO) was trusted with developing a model for transactions between Taiwan's chipmakers and equipment suppliers to foster procurement of locally-produced equipment.[145]

Nankang Integrated Chip Design Science Park

In 2003, the government opened the Nankang Integrated Chip Design Science Park in Taipei, emphasizing the design of systems-on-a-chip (SoC). The park featured sites for IC design firms, an incubation center for start-ups, an open lab and a service and management section. The principle objective of the new park was to incubate IC design start-ups with up to 35 employees.[146] The park was part of a broader national effort to promote indigenous silicon intellectual properties (SiP) in Taiwan, to streamline electronic design automation (EDA) software, and to promote the development of a domestic IC design industry.[147]

Taiwan's Emerging Competitive Edge

By 2007, Taiwan's semiconductor industry led the world in a number of key technology areas including foundry manufacturing and testing. Taiwan had more state-of-the-art 12 inch wafer fabrication facilities than any other country in the world, 11 fabs.[148] An ITRI official declared in 2007 that Taiwan's semiconductor industry ranked "first in the world in competitiveness, with no other countries expected to rival it over the next few years."[149]

TABLE APP-A3-8 Taiwan's Competitive Position in Semiconductors, 2007.

Category	Global Rank	Global Market Share (Percent)
Foundry	1	66.6
Mask ROM	1	92.9
IC Packaging	1	44.4
IC testing	1	63.0
Large wafer size (10"+)	1	46.4
IC design	2	23.9
IC substrate	2	26.4
DRAM	2	22.4

SOURCE: Industry and Technology Intelligence Services (IT IS) cited in Hwa Meei Liou, "Overview of the Photovoltaic Technology Status and Perspective in Taiwan," *Renewable and Sustainable Energy Reviews* (2010).

---

[145] "Taiwan Gov't Persuades Local Chipmakers to Buy Home-made Equipment," *Taiwan Economic News* (November 27, 2007).
[146] "World's First System on Chip Design Center Will Open in Taiwan Industrial Park," *China Post* (April 23, 2003).
[147] "Taiwan Aiming to be Global SoC Design Center," *China Post* (December 12, 2003).
[148] "Taiwan a World Leader in 12-inch Wafer Foundries," *China Post* (September 26, 2006).
[149] "Taiwan Chip Industry Still Leads World in Competitiveness," *Asia in Focus* (April 22, 2007).

ITRI is commonly said to have ended its support for the semiconductor industry because of the industry's commercial success, but in fact, ITRI is engaged in significant developmental work in microelectronics. In 2008, ITRI organized the Advanced Stacked System Technology and Application Consortium to develop technology for 3-dimensional integrated circuits, based on the stacking of multiple thinned ICs with through-die area array interconnects between them.[150] SUSS Microtec, a German semiconductor equipment supplier and developer of 3D integration technology joined the consortium in 2009.[151] The leading U.S. semiconductor equipment vendor, Applied Materials, joined the consortium in 2009. The consortium established a 300 mm demonstration production line at ITRI using state-of-the-art equipment and materials to produce "through-silicon vias" (TSVS), a technique for making stacked memory and logic chips.[152] Rambus, one of the world's leading technology licensing companies, joined the 3D IC consortium in 2011.[153] In early 2012, ITRI was reportedly in discussions with a fabless U.S. semiconductor company about a collaboration in which ITRI would manufacture the U.S. firm's 3D-1C design on a pilot line.[154]

In 2010, ITRI was reportedly working with a team of "dozens" at National Tsing Hua University to develop architecture for a high-density 3-dimensional nonvolatile resistive random access memory (RRAM) device that would consume less power and yield higher speeds than existing memory chips.[155] In late 2011, ITRI exhibited a prototype high speed non-volatile RRAM which it had developed. The chip has up to 20 times the speed of a flash memory device but consumes only 20 percent of the power and can survive up to 10 years at a temperature of 200 degrees Celsius.[156] As of February 2012 ITRI believed that it was "ahead of everybody" in RRAM technology, which it expected to license to Taiwanese companies.[157]

In December 2011, Intel Corporation announced a five-year collaboration with ITRI to develop ultra-fast and energy saving semiconductor memory products. The collaboration will focus on producing memory devices utilizing 3D IC stacking technology for applications in cloud computing data

---

[150]The idea of stacking arose out the difficulty in moving lithography technology forward from the 32 nanometer line width node. Stacking is an alternative to further miniaturization of 2D IC devices. "Organizations Establish Taiwan's 3D IC Consortium," *Taiwan Economic News* (July 26, 2008).

[151]"SUSS Microtec Cooperates with Research Institute ITRI on Technology Development in 3D Integration," *Business Wire* (September 28, 2009).

[152]"ITRI and Applied Materials Collaborate to Advance 3D IC Technology," *Business Wire* (October 15, 2009).

[153]"Rambus and ITRI Collaborate to Develop Interconnect and Advanced 3D Packaging Technologies," *Business Wire.*

[154]Interview with Dr. Yi-Jen Chan, Director, ITRI Electronics & Optoelectronics Research Laboratories, Hsinchu, Taiwan, February 14, 2012.

[155]"Taiwan's Tiny RRAM," *IEEE Spectrum* (November 2010).

[156]"ITRI Shows Off Taiwan's Latest Technology," *Taiwan Today* (October 6, 2011).

[157]Interview with Dr. Yi-Jen Chan, Director, ITRI Electronics and Optoelectronics Research Laboratories, Hsinchu, Taiwan, February 14, 2012.

centers and next-generation mobile devices.[158] MOEA pledged to provide $5 million in funding for this project, which the General Director of ITRI's Information and Communications Research Laboratory, Wu Cheng-wen, said would "enable Taiwanese DRAM chipmakers to produce more value-added chips utilizing existing technologies" not requiring a major investment in new equipment.[159]

Photovoltaics

ITRI has been working to establish a photovoltaic solar equipment industry in Taiwan for over two decades, and "in the 2000s one can see the solar photovoltaic (PV) industry emerging as the 'third pillar' of Taiwan's high-tech industrialization efforts" (the first two being semiconductors and displays.[160] Taiwan's existing competency in semiconductors is a major asset in the PV effort.[161] Major Taiwanese companies like TSMC, AuO Optronics, Delta Electronics and possibly Formosa Plastics are investing in the industry, moves which "promise to transform the Taiwan industry and make it one of the world's leading global players, alongside China, Germany, Japan and the USA."[162]

In 1987, ITRI's Energy and Mining Research Division began R&D on monocrystalline and amorphous silicon, materials used to manufacture solar photovoltaic cells. In 1988, Taiwan's first solar cell company was established, Sinonar Amorphous Company, using technology developed by ITRI and founded by two former ITRI staffers. The new company focused on indoor solar cells using amorphous silicon. In 2004, ITRI spun off a solar cell company, DelPoint, a joint venture with Delta Electronics, which received technology, capital and staff from ITRI.[163] The following year Delta began production of solar cells, and a large number of Taiwanese entered the solar cell companies business, some of which "[stole] talents from the government-backed Industrial Technology Research Institute (ITRI), some of whose laboratories

---

[158] "Intel, ITRI Team Up on Memory R&D," *China Post* (December 7, 2011).

[159] "Government Makes Right Policy Call," *Taipei Times Online* (December 12, 2011).

[160] Mathews, *Hsinchu Model* (2010) op. cit., p. 11.

[161] Morris Chang, the founder of TSMC, Taiwan's largest semiconductor maker and an ITRI spin-off, was asked in a 2009 interview about his company's entry into the light-emitting diode and solar cell industries. Interviewer: "You say TSMC wants to go into LEDs and solar cells. What core competencies does TSMC have in these areas?" Chang: "I won't tell you outright that we have this capability. But LEDs and solar cells are semiconductors. When I was working on my Ph.D., my doctoral thesis was related to II-V compounds and LEDs were considered part of my field. The physics of LEDs are the same as they were for semiconductors." "I'm Willing to Start from Scratch," *Commonwealth* (June 18, 2009).

[162] Mathews, Hu and Wu, "Fast Followers Dynamics." (2011) op. cit. p. 190.

[163] John A. Mathews, Mei-Chi Hu and Ching-Uan Wu, "Fast Follower Industrial Dynamics: The Case of Taiwan's Emergent Solar Voltaic Industry," *Industry and Innovation* (February 2011), pp. 186, 197.

developed the technology."[164] The government implemented a series of policy measures to foster the grown of the photovoltaic industry, providing subsidies for installation of photovoltaic power generation systems, including installation subsidies to households for up to 50 percent of the total cost of installing rooftop PV systems.[165]

In 2006, MOEA established the Taiwan Silicon Conference to develop a strategy for creating an industry supply chain in photovoltaics, an initiative that led to the establishment of Taiwan's first polysilicon manufacturing company to provide upstream materials for PV cell makers. Photovoltaic supply chains now exist at three Taiwanese Science Parks.

ITRI has emerged as Taiwan's foremost source of expertise and research activity with respect to solar cells. Between 1984 and 2008 ITRI had secured or was pursuing applications for 1,940 patents in solar cell technology. In 2006, ITRI opened the PV Technology Center.[166] ITRI-trained scientists and engineers have played an important role in the development of Taiwan's solar

TABLE APP-A3-9 Creating the PV Supply Chain

	Materials	Cells	Modules	Systems
Hsinchu Science Park				
	WW Corp	Gintech	Gintech	Delta
	SAS	DelSolar	DelSolar	
		Neo Solar		
Central Taiwan Science Park				
	Setek	AUO	AUO	AUO
	SR Solar	BMC	Delta	Delta
		Nextpower	Nextpower	
Southern Taiwan Science Park				
	Motech	Motech	Delta	Delta
		E-Ton		
		Nanowin		

SOURCE: Dr. Weileun Fang, National Tsing Hua University, *Applications of Micro Tech for Renewable Energy at Taiwan: Progress and Prospects* (2011).

164"Taiwanese Enterprises Rush to Pan Gold in Solar Cell Business," *Taiwan Economic News* (July 6, 2005); "Delta Electronics to Mass Produce Solar Cells in Q4," *Taiwan Economic News* (August 22, 2005).

165In 2007, the Executive Yuan's Science and Technology Review Board recommended that Taiwan take advantage of its strong position in semiconductor and flat panel displays and its experience in developing strategic alliances with multinationals "to actively develop the photovoltaic industry." In 2009, Taiwan enacted the *Renewable Energy Development Act,* which provided incentives for installation of renewable energy power generation, set up a fund for renewable energy, authorized subsidies to support the use of renewable-generated electricity, and made provision for creation of feed-in tariff (FIT) pricing schemes to make renewable energy competitive with fossil-fuel generated electricity. The government's various incentive programs are summarized in Hwa Meei Liou, "Overview of the Photovoltaic Technology Status and Perspective in Taiwan," *Renewable and Sustainable Energy Reviews* (2010).

166Mathews, Hu and Wu (2011) op. cit. p. 197.

cell industry.[167] In 2007, ITRI decided to release 233 patents in the fields of solar energy and energy storage and efficiency to local Taiwanese companies. Most of the patents covered solar power generation, and were released in order to "[upgrade] local technology levels and assist … industries to expand their global market share in the wake of surging demand for green and clean energy."[168] In January 2010, ITRI spinoff TSMC acquired a 20 percent stake in Motech Industries, Taiwan's largest maker of solar cells, for $196 million.[169]

Taiwan's semiconductor manufacturers, including two ITRI spinoffs, are playing a significant role in the development of Taiwan's thin film solar cell industry, a technology area holding great promise for solar power generation.[170] In 2008, a UMC subsidiary, NextPower Technology Inc., was the first Taiwanese firm to begin volume production of thin film solar cells.[171] In 2010, TSMC began construction of a $258 million Thin Film Solar R&D Center and fab in Central Taiwan Science Park.[172] Concurrently TSMC acquired a $50 million equity stake in California-based Stion, a start-up conducting R&D in CIG5 technology.[173] ITRI has worked with several Taiwanese universities to develop modification in the thin-film PV CIGS manufacturing process that do not utilize toxic selenide.[174] Taiwan's MOEA has formed the CIGS Alliance, an ITRI technology-diffusion alliance which includes a number of small TF companies as well as large firms like TSMC and AU Optronics.[175]

By 2011, Taiwan's solar cell industry ranked number two worldwide by production value ($4.3 billion), having surpassed Japan and Germany. Taiwan currently holds 20 percent of the world market for solar cells, and while

---

[167]Dr. Kuo En Chang worked on battery research as an ITRI staffer in the 1990s. He joined Motech, one of Taiwan's pioneering solar cell companies in 1999 as Chief Technology Officer, becoming President of Motech's Solar Division in 2008. Dr. Sam Hung, President and CEO of Neo Solar Power, previously served as ITRI's Research Director, Solar Energy Division. He also served as Vice President and Plant Director of Taiwan's first amorphous silicon manufacturer, Sinonar Amorphous Silicon Solar Cell Co. "Editorial Advisory Board, PVTech, *<http://www.pv-tech.org/about/editorial_advisory_board>*.

[168]"Taiwan's ITRI Releases Patents for Solar Power Technology," *Asia in Focus* (September 27, 2007).

[169]Motech's CEO was a former TSMC executive responsible for materials and risk management at TSMC. TSMC CEO Morris Chang reportedly plans to back Motech's efforts "to expand into a solar conglomerate from a solar cell maker." "Solar Newcomer Makes an Impact as CEO of Motech," *Taipei Times Online* (May 17, 2010).

[170]So-called "thin film" (TF) photovoltaic cells hold great promise in the field of solar energy. TF cells are an alternative to full crystalline silicon ("thick film") cells, and utilize a thin film of copper indium gallium selenide (CIGS) semiconductor materials on glass, permitting a substantial reduction in the cost of materials.

[171]"NextPower Vows to Top Taiwan's Thin Film Solar Cell Industry," *Taiwan Economic News* (September 10, 2008).

[172]The fab will be operated by a subsidiary, BMC Solar, with the first commercial shipments expected in 2012. "TSMC Solar's First SIGSSe '5-Fab' Prping for Commercial Production Ramp," *PV Tech* (October 14, 2011).

[173]"TSMC Sinks $50 Million Into Stealthy Thin-Film solar Startup Stion," *Venturebeat* (June 16, 2010).

[174]"ITRI Develops 3.5G Turnkey Solution for Thin-Film PV Production." *Digitimes* (June 12, 2009).

[175]Mathews, "Fast Follower Industrial Dynamics" (2011) op. cit. p. 196.

its revenues rank behind those of China, Taiwan is more technologically advanced than China in this field.[176]

Biotechnology

Taiwan has been promoting the development of a biotechnology industry since the early 1980s but the results to date have been mixed at best, despite a massive deployment of government resources.[177] In contrast to Taiwan's developmental effort in microelectronics, in which ITRI was the governmental entity responsible for promoting the industry, in biotechnology, Taiwan's "agencies and units involved are so numerous and diverse that a serious coordination problem has emerged."[178] Instead of a single main technology cluster (Hsinchu) which characterized the formative years of the semiconductor industry and continues to account for most of the semiconductor industry's output, Taiwan has already established six biotechnology parks with more planned.[179] The Development Center for Biotechnology (DCB), a government research organization formed to play an ITRI-like role as a technology intermediary—turning basic research into commercial products—has drifted into an emphasis on basic research.[180]

ITRI has long taken the position that Taiwan's developmental efforts in biotechnology should be limited areas where the country could leverage existing strong competencies rather than risky, leaps into blue-sky areas with uncertain prospects. ITRI has argued that Taiwan should seek to integrate its medical research infrastructure with its strengths in electronics and the information and communications technologies. ITRI also favors establishment of a few centers of excellence in the handful of medical areas in which Taiwan is a world leader—most notably liver diseases—in order to attract multinational biopharmaceutical companies to the island for R&D.[181] ITRI has concentrated

---

[176]Taiwan Strives to Build International Brand Names," *Taiwan Insights* (July 19, 2011).
[177]"Taiwan's Technology Success Underappreciated: Canadian Scientists," *Focus Taiwan* (July 24, 2010); Dodgson, et al, "Taiwan's National Innovation System" (2008) op. cit. p. 431.
[178]Yu-Shan Wu, Academia Sinica, "Taiwan's Developmental State: After the Economic and Political Turmoil," paper prepared for delivery at the Conference on A Decade After the Asian Financial Crisis, Thammasat University, Bangkok, February 23-24, 2007.
[179]Mark Dodgson, John Mathews, Tim Kartelles and Mei-Chih Hu, "The Evolving Nature of Taiwan's National Innovation System: The Case of Biotechnology Innovation Networks," *Research Policy* (2008), pp. 436-7. Despite the existence of erstwhile clusters, "The research institutes and firms are not effectively linked, nor are the science community and the ministries and agencies." Yu-Shan Wa, "Taiwan's Developmental State" (2007) op. cit. p. 23.
[180]The DCB was founded in 1984 to close the gap between academia and industry in biotechnology. However, DCB has many more basic researchers on its staff than translation researchers. Wong, Chi-huey, President of Academia Sinica and a biochemist, commented that "what you're seeing is that institutions in different roles have hired people from the same background who end up doing the same things. I think there should be more coordination." Andrea Yung, "A Long Haul for Biotech," *Topics* (October 2009).
[181]"ITRI Advices Using Strengths in ICT for Biotech Industry," *Taipei Times* (July 22, 2008). Reflecting the fact that hepatitis is one of Taiwan's most prevalent diseases, ITRI has prioritized the

its biotechnology efforts in areas of strength, and can point to some market success stories. One analyst observed in 2007 that "only in areas where biotechnology and microelectronics overlap can one find some prospect of success."[182] ITRI's Medical Electronics and Device Technology Center (MED) is driving this effort.[183] Taiwanese medical device makers have already made significant inroads in some global markets.

ITRI established a Biomedical Engineering Center in 1999 (BMEC), which was split in 2006 to form the Biomedical Engineering Research Laboratories (BEL) and the Medical Electronics and Device Technology Center (MED). BMEC's Biochip Program combined research efforts of five ITRI laboratories to develop DNA microarray and microfluidics technology, which is essential to the rapid unraveling of the human genome. Sixteen patents from this effort were transferred from ITRI to a spin-off company, Phalanx Biotechnology Group, established in 2003. The research effort organized by ITRI drew in the Canadian Genetic Diseases Network (CGDN) and the Information System for Biotechnology (ISB), a U.S.-based research organization. Using ITRI's technology, Phalanx was able to drop the price of one gene chip from about $1,000 down to $50-80 per slide. A CGDN executive commented that "we have looked at technologies from around the world and we are sure this is the best tech available [in the field of gene chips]."[184]

In 2002 ITRI spun off CESCO Bioengineering Co., Ltd., comprised of eight team members from ITRI's Tissue Engineering and Biomaterial Laboratory, to commercialize high cell density culture technology.[185] CESCO developed a novel disposable pact bed contractile (DPBC) bioreactor suitable

TABLE APP-A3-10 Global Share for Taiwanese Medical Devices

Item	2009 Global Market Share (Percent)
Digital blood pressure monitor	42
Electric wheelchair/scooter	30
Electric ear thermometer	30
Electronic thermometer	60
Pre-filled IV auto injector	50

SOURCE: Dr. Chei-Hsiang Chen, Director, Biotech & Pharmaceutical Industries Program office, MOEA *Taiwan's Biotech Industry Overview,* October 21, 2010.

study of liver diseases. It has developed the world's most comprehensive liver proteome database as well as projects with applications in fighting liver ailments, including high density microarray gene chips, new herbal medicines to combat liver disease, tissue engineering for artificial livers, and non-invasive liver disease medical instruments. "Taiwan Strengthens Liver Disease Research," *BioSpectrum Asia Edition* (July 31, 2007).

[182]Yu-Shan Wu, Taiwan's Developmental State" (2007) op. cit. p. 23.

[183]"ITRI and NCKU Collaborate on Biomedical Device Related Technology," *Business Wire* (May 7, 2009).

[184]"Gene Chip Venture Phalanx Inaugurated," *China Post* (January 24, 2003).

[185]*<http://www.cescobio.com.tw/history.php>*.

for producing various proteins and viruses and non-adherent cell cultures including embryonic stem cells.[186] ITRI spun off DailyCare Biomedical in 2004 to commercialize low cost, portable medical devices for home core users. The company's new CEO was K.P. Lin, the former chief of ITRI's Biomedical Engineering Center.[187] The care technology underlying one of these products, ReadMyHeart, a portable electrocardiogram, was developed by ITRI's BEC.[188] This product was approved for use in Japan in 2007, representing the first non-Japanese company to obtain a class-II medical device license under Japan's Pharmaceutical Affairs Law, which took effect in 2005.[189]

FUTURE DIRECTIONS

The Statute for Industrial Innovation (SII) enacted in 2010 is the most recent of the promotional statutes that have provided the institutional framework for Taiwan's industrial development. SII represents a departure from the previous laws in that it provides not only for the promotion of manufacturing but also agriculture and services, and seeks to redirect Taiwan from its focus on manufacturing toward innovation, "soft power," development of national brands, and advanced logistics.[190] ITRI currently singles out 20 sectors which represent developmental priorities.[191]

Current ITRI developmental projects are underway addressing themes relevant to all of the smart industries noted above as well as biotechnology,

TABLE APP-A3-11 Taiwan's Developmental Priorities

Emerging Industries	Smart Industries	Service Industries
Health Care	Cloud computing	WiMAX
Biotechnology	Smart electronic vehicles	International medical
Quality agriculture	Smart green buildings	High tech finance
Tourism	Patent industrialization	Music/digital content
Cultural/creative		Exhibitions
Green energy		International logistics
		Urban renewal
		International cuisine
		Chinese e-commerce
		High education output

[186]Hera Andrade-Zaldivar, Leticia Santos and Antonio De Leon Rodriguez, "Expansion of Human Hemapoietic Stem Cells for Transplantation: Trends and Perspectives," *Cytotechnology* (March 2008).
[187]"DailyCare Announces Telehealthcare System," *Taiwan Economic News* (November 28, 2005).
[188]Dodgson, et al., "Biotechnology Innovation Networks" (2008) op. cit. p. 440.
[189]"DailyCare's Handheld ECG Penetrates Japan's Medical Device Market," *Taiwan Economic News* (May 15, 2007).
[190]SII established a uniform 17 percent corporate tax and eliminated tax incentives except those associated with R&D.
[191]ITRI International Center, "Innovation Industrialization Corporation (IIC) Program Proposed for Bridging the Cooperation Between the United States and Taiwan" (February 14, 2012).

health care, quality agriculture, green energy, and WiMAX. In most cases ITRI has formed industry alliances to prepare for the production of the new technologies.

ITRI is also developing what it characterizes as the latest in a series of related electronic hardware technologies which incorporate semiconductor technology, including solid state lighting (LED), resistive random access memory (RRAM, high-speed, low power consumption semiconductors) 3D television, flexible electronics, "floating image," and optoelectronic medical devices.[192]

ITRI is currently seeking to strengthen its capability in "upstream" pre-competitive R&D, to enhance the interdisciplinary character of its work, and to shift from an emphasis on the manufacture of components to development of systems, services and applications.[193]

CONCLUDING PERSPECTIVE

Taiwan's economic development and ITRI's role in creating advanced industries have been widely studied, but the extent to which the ITRI model is adaptable elsewhere is unclear. The circumstances underlying Taiwan's success are not necessarily present in other countries and for that matter may not always be present in every case in Taiwan itself. The creation of ITRI brought together cosmopolitan elite commanding a vast pool of knowledge absorbed from the world's finest research universities and the most technologically advanced international companies. This remarkable founders' generation was backed in its efforts by a government in which the ruling party held a monopoly on political power and an abiding belief that economic policy should be conducted by skilled professionals shielded from political pressure. Government funding

---

[192]In January 2012, ITRI reported a breakthrough in LED technology which overcomes the longstanding problem of a LED bulb's narrow beam angle. The prototype LED weighs less than half of a normal LED bulb, is unbreakable, can be manufactured at low cost, and has a beam angle of 330 degrees. "ITRI Develops Wide Beam Angle LED Bulb." *Central News Agency* (January 31, 2012). "Floating image" refers to a technology in which a device projects an image I the air in front of a person. In some cases the image may consist of a keypad which can be operated by poking a finger through the images of keys, motion which the device detects with a sensor. The device can be operated in this manner without human touch. Interview with Dr. Yi-Jen Chan, Director, ITRI Electronics and Optoelectronics Research Laboratory, Hsinchu, Taiwan, February 14, 2012. MOEA indicated in April 2012 that the government was planning to invest $339 million between 2012 and 2015 in the development of smart handheld devices and an associated supply chain. "Over NT $10 Bil. To be Pumped Into Smart Device Sector: MOEA," *China Post Online* (April 7, 2012). MOEA reportedly plans to task ITRI to develop core components technology and to explore customer-oriented applications for smart handheld devices. "Taiwan Smart Handhelds to Top US $43.8 billion," *Taiwan Today* (September 13, 2011).

[193]Interview with Dr. Yi-Jen Chan, Director, ITRI Electronics and Optoelectronics Research Laboratory, Hsinchu, Taiwan, February 14, 2012.

for ITRI was substantial and sustained.[194] Ancillary government laws and policies were implemented to support ITRI in areas such as human resources, venture capital, and creation of science parks, intellectual property protection, and start-up of new companies. Replicating this mix of factors would be a challenge for any government.

ITRI has remained relevant by continual focus on its original, carefully-defined mission—not that of a research laboratory, but of a bridge connecting the international research community to small and medium Taiwanese businesses which—although entrepreneurial and highly motivated—lack the resources to conduct research and stay abreast of international technology trends. ITRI has developed institutional practices and policies which serve this mission:

- Its policy of encouraging its own personnel to leave after four to five years continually seeds Taiwanese industry with experienced and technology-savvy leaders.
- Its technology integration centers perform an internal interdisciplinary silo-breaking function which in its later stages is passed along to the private sector in the form of complete industry supply chains developed in ITRI-led industry alliances and research projects.
- The physical proximity of researchers, labs and companies in the Hsinchu technology cluster fosters continual knowledge spillovers from the research community to private companies.
- ITRI's incubation center provides innovative start-ups with technology with commercial potential, business and legal advice, funding opportunities, access to research facilities, and introductions to potential customers and investors.
- ITRI's international outreach efforts acquire technology, but are also leading to foreign research efforts located in Taiwan itself with participation of Taiwanese companies.

Significantly, ITRI has not veered off course to pursue Nobel prizes or engage in industrial empire-building like Japan's Ministry of Economic Trade and Industry in the 1960s and 1970s. It has maintained its focus on its core mission of technology intermediation between the global research community and Taiwanese companies.[195]

---

[194]ITRI's annual budget from MOEA passes "legislative reviews without questions, which shows that the institute's achievements have won approval from all political parties." "ITRI Has Contributed Greatly to Taiwan's Premier," *Asia Pulse* (May 23, 2006).

[195]Dr. John Chen, Director of ITRI's Display Technology Center, comments that the 50-50 split in ITRI's funding between the government and the private sector forces ITRI to stay on-mission to remain relevant to industry. Half of its income derives from the sale of IP, technology licensing and the provision of knowledge-based services to industry. Interview with John Chen, Hsinchu, Taiwan, February 13, 2012.

ITRI's successes have been based, above all, on the excellence of its scientists, engineers and managers. Originally ITRI was able to draw upon a large number of U.S.-educated Taiwanese with relevant work experience in multinational corporations. Today it increasingly recruits graduates from domestic universities with strong and improving programs in engineering and the sciences. The competitiveness of the industries in the Hsinchu technology cluster "derives from the availability of a significant pool of well-trained workers and technical and managerial personnel [from ITRI and Hsinchu's two science-based universities] who now stand in comparison with the concentration s of talent in other [global] R&D intensive locations."[196] *Business* Week terms this pool "one of the deepest reserves of high tech talent in the world."[197] Its existence reflects decades of public investment in education, both in promotion of study abroad and in upgrading the domestic educational infrastructure.[198]

---

[196] Mathews and Hu, "Enhancing the Role of Universities," (2007) op. cit. p. 1006.
[197] "Why Taiwan Matters," *Business Week* (May 16, 2005).
[198] Soren Eriksson, "Innovation Policies in South Korea and Taiwan," Vinnova Analysis Va 2005:03 (2005) p. 31.

Appendix A4

Catapult: Britian's New Initiative in Applied Research

The global financial crisis that began in 2008 served to dispel the widely-held view in Britain that the effects of long erosion of the country's manufacturing sector would be offset by a dynamic and growing financial services sector. With banks collapsing and property values plummeting, across the political party spectrum, British policymakers called for a national revival in manufacturing. The renaissance of British manufacturing would be driven by innovation derived from the country's scientific research base, regarded as the second-best in the world after that of the United States. The centerpiece of this new national effort is a network of "Technology and Innovation Centres" (TICs) for applied research that, for branding purposes, have been dubbed "Catapults." Their mission is to serve as a bridge between the country's research base and British industry.

The Catapults are intended to bring some coherence to the bewildering array of public and private organizations in the United Kingdom that are engaged in technology transfer from universities and other basic research organizations to industry. The Catapult program will take some of the best of these organizations, provide "core' funding from the government over a five-year horizon, establish governance arrangements, and align each center with national strategy. Each Catapult will focus on a technology area in which the United Kingdom enjoys a strong manufacturing presence already. Total government funding for this effort has been set at £200 million over a five year period. Overall direction and core funding for the Catapult network is being provided by the Technology Strategy Board, a non-departmental public body largely staffed by individuals with backgrounds in business and industry.

Currently seven Catapults have been or are being established, with the expectation they will all be operating by the end of 2013.[1]

BRITAIN'S CHALLENGE: CAPTURING VALUE FROM SCIENTIFIC RESEARCH

Britain's leaders have been struggling to translate research from the nation's excellent science base into technologies with industrial and commercial impact for a century, but with a few exceptions such as pharmaceuticals, the challenge has proven intractable. Perhaps the biggest obstacle has been embedded attitudes in the science and business community that inhibit close collaboration and a reluctance by successive governments to become too deeply involved in "industrial policy." Additional challenges are presented by the erosion of the manufacturing base, a shrinking industrial research infrastructure, and a work force that often lacks the skills needed by technology-intensive companies. Globalization and the rise of new low-cost industrial powers like China, Korea, and Brazil have given this longstanding British dilemma a new urgency. "Bridging the gap between academia and industry to create cutting-edge technologies in manufacturing is important if the United Kingdom is to remain competitive in the global economy."[2]

Numerous studies of British international competitiveness have concluded that while the United Kingdom has a basic science capability second to only that of the United States, "it falls short on translating scientific leads into leading positions in new industries."[3] In 1919, the economist Alfred Marshall observed that "the small band of British scientific men have made revolutionary discoveries in science, but yet the chief fruits of their work have been reaped by

---

[1]The main inspiration for the Catapult initiative is the Fraunhofer-Gesellschaft, Germany's successful network of parapublic applied research institutes, and some elements of the Fraunhofer model have been replicated in the current British effort. Like the Fraunhofer, each Catapult will receive "core" funding from the government over a comparatively long time frame (five years) which it is expected to augment with revenue from contract research for industry and public bodies. The designation "Catapult" represents an effort to establish a brand with a reputation for excellence, as the Fraunhofer has succeeded in doing on a global basis. Like the Fraunhofer, the Catapults are not mere conduits for research results from universities to companies but will operate their own well-equipped facilities that perform on-site research to develop, test, and prove innovative products and processes for industrial consumers. The expectation is that the Catapult initiative will translate into new manufacturing jobs in Britain.

[2]*Professional Engineering.* 2012. "Fast Growth at Technology Centre Bodes Well for Manufacturing." June 14.

[3]Hauser. Hermann. 2010. *The Current and Future Role of Technology and Innovation Centres in the UK.* pp. 1. A 2009 Cambridge University study observed that "The perceived failure of the UK to exploit effectively its science and technology base has been the subject of hand-wringing by politicians and policy specialists for nearly a century and government efforts to remedy this have been redoubled over the past 15 years." Mina, Andrea, David Connell and Alan Hughes. 2009. *Models of Technology Development in Intermediate Research Organizations.* University of Cambridge: Center for Business Research. Working Paper No. 396. December. p. 3.

businesses in Germany and other countries where industry and science have been in close touch with one another."[4] In fact, many British innovations have led to commercial successes, but often the commercial benefit has inured to countries other than Britain.[5]

A Commanding Position in Basic Science

Britain enjoys an extremely strong global position in basic science, ranking second only to the United States according to some commonly-utilized benchmarking metrics.[6] With respect to highly-cited published scientific papers, Britain is "solidly second among all countries," ranking second or third across a broad range of scientific disciplines.[7] "The UK science base is one of the most productive and influential systems of publicly funded research in the world."[8] British researchers are the most efficient and productive in the world in terms of output per researcher and per unit research spending.[9] British universities

---

[4]Cited in a speech by University and Science Minister, David Willetts at Policy Exchange, January 4, 2012.

[5]Although there is scant quantitative information about how widespread this phenomenon has become, Parliament was regaled in 2012 with anecdotal examples of British inventions exploited outside of Britain and yielding little if any value-added in Britain. The University of Edinburgh developed a novel remediation technology for the removal of hazardous subsurface contamination but no UK-based licensing partner could be established, so the technology was licensed to a Canadian firm and is undergoing successful testing in the United States. House of Commons, Science and Technology Committee, Technology and Innovation Centre Enquiry, Written evidence submitted by UK Deans of Science. April 20, 2012. Although the United Kingdom has done pioneering research in carbon fibers, all major carbon fiber production is overseas. *Bridging the Valley.* 2012. Op. Cit. Written Evidence submitted by the Association of Independent Research and Technology Organizations (V45). Cambridge-based Raspberry Pi revised plans to manufacture low cost computers in the United Kingdom, choosing Taiwan and China instead, because of "prohibitive taxation" and "lack of UK competitiveness." Plastic Logic, a spinoff from Cambridge University's Cavendish Laboratory, decided to manufacture its plastic electronics products in Dresden, citing the "high skills base of the work force, the presence of Fraunhofer facilities and the difficulties associated with planning and construction timescales in the UK."

[6]As of 2010, the United Kingdom had produced 116 Nobel Prize winners, second only to the United States' 320 winners from a population around five times as large as that of the United Kingdom. Recent British Nobel Laureates include Sir Robert G. Edwards (2010), the pioneer of in-vitro fertilization; Sir Martin Evans and Matthew Kaufman (2007), the first individuals to clone embryonic mice stem cells; Oliver Smithies (2007), the inventor of gel electrophoresis and a pioneer in the development of techniques for altering animal genomes; Sir Andre Geim and Konstantin Novoselov (2010), who conducted groundbreaking research on graphenes; and Sir Anthony James Leggett (2003) who made major contributions to the theory of superconductors and superfluids. Dyson. 2010. *Ingenious Britain.* Op cit. p. 35.

[7]Weinberg, Bruce A. 2009. "An Assessment of British Science Over the Twentieth Century." *The Economic Journal.* June. The rankings were based on data from the Institute for Science Information's Highly Cited. Ibid. pp. F261-62.

[8]HM Treasury. 2004. *Science and Innovation Investment Framework 2004-2014.* July. p. 8.

[9]Department of Business, Innovation, and Skills. 2011. *International Comparative Performances of the UK Research Base.*

consistently rank among the global leaders across the scientific disciplines in education and research, and in the field of clinical/pre-clinical health, three of the top five ranked institutions are British, including number one, Oxford.[10]

Given Britain's rich scientific heritage, it is a question of central importance to the nation why the country's scientific research and discoveries have not had greater industrial and commercial impact. A number of interrelated factors appear to be involved, including cultural attitudes, an eroded manufacturing base, the decline of the public and private research infrastructure, an opaque clutter of intermediate research organizations, and a work force that too often lacks the skills for technology-intensive manufacturing.[11]

The Cultural Challenge

British values, ideas and its rich intellectual tradition are celebrated at home and admired around the world. However, certain aspects of British culture and economic ideology have proven serious impediments to the forging of close working relationships between the science community, government, and private industry. They are sometimes noted as key reasons that adoption of the German model of innovation would be difficult in the United Kingdom. A number of recent studies of Britain's innovation system have urged a major national effort to modify traditional national biases and perspectives[12].

In countries like Germany and Taiwan, which have proven adept at translating scientific knowledge into competitive products, scientists, engineers, technicians, and skilled factory workers are admired throughout the broader society. That is less so in Britain, a fact that is reflected in phenomena such as the lower priority historically accorded science and engineering curricula at some universities, the tendency of young people to avoid careers in these fields, and ultimately, shortages of personnel with the education and skills needed by technology intensive domestic companies.[13] Sir Peter Williams, Chairman of the

---

[10]Times Higher Education World University Rankings 2011-12.

[11]Goodwin, Tom and Nick Matthews. 1998. *Knowledge Transfer: A UK Competitive Weakness.* London: Institute for Public Policy Research.

[12]Dyson. James. 2010. *Ingenious Britain: Making the UK the Leading High Tech Exporter in Europe..* p. 13-22.

[13]"Successive governments have been concerned that negative public perceptions are acting as a barrier to growth in manufacturing…a fifth of respondents [in a recent Institute for Manufacturing survey] would encourage their children to follow a career in manufacturing, compared to a third in the U.S." Parliamentary Office of Science and Technology. 2012. *Advanced Manufacturing.* Postnote Number 420. September. p. 4 In 1991, *New Scientist* lamented a common national "ingrained belief that making things is a second-class occupation. *The New Scientist.* 1991. "Last Chance for British Industry." November 30. A 2010 report to the Conservative Party by entrepreneur James Dyson observed with respect to British national attitudes, "Young people's perception of engineers and scientists would be comical if it were not tragic. Look at the national stereotypes. Scientists are egghead lab-coated geeks; engineers are metal-bashing factory workers or mechanics fixing broken appliances. It's no wonder careers in science and technology are deemed

United Kingdom's National Physical Laboratory, told Parliament in 2010 that "the technician class is a forgotten, underrated, and undervalued one in this country and has been endemically."[14]

The German and Taiwanese innovation models relentlessly drive findings derived from basic research into the factory and marketplace. Given Britain's formidable position in basic science, a comparable attention to commercialization of research would have undoubtedly had a dramatic effect on British international competitiveness. However, a longstanding attitude commonly found in the United Kingdom's science establishment has been indifference, if not disdain, for the practical application of research results. Britain's greatest electronics research team of the Twentieth Century reportedly used to toast their innovations with the cry "And may it never be of use!"[15] In contrast to Germany, where close interaction between universities and industry can be traced back to the Eighteenth Century, traditionally in Britain, "the university and the business and industrial community always treat[ed] each other with indifference, if not distrust and hostility."[16]

---

unappealing by both parents and their children. By contrast, countries like the USA, Germany and France hold these careers in much higher esteem." Dyson. 2010. *Ingenious Britain.* Op Cit p. 13.

[14]Williams complained that "in the 1990s we had this obsession with turning everybody in the country into a university graduate without any thought whether the market and demand was there for such people, and at the same time and in the same process, we therefore implicitly devalued what the Germans might term the product of the *Fachhochschule*, rather than the university. The net result is that…you reap what you sow…the technician class for big and growing businesses is sorely in need of more technicians." House of Commons, Science and Technology Select Committee. "Bridging the Valley of Death." Improving Commercialisation of Research." Oral Evidence. Q112. The American historian Martin Wiener argues that despite the fact that Britain was the first country to industrialize, a persistent strain exists in British culture, particularly in the upper classes, which disdains industry and business, an attitude that Margaret Thatcher's government acknowledged and fought to change. Mrs. Thatcher's closest political ally, Sir Keith Joseph, distributed to every Minister a copy of Wiener's book, *English Culture and the Decline of the Industrial Spirit, 1850-1980. The Economist.* 2010. "Empty Shelves." April 27.

[15]Toast at a dinner held for the Rutherford team of Cavendish Laboratory, Cambridge University. The work of this team contributed substantially to the development of the electronics industry and atomic energy. Oakley, Brian and Kevin Owen. 1989. *Alvey: Britain's Strategic Computing Industry.* Cambridge MA: MIT. p. 267. Professor Keith Ridgeway, Research Director of the Advanced Manufacturing Research Centre, told Parliament in 2010 that part of Britain's problem in translating scientific knowledge into commercial products and industrial processes was that "the universities have had a system of appraisal and monitoring that has been a bit adverse to work in the industrial sector. It rewarded publication and pure research rather than work within industry. That has been a factor." House of Commons, Science, and Technology Committee, Technical and Innovation Centres enquiry. Oral evidence. January 20 2010. Q42 Ev 12.

[16]Chandler, Alfred. 1990. *Scale and Scope: The Dynamics of Industrial Capitalism*. Cambridge MA; Harvard University Press. P. 293. Chandler, an economic historian, comments that British production managers and engineers agreed that "practical work on the job was far more useful than spending the same amount of time at a university." At the universities, training in science was a proper function of the university; vocational training was not. As a result of this belief, the critical linkage between higher education and industry, so essential to the development of long-term

Today, generations after this dynamic was observable, it remains an issue in the U.K.[17] A 2009 study by the University of Cambridge commented that "the first and foremost point to make is that the evidence overwhelmingly shows that as direct contributors to the development of commercial innovation processes higher education institutions are way down the list of sources of knowledge for information that UK businesses cite."[18] "Although British science is frequently lauded as being second only to the United States in capability, the gap between the research carried out in academia and its successful application commercially has vexed successive governments."[19]

In the historical accounts of how Germany overtook Britain in industrialization in the late Nineteenth and early Twentieth Centuries, British manufacturers are faulted for the fact that "the British did not innovate as rapidly or effectively" as their competitors, and were slow to introduce new technologies in their operations.[20] British economic and military historian Corelli Barnett cites Britain's "cult of the 'practical man' [which] led to a positive distrust [in industry] of the application of intellectual study or scientific research to industrial problems."[21]Alfred Chandler chronicles how British industry failed to exploit the new production technologies of the "Second Industrial Revolution," capital-intensive, continuous process machinery that achieved high volume throughput and economies of scale.[22] In postwar Britain, the manufacturing sector was criticized to the effect that there was not enough technical training on the shop floor, managers were poorly trained, and companies failed to attract talented young people. A U.S.-based management expert, Dr. David Granick, characterized Britain as the "home of the amateur." British managers "did not think there was any special experience that was useful for a future managing director. Granick observed that "professionalism was a serious charge against an individual in British industry."[23]

In countries like Germany, the United States and Taiwan, which are innovation leaders, the government plays a supportive role in coordinating and sometimes funding collaborations between the research base and industry. For

---

industrial capabilities in Germany and the United States, remained tenuous in Britain before 1914 and continued to be so during the interwar years.

[17]For specific examples of industry-university applied research collaborations that left the industrial side dissatisfied, see, "The Power of Knowledge," *The Manufacturer* (April 18, 2008).

[18]Mina, Andrea, David Connell and Alan Hughes. 2009. *Models of Technology Development in Intermediate Research Organizations.* University of Cambridge: Center for Business Research. Working Paper No. 396. December. p. 3

[19]*Professional Engineering.* 2012. "Catapults to Fame and Fortune." July 31.

[20]Friedberg, Aaron L. 1988. *The Weary Titan: Britain and the Experience of Relative Decline 1895-1905.* Princeton: Princeton University Press. p. 85.

[21]Barnett, Corelli. 1972. *The Collapse of British Power.* Atlantic Highland, NJ: Humanities Press International. p. 94.

[22]Chandler, Alfred. 1990. *Scale and Scope: The Dynamics of Industrial Capitalism.* Cambridge MA; Harvard University Press.

[23]TUC. 2011. *German Lessons: Developing Industrial Policy in the UK.* December 8. p. 26.

many years, representatives of British business have urged the government to put in place a comparable long-term industrial strategy.[24] However, the laissez-faire policies introduced by Margaret Thatcher beginning in 1979 were carried forward by successive governments, including New Labour under Tony Blair.[25]

Such government action has been inhibited, albeit not foreclosed, by a longstanding and commonly-held view that the government should not intervene in the economy to promote particular sectors and companies.[26] This attitude has not prevented successive British governments from implementing industrial promotion initiatives, but it has inhibited such efforts in terms of funding, duration, and scale.[27]

Until recently, many British policymakers were dismissive of the erosion of the manufacturing sector, arguing that the national economy was undergoing a structural transformation in which a dynamic financial services industry would offset job losses in manufacturing, an "evolutionary" perspective

---

[24]A 2012 editorial in *Professional Engineering* commented that Business Secretary Vince Cable "has his work cut out putting his goodwill and fine words into action. The first step to achieving this would be the development of a national industrial strategy. It's hard to see how manufacturers can make long-term investment decisions when they don't know the long-term plans of those in power." *Professional Engineering.* 2012. "Fewer Fine Words and More Action to Support Industry are Needed from Vince Cable." July 31.

[25]In 2009, the wake of the financial collapse in 2008, Business Secretary Lord Peter Mandelson introduced a "New Industry, New Jobs" program featuring a more activist role for the government in industry than any policy since the 1970s. Among other things the program provided for government review of whether long-term government capital investment should be undertaken in innovative small and medium companies. HM Government. 2009. *New Industry, New Jobs.* April.

[26]"In Britain, the tension between the ruling party's free market ideology and the patent need to adjust and upgrade industry resulted in inconsistent and underfunded programmes." Jacobsen, Kurt. 1992. "Microchips and Public Policy: The Political Economy of High Technology." *British Journal of Political Science.* October. p. 506.

[27]Paul Jackson, the Chief Executive of EngineeringUK, a not-for profit which promoted engineering, said in 2012 that "We desperately need an industrial policy that sets the vision and framework for rebalancing the economy. The government has announced many useful initiatives to support engineering and manufacturing. But it has gone about things in a piecemeal fashion. That means the initiatives haven't had the desired overall effect." *Professional Engineering.* 2012. "Engineers Call on Government to Develop Nationwide Industrial Policy." May 11. The Alvey Programme, launched in 1983 by Margaret Thatcher's government to promote national competencies in computer science, received funding from the government and industry and involved industry-university research collaboration. In 2012, Sir Peter Williams recalled for Parliament the outcome of the Alvey Project: "In the typical British style of risk aversion, because it was then a large sum of money, Alvey was scattered in small pieces so that nobody could blow a large sum of money that was high risk. The net result is that I suspect for many in this room the name Alvey is entirely new. I commend that you go back and look at the evaluation reports on Alvey. You will see that it made no difference whatsoever to the electronics and IT industries in the United Kingdom in subsequent decades." House of Commons, Science and Technology Select Committee. *Bridging the Valley of Death.* Oral Evidence. Q116.

that was also espoused by many economists. That perspective has been partially if not wholly discredited by the events of the past five years.[28]

> *Not so long ago, it was fashionable in Whitehall and Westminster to claim that this relative industrial decline didn't matter. Then came the banking crisis, which showed what happened when you bet the lot on the City—and the bet didn't come off.*[29]

British attitudes that have hampered innovation, are beginning to change. In government, all major parties recognize the need for a strategy to enhance the competitiveness of the manufacturing sector, although a reluctance remains to pick "winners and losers"[30]. The Chancellor of the Exchequer closed his March 2011 budget statement with the following statement:

> *We want the words: "Made in Britain," "Created in Britain," "Designed in Britain," and "Invented in Britain" to drive our nation forward—a Britain carried aloft by the march of the makers. That is how we will create jobs and support families.*[31]

With respect to the research community, high tech entrepreneur Hermann Hauser commented in 2010 that "when I first talked to our scientists about how to engineer their research into a real product rather than a scientific idea, I got brushed off. I find that practically none of that attitude remains."[32]Lord Sainsbury of Turville, author of a 2007 study of the British government's science and innovation policies, said in 2010 that he believed the country had "turned the corner" in terms of young people going into science and engineering, with increases in the number of entrants into engineering courses

---

[28]"The banking crisis and subsequent recession showed that the UK had become over-dependent on financial services and property. Even worse, it is now clear that the banking and financial services sectors, taken as a whole, did not generate as much added value as has been supposed. Instead, paper profits were reported which were based on leveraging the price rise of financial assets." Dyson. 2010. *Ingenious Britain.* Op Cit. p. 7.

[29]*The Guardian.* 2010. "David Cameron's Talk About Reviving British Industry is Nonsense." November 1.

[30]According to the Trades Union Congress, the current UK "Coalition government is not in favour of an activist approach to industrial policy. Whether this is because of its focus on deficit reduction, which limits the amount of money available to support such an approach, or is the result of an ideological attitude to government intervention, is unclear. In truth, it is probably a mixture of the two." TUC. 2011. *German Lessons: Developing Industrial Policy in the UK.* December 8. p. 26.

[31]HC Deb 23 March 2011 c966.

[32]*The Engineer.* 2010. "Centre Point: Hermann Hauser Discusses Commercializing Technology." September 6.

and the number of young people taking apprenticeships.[33] The need to "rebalance" Britain's economy away from financial services and toward manufacturing has "been spouted by politicians of all stripes since the recession [including] Lord Mandelson, the Chancellor John Osborne, business secretary Vince Cable, and David Cameron."[34]

A Vulnerable Manufacturing Base

At the end of three decades of free market fundamentalism under successive Conservative and New Labour governments, Britain is now looking back at "one of the biggest industrial declines observable in postwar Western Europe"[35]. At the end of the 1970s, manufacturing accounted for nearly 30 percent of national GDP and employed 6.8 million people, but by 2010, its share of GDP had shrunk to 11 percent of the economy and the sector employed about 2.5 million people, an attrition rate of nearly two-thirds[36]. The composition of the British manufacturing sector has also undergone a transformation, with large and medium-sized firms at the top of the manufacturing chain headquartered in the United Kingdom having become "a nearly extinct species." ICI, GCE, Lucas, "and all the rest have all been broken up and sold off."

> *The big British-owned factories of the 1970s are mostly closed or sold off either because of shareholder value demands for profit which encouraged retreat, or as a result of inept privatization which destroyed supply chains.*[37]

The closure of the factories doomed many of the assembly and process companies that supported them: "When the big factories closed, the supporting infrastructure decayed. Import dependency is the legacy…British manufacturing has downsized into workshops, as it loses its industrial districts."[38]

The proportion of British companies with 10 or fewer employees grew from 52 percent in 1983 to 76 percent in 2010, while the proportion of firms with 200 or more employees shrank from 4 percent to 1 percent during the same period.[39] Manchester University's Center for Research on Socio-Cultural

---

[33]*Professional Engineering.* 2010. "Gordon Brown's Record on Science, Engineering and Technology Defended by Lord Sainsbury." May 14.
[34]*Professional Engineering.* 2012. "Getting the Balance Right." May 1.
[35]Chakrabortty, Aditya. *The Guardian.* 2011. "Why Doesn't Britain Make Things Any More?" November 16.
[36]Ibid.
[37]Centre for Research on Socio-Cultural Change. 2011. *Rebalancing the Economy (or Buyer's Remorse).* Working Paper No. 87. Pp. 29-30.
[38]CRESC. 2011. *Rebalancing the Economy.* Op. Cit. p. 33.
[39]*The Guardian.* 2010. "David Cameron's Talk About Reviving British Industry is Nonsense." November 1.

Change (CRESC) commented in 2010 that "the legacy of Thatcherism and New Labour is a British manufacturing sector dominated by small workshops."[40]

CRESC released a study in 2011 which concluded that thirty years of welcoming inward foreign investment, coupled with the advent of a manufacturing sector dominated by small "workshops," would limit Britain's ability to use innovative technology to capture high levels of UK-content in manufacturing. Companies at the top of the United Kingdom's supply chains are typically foreign-owned branch assembly plants, such as the Japanese automakers, "limited by their role in a global division of labour established by a corporate parent for whom the United Kingdom is an important market but a relatively high wage cost production base whose export profitability is complicated by currency fluctuations against the Euro." Small workshops firms are often cogs in global supply chains which are largely non-British.[41]

A number of observers point out that the dramatic shrinkage of British manufacturing overshadows areas of abiding strength. The United Kingdom is still the ninth largest manufacturing nation on earth. Although the sector has contracted in terms of jobs and percentage of GDP, output is actually 25 percent higher than it was in 1970, reflecting the rapid improvement of labor productivity.[42] The British machine tool industry—critical to the manufacturing sector—recovered quickly from the recession and enjoyed dramatic growth after 2010, reflecting equipment orders from automotive and aerospace firms.[43] Rolls-Royce is a world class producer of technology-intensive products and is the second-largest maker of jet engines in the world. The United Kingdom has attracted investment by Japanese automobile producers and India's Tata group and retains the third largest auto industry in Europe. Two of the world's largest

---

[40]CRESC. 2011. *Rebalancing the Economy.* Op. Cit. p. 30. A technological divide also exists which, in effect, creates a "tale of two sectors" in manufacturing. "There is a cluster of highly successful firms in high-tech, innovative market sectors. But there is also a long tail of low value-added manufacturing firms that compete largely on price." *Engineering UK 2012*. Part 1.3.1.

[41]A 2009 Cambridge University study comments that because so much of the value chain in many sectors is outside the UK, "partnering with major multinational companies, often promoted as a reason for encouraging academic-industrial collaboration at the research stage, is quite unlikely to lead to significant added value, and jobs, in the UK." Mina, Connell and Hughes. 2009. *Intermediate Research Organizations.* Op Cit. Pp. 7. Tim Crocker, a representative of the UK's SME Innovation Alliance, told Parliament in 2012 that "I can promise you that there is only one business model in town for small companies: to get to £20 million and sell out. There has been a lot of research on building what has now become recognised as the equivalent of the German *Mittelstand* and how we grow medium-size companies from a sell-out value of £20 million to something with a revenue of maybe £100 million, which we are really short of in this country. To do that, you have to have longer-term finance that will buy out the VCs and keep the companies in this country. It may be in partnership with a UK company, but the statistics show that most of these companies are sold overseas because the Americans, Indians and Chinese all have their chequebooks open." House of Commons, Science and Technology Select Committee. *Bridging the Valley of Death. Improving Commercialisation of Research.* Oral Evidence. Q219.

[42]Engineering UK 2012. Part 1.3.1.

[43]*Professional Engineering.* 2012. "Sharpening Up." April 3.

and most successful pharmaceutical companies, AstraZeneca and GlaxoSmithKline, are UK-based. The United Kingdom is one of the world's leading countries in the development and manufacture of advanced composite materials. But "the erosion of the UK's manufacturing base over the last 30 years means that it is now highly skewed with some strong R&D intensive sectors such as pharmaceuticals and defense and aerospace and some, including electronic devices, ITC and materials, with weak absorptive capacity."[44]

A Shrinking Industrial Research Infrastructure

Structural changes in the British manufacturing sector have affected the industrial research infrastructure supporting manufacturing. Since the 1980s, a significant number of British companies have shut down their in-house laboratories, replaced by an approach exemplified by Cisco Systems—the acquisition of promising start-ups and/or their intellectual property[45]. AT&T Laboratories Cambridge, for three years Europe's leading communications engineering research laboratory, was closed in 2002[46]. Public research institutions supporting British industry have also been downsized. Rolls-Royce told Parliament in 2010 that in contrast to Germany:

> *The UK has over the past 25 years consistently dismantled its national research centre base with the loss of the Royal Aircraft Establishment, Marchwood and Leatherhead CEGB Centres and many others. It has also significantly reduced the scale and scope of remaining centres such as Culham, the National Physical Laboratory and the National Engineering Laboratory.*[47]

---

[44]Mina, Connell and Hughes. 2009. *Intermediate Research Organizations.* Op Cit. Pp. 7. A recent economic model commissioned by the BBC and developed by the Office of National Statistics and Cambridge Econometrics forecasts that Britain's manufacturing sector will continue to decline steadily through 2020. *Professional Engineering.* 2012. "Getting the Balance Right." May 1.

[45]Corporate laboratories closing included British Telecom's Martlesham laboratory and GEC Marconi's central research. *The Engineer:* 2010. "Centre Point: Hermann Hauser Discusses Commercializing Technology." September, 2010.

[46]*The Guardian* commented on the closing of the AT&T facility that "a great research laboratory is a very delicate organism - rather like a great symphony orchestra. It takes years to create, and very skilled management to keep it vibrant. Once broken up, it is impossible to put it back together again. Sometimes these strange social organisms run out of creative steam and deserve to die. But the AT&T lab was not one of these. In fact, in the last years of its life it invented what many of us thought would be one of the defining communications devices of this century (and, incidentally, the salvation of its telco owner) - the broadband phone." *The Guardian.* 2002. "Wrangling Money Men Shut Down the Future." April 20.

[47]House of Commons, Science and Technology Committee, Technology and Innovation Centre Enquiry, Written evidence submitted by Rolls-Royce. TIC 82. Ev. 68.

The small firms that increasingly comprise the British manufacturing sector are not well positioned to fill the research and training gaps left by the retreat of the big industrial labs.[48]

A major deterrent to R&D by small companies is the fact that even when a patent is secured on a promising technology, "you cannot exercise the rights of those patents unless maybe you have a fighting fund of half a million quid; that would be a minimum sort of fighting fund, as lawyers would advise you."[49]

Weaknesses in the Existing Knowledge Transfer Ecosystem

When the British government began examining the challenge of knowledge transfer from basic science to industry, it was frequently pointed out that the United Kingdom already had an extensive system of public and private organizations engaged in knowledge transfer, some of them excellent.[50] However, Britain's network of knowledge transfer organizations is a confusing jumble of entities that has been established over time with widely varying objectives and methods, not operating pursuant to a coherent national strategy. In 1992, an observer characterized this landscape as "an individually impressive but motley array of research institutions, science parks, training councils, and whatnot" in need of government direction, a description that is still apt today.[51] Ian Gray, the Chief Executive of the Technology Strategy Board, told Parliament in 2011 that "there are some very good centres operating within the regions today. A much bigger gap is the fact that they are sub-scale and that they have not been operating in a joined-up way, focused around UK-wide national

---

[48] *The Guardian*. 2010. "David Cameron's Talk About Reviving British Industry is Nonsense." November 1. Clive Hickman, CEO of the Coventry-based Manufacturing Technology Centre (MTC), observed in 2012 that in the current economic climate, "we know that all the small firms that are local to our area would love to take on apprentices but they can't see four years ahead of their order book so they won't take the risk. *Professional Engineering*. 2012. "Fast Growth at Technology Centre Bodes Well for Manufacturing." June 14. Aditya Chakrabortty, economics lead writer for *The Guardian*, commented in 2010: "Here's the problem with workshop Britain: compared with big companies, tiny establishments are less able to cope when recession hits: they do less research and development, and they are badly placed when it comes to exporting. In other words, they can't deliver on all those things politicians dream of. One- or two-man bands also don't tend to offer the apprenticeships so important to a career in engineering."

[49] Testimony of Time Crocker. House of Commons, Science and Technology Select Committee. "Bridging the Valley of Death." Improving Commercialisation of Research." Oral Evidence. Q219.

[50] House of Commons, Science and Technology Committee, Technology and Innovation Centre Enquiry, Written Evidence Submitted by the Universities UK-AURIL (TIC 77).

[51] *British Journal of Political Science*. 1992. "Microchips and Public Policy—The Political Economy of High Technology." October. p. 503.

priorities."[52] Thus, companies seeking technical support confront a bewildering task.[53]

The Association of Independent Research and Technology Organisations (AIRTO Ltd.) observed that—

> *The current UK approach has often resulted in sub-optimal and dispersed investments with the lack of long-term funding certainly damaging the ability of the established centres to: engage with business; realise the full potential of their assets; invest in long-term capability; recruit and retain the best staff; and commercialise leading edge research.*[54]

Work Force Issues

Various surveys of British industrial competitiveness have warned that the country faces a skills deficit in manufacturing. The 2010 Dyson report noted

[52]House of Commons, Science, and Technology Committee, Technical and Innovation Centres enquiry. Oral evidence. January 12 2011. Q84 Ev 24.

[53]The 2011 House of Commons enquiry into proposed Technology and Innovation Centres revealed that "nobody" had a clear understanding of what technology intermediary organizations already existed in the United Kingdom. Dr. Tim Bradshaw, Head of Enterprise and Innovation at the Confederation of British Industry, testified that because no one "really knows" what already existed in the United Kingdom, small and medium enterprises, in particular did not know where to look for research support. "SMEs, in particular, do not know where the best facilities are that they could go and tap into, where there is research going on from which they can benefit and where there is best practice from which they could learn. Part of setting up a TIC model ought to be to try to work out a better model of getting that information out to a community that might actually want to use that. House of Commons, Science and Technology Committee, Technology and Innovation Centre Enquiry, Oral Evidence. December 15, 2010, Q27.

[54]House of Commons, Science and Technology Committee, Technology and Innovation Centre Enquiry, Written Evidence of Association of Independent Research and Technology Associations (AIRTO Ltd.) (TIC 12). Ev 38. The UK's network of "Micro and Nanotechnology Centres" (MTN) is sometimes cited as an example of the weakness of the country's intermediary research infrastructure. The MTNs were widely-dispersed research centers established in 2004-5 funded by the British government, Regional Development Agencies (RDAs), and the Devolved Administrations of Wales and Scotland. A review of the MTNs commissioned by the TSB, the RDAs and the Devolved Administrations concluded that "while the development of open access facilities for business was undoubtedly beneficial, the investment was thinly spread across a number of centres, resulting in "sub-critical" activities that compromised the abilities of the centres to achieve world-class capabilities." Professor Ric Parker, Director of Research and Technology at Rolls-Royce, was more unsparing: "If you look at the £50 million the UK invested in nanotechnology centres, we tried to create 32 centres with £50 million. Frankly, it was a bit of a disaster, because you are not going to create any critical mass of activity if you spread it that thinly." House of Commons, Science and Technology Committee, Technology and Innovation Centre Enquiry, Oral Evidence. December 15, 2010, Q4 Ev 2.

that 43 percent of the country's engineering companies were experiencing difficulty in securing the right graduate recruits.[55]

At present, the United Kingdom has one of the lowest literacy rates in Europe. Its secondary educational system does not sufficiently foster education in the sciences. That fact makes it harder for British universities to assemble students who are capable of pursuing degree programs in areas such as mathematics, chemistry, the physical sciences, and biology.[56] In the universities, "in many cases, courses in the applied sciences are too theory-oriented and lack practical application in a workplace setting."[57]

The failure of Britain's education system to prepare young people for industrial careers has been the subject of public discussion for well over a century.[58] Throughout the Twentieth Century, Britain trailed far behind Germany in the number and quality of technical schools and polytechnics. Industrial apprenticeship became "entangled in conflicts between employers and craft unions" and in many cases devolved into mere exploitation of young people as cheap laborers.[59]

---

[55]Dyson. 2010. *Ingenious Britain.* Op cit. pp. 21. Engineering UK, a not-for-profit corporation promoting engineering, observed in its 2012 annual report that "if the engineering and manufacturing sector is to rebalance the economy the UK needs to fundamentally rebalance its skills capacity...[T]he education system is not giving young people the skills that businesses need. This skills shortage is particularly acute in the manufacturing sector. In a recent survey, 20 percent of manufacturers reported skills gaps, while 31 percent of high-tech manufacturing firms had recruited people from outside the UK owing to a lack of suitably qualified people from within the UK." Engineering UK, *Engineering UK 2012.*

[56]Interview with Professor Nigel Titchener-Hooker. Head, Department of Biochemical Engineering. University College London. June 11, 2012.

[57]CRESC. 2011. *Rebalancing the Economy.* Op. Cit. p. 44-45

[58]In 1902, during debates over what would become the country's first legislation establishing a national education system, the eminent historian and educator G.M. Trevelyan commented that "This bill...is demanded by the people from a sense of shame in our possessing the worst industrial peasantry in the west of Europe, a fear on the part of our industrial population that we shall not be able to meet commercial competition." Lowndes, G.A.N. 1937. The Silent Social Revolution: An Account of the Expansion of Public Education in England and Wales. Oxford University Press. p. 91.

[59]Thelen, Kathleen, 2004. How Institutions Evolve: The Political Economy of Skills in Germany, Britain, the United States, and Japan. Cambridge: Cambridge University Press. Pp. 92-147, 283. In 1994 the government initiated the Modern Apprenticeships program, managed by regional Training and Enterprise Councils (TECs), providing for employer training of 16-25 year olds and certification of skills on a sectoral basis. A study of the program published in 1998 observed that it was "somewhat fragmented," that without exception all the TECs in the study stated that employers do not have the training resources to take on modern apprentices," and that the TECs felt "most of those leaving school at 16 to work full-time do not have the capacity to undertake a Modern Apprenticeship...a sad commentary on the current pre-16 education system." The study concluded that "unless that is both robust financial commitment to apprenticeship training and a scheme which is national rather than local in focus, it is unlikely that the United Kingdom will produce a labour force capable of competing with our European trading partners." Gray, David and Mark Morgan. 1998. "Modern Apprenticeships: Filling the Skills Gap?" *Journal of Vocational Education and Training.* Vol. 50, No. 1. pp. 123, 129, 131.

During the past two decades, the British government has devoted a substantial effort to upgrading its apprenticeship programs, and between 2001-2 and 2008 the number of apprentices completing their programs grew from 40,000 per year to 100,000 annually, a trend that the government characterized as a "renaissance."[60] Effective in 2012, small and medium enterprises hiring their first apprentice were entitled to a £1500 government incentive.[61] In 2012, BIS launched a review of government-sponsored apprenticeships to "build on the record success of recent years," maximize the value of the government's investment, ensure that each apprenticeship delivers high quality training. The review is headed by Doug Richards, a serial entrepreneur and founder of the School for Startups.[62] Despite the considerable progress that has been achieved in reforming Britain's apprenticeship system in the past decade, daunting challenges remain. A 2012 study found that SMEs—

> *expressed concern about the lack of preparedness of some 16-18 year olds for the realities of apprenticeship. This ranges from difficulty in getting to work on time, unrealistic expectations of what the role entails and a lack of basic English and Maths. It seems that schools no longer provide this support.*[63]

---

[60]Department for Innovation, Universities and Skills. 2008. *Would Class Apprenticeships: Unlocking Talent, Building Skills for All.* January. P. 3,5. "Apprenticeship Frameworks" are drawn up by Sector Skills Councils (independent, employer-led organizations) outlining a program of learning and training against which the government provides apprenticeship funding. The Frameworks must conform to guidelines set forth in the government's National Apprenticeship's blueprint. Apprentices completing such programs receive a nationally consistent completion certificate. In the apprenticeship programs, the employer offers a paid job (at a reduced wage)to the employee, plus training and support. The government fully funds apprenticeships for employees aged 16-18, and co-pays half the cost (with the employer providing the other half, for individuals between 19 and 24 years old. Government support is reduced for individuals who are 25 or older, those at large employers, and those with prior skills. Beginning in 2013-14, government loans will be available to individuals 24+ years old to finance half the cost of the apprenticeship. BIS. 2012. *The Richard Review of Apprenticeships: Background Evidence.* p. 7.

[61]Holt, Jason. 2012. *Making Apprenticeships More Accessible to Small and Medium Enterprises.* May. p. 30.

[62]*Apprenticeships.* 2012. "Doug Richards to Lead Government Apprenticeships Review." June 12.

[63]Holt, Jason. 2012. *Making Apprenticeships More Accessible to Small and Medium Enterprises.* May. p. 29. Another challenge is an acute shortage of employers willing to take on apprentices. CRESC observed in 2011 that "Thirty years ago, the large British firms in every major town accepted the costs of apprenticeship and technical education as a kind of social overhead which they willingly paid; now it's 'can't pay and won't pay' from both the small firms and those that are large. In 2011, the number of applications in the United Kingdom for apprenticeships exceeded available positions at companies by a ratio of about 10 to 1. BIS. 2011. *Equality Impact Assessment: Changes to the Apprenticeship Programme.* November 25.

GOVERNMENT SUPPORT FOR APPLIED RESEARCH

The British government has been grappling with the challenge of applied research for decades. The government institutions directing this effort have undergone reorganizations with a frequency which contrasts sharply with the relative stasis of the German government's organizational structures for fostering innovation. The key British government organizations involved in supporting applied research and innovation are summarized below.

Technology Strategy Board

The Technology Strategy Board (TSB) was set up in 2004 as an advisory organization in the UK Department of Trade and Industry (DTI). In 2007, under Prime Minister Gordon Brown, DTI was reorganized into the Department of Innovation, Universities, and Skills (DIUS) and the Department for Business, Enterprise, and Regulatory Reform (BERR). The TSB was spun off as an independent public organization ("non departmental public body") with a mandate to operate at arm's length from the government. DIUS was merged with BERR in 2009 to create the Department for Business, Innovation and Skills (BIS). The TSB is funded by BIS with a mission of investing in "market pull" collaborative R&D, promoting university-industry interaction, and promoting the exploitation of emerging technologies. It funds individual companies and institutions based on competitive peer review. The TSB is comprised largely of individuals with business experience.[64]

Research Councils

The United Kingdom's seven Research Councils fund basic research and postgraduate training. They are funded from the government's science budget through the Department for Business, Innovation and Skills (BIS). Funds are allocated to universities on the basis of competitive peer review.

---

[64]The TSB is staffed by "industry R&D types and a few former civil servants…no kids out of school." Interview with David Way, Director of Knowledge Exchange and Special Projects, Technology Strategy Board, London, June 12, 2012. The TSB does not support basic research or "right-before-market" research, but concentrates on the broad area in between the two. It "does not pick winners, but winning areas." It concentrates on technology areas in which the United Kingdom already possesses world class science and in which there has been "clear market failure" with respect to the commercial exploitation of that science. It carries projects "across the valley of death" to commercialization, with typical levels of support around £2 million. The TSB is currently investing £350 million per year in research projects. Interview with Professor Chris Maran, Head of Regenerative Medicine Bioprocess Group, Advanced Centre for Biochemical Engineering, University College London, June 11, 2012.

Higher Education Funding Councils

The Higher Education Funding Councils (England, Wales, Scotland, and Northern Ireland) are overseen by BIS and provide grants to universities supporting research and research infrastructure. The Higher Education Innovation Fund (HEIF) is specifically targeted at knowledge transfer from universities to industry.

Regional Development Agencies

Until 2012, Regional Development Agencies (RDAs) were business-based organizations aimed at promoting regional economic development, including innovation and science. The RDAs supported the establishment of innovation clusters adjacent to universities, science parks, and incubators for start-ups.

In June 2010, the British government indicated that it would abolish the RDAs in 2012, with future economic development relegated to Local Enterprise Partnerships (LEPs), which did not receive any funding from the central government.[65]

Small Business Research Initiative (SBRI)

The SBRI program uses government procurement to provide business opportunities for small businesses by funding research challenges facing government organizations. The individual projects are funded by the organizations seeking the research, with companies selected in competitive process administered by the TSB (with the exception of Health and Defense).

Manufacturing Advisory Service

The Manufacturing Advisory Service (MAS) is a government agency established in 2011 by the Department of Trade and Industry to provide strategic and technical advice to small and medium enterprises in the United Kingdom. MAS was originally comprised of regional Centres of Manufacturing Excellence in various parts of the United Kingdom, funded through the regional development agencies. In 2011, with the abolition of the RDAs looming, MAS was reorganized as a national level agency with a 25 percent lower budget.[66]

---

[65]LEPs are partnerships between local governments and businesses to help identify local economic priorities and opportunities for economic growth and job creation.
[66]Interview with Manufacturing Advisory Service. London. June 11, 2012.

Pitfalls of the Fraunhofer Model

British policymakers have long looked with admiration at Germany's Fraunhofer-Gesellschaft and debated the possibility of adapting the Fraunhofer model to improve Britain's manufacturing competitiveness. Experience has demonstrated, however, that the complexity of the German system of innovation raises the risk that the wrong lessons can be drawn from it with unfortunate consequences.

In 1990, with the United Kingdom in recession, the Thatcher government began considering measures to bolster the British manufacturing sector.[67] In 1991 the House of Lords Select Committee on Science and Technology issued a report with castigated Britain's "antipathy to the manufacturing industry," citing the "great strength" of countries like Germany and Japan as arising from "well-tried systems for transferring technology from academic institutions to industry and from one company to another."[68]

In 1992 the Prince of Wales, who was interested in innovation policy, asked Sir John Fairclough, a computer engineer and former advisor to the Thatcher government, to head a working group to explore how to improve technology transfer. The working group published an interim report in 1992 recommending that the United Kingdom set up a network of "Faraday Centres," modeled on Germany's Fraunhofer institutes, to foster innovation.[69] These institutions would concentrate on transferring technology by transferring people.[70] The Conservative government endorsed this proposal, with the caveat that the Faraday Centres should be created out of existing institutions and funded by private industry. The Labour Party declared that it, too, would open technology transfer centers if it won upcoming elections although it would call them "Newton Centres" and build new institutes funded in part by unspent funds already in DTI's budget.[71] The Liberal Democrats endorsed a similar concept in 1991. Britain's major parties entered the General Election of 1992 in the midst of "Fraunhofer Fever," with all three major parties advancing proposals to

---

[67] *New Scientist*. 1990. "Comment: Manufacturing Disaster." November 30; *The New Statesman*. 1992. "The Revolt of the Industrialists." October 31; *New Scientist*. 1991. "Last Chance for British Industry." November 30.

[68] House of Lords Select Committee on Science and Technology. 1991. *Innovation in Manufacturing History*. HM Stationery Office; *New Scientist*. 1991. "Last Chance for British Industry." November 30.

[69] Wood, Audrey. 2001. *The Magnetic Venture: The Story of Oxford Instruments*. Oxford: Oxford University Press. Pp. 338; *The Independent*. 1992. "Germany Shows Path to Industrial Recovery." October 14.

[70] "The centres will take on graduate engineers, train them to PhD level in commercial surroundings and allow them to move to industry." DTI Secretary Lilley said that "nine times out of ten it is people that transfers technology. These students will go into industry with that much greater experience." *New Scientist*. 1992. "Faraday or Newton to Bridge the Innovation Gap." February 29.

[71] *New Scientist*. 1992. "Faraday or Newton to Bridge the Innovation Gap." February 29.

emulate the Fraunhofer model.[72] The Conservatives won the election and subsequently ruled out the literal replication of Fraunhofers in the United Kingdom "in the different circumstances of this country." There would be "no new buildings." Instead, DRI would seek to reproduce the Fraunhofers' success by "arranging marriages between higher-education institutions, including universities and polytechnics, and industrial research organizations."[73] In October 2002 the first 50 students enrolled in five pilot projects constituting prototype Faraday Centres—"Postgraduate Training Partnerships."[74]

Critics of the Faraday Centres pointed out that the government "fundamentally misunderstood" the Fraunhofer model. They noted that the Prince of Wales' working group had concluded that the Fraunhofers received "a large slice of their funding from industry." In fact, although Fraunhofer was at the time receiving about 70 percent of its income from contract research, over half of this represented government funds used for public research contracts. Industry was contracting with the institute for the remainder but much of this research was also co-funded by the German research ministry, BMFT, and other EU and German government entities. "So perhaps as little as 20 percent of a Fraunhofer's income comes from outside government."[75]

The government's belief that "most of the technology transferred to industry by the Fraunhofer takes place through 'people transfer'" was also wrong. It was noted that according to Dirk-Meintz Polter, Deputy Director of the Fraunhofer, the main vehicle for technology transfer was contract research for industry:

> *Technology transfer through people is a secondary thing. It is something we want, and encourage, but it is not a primary goal. It is not something we measure.*[76]

---

[72]*Physics World.* 1992. "Fraunhofer Fever Hits the UK." March.
[73]*New Scientist.* 1992. "British Innovation, German Style." March 21.
[74]In these projects the government gave grants to the students collaborating with universities or contract research organizations with the expectation that the students would gain their PhDs performing contract research for industry. The government paid the students' tuition at participating universities, and the contract research organizations received funds from DTI to cover associated costs. There were "high hopes that the Faradays [would] become first-class training centres for engineers."*New Scientist.* 1992. "German Innovation, British Innovation." November 21.
[75]Even this conclusion was misleading because the term "income" was used to refer to running costs only. "All expenditure for building and equipment comes from either the BMFT or state governments…In contrast to Lilley's belief that the Fraunhofers do not rely on public expenditure, well over 80 percent of their money comes from taxpayers." Ibid.
[76]An editorial in the New Scientist in 1992 commented that "Although they have been touted as a way of helping Britain emulate German success in transferring technology from academia to industry, the first five pilot centres for the Faraday programme are merely a new way of training postgraduate engineers." New Scientist. 1992. "German Lessons Half Learnt." November 21.

The Faraday Centres also lacked the political and popular support that Fraunhofer enjoys throughout German society. Michael Heseltine, President of the Board of Trade and the House of Lords Select Committee on Science and Technology, rejected the whole concept, declaring that "I do not think there is a case for trying to set up a new organization called Faraday Centres."[77] The Lords stated that "the programme is not necessary to give scientists experience of industrially relevant research, nor for providing technology transfer."[78] Crucially, the Department of Trade and Industry "did not find the budget to provide what would be the core funding."[79] The Faradays were forced to apply for peer-reviewed grant funds "along with everybody else." As the Faradays pursued industry funding by moving research toward the applied end of the spectrum, their scores in the peer review process worsened.[80] In 2004 the surviving Faraday Centres were merged or replaced by so-called Knowledge Transfer Networks (KTNs) administered by the Technology Strategy Board.[81] Professor Ric Parker of Rolls-Royce commented before Parliament in 2010: "How many of the Faraday Centres that we set up in the 90s still exist? There is only one in the guise of the Begbroke Science Park at the Materials Centre there."[82]

THE TIC INITIATIVE

The failure of the Faraday Centres was followed by a succession of major studies by the government, the political parties, and public organizations hammering on the same theme—Britain urgently need to put its science to work for industry[83]. In March 2010, the United Kingdom's two leading political parties each released a study examining the potential for utilizing innovation to improve the global competitiveness of British industry. Although not identical in

---

[77]*Nature.*1993. "Faraday Centres Appear Doomed After Criticism by Lords' Panel." (Issue 6412).
[78]*New Scientist.* 1993. "Spark of Life for Faraday Centres." March 6.
[79]DTI "had major reservations about providing any additional resources for funding new mechanisms, particularly ones which smacked of state subsidy for what should be market-driven innovation." Webster, Andrew. 1994. "Bridging Institutions: The Role of Contract Research Organizations in Technology Transfer." *Science and Public Policy.* April. Pp. 95.
[80]House of Commons, Science and Technology Committee. *Technology and Innovation Centres.* Second Report of Session 2010-11. Pp. 8.
[81]"When the TSB was formed these centres were folded into its remit, and most have now become KTNs that bring scholars and businesses closer together." *Times Higher Education.* 2010. "Research Intelligence: Getting More 'D' Out of R&D." April 15.
[82]House of Commons, Science and Technology Committee. Enquiry about Technology and Innovation Centres. December 15, 2010. Q12 Ev 5.
[83]HMG. 2003. *Lambert Review of Business-University Collaboration.* HM Stationary Office; Lord Sainsbury of Turville. *The Race to the Top: A Review of Government's Science and Innovation Policies.* Dyson, James. 2010. *Ingenious Britain: Making the UK the Leading High Tech Exporter in Europe.* Hauser, Hermann. 2010. *The Current and Future Role of Technology and Innovation Centres in the U.K.*

their findings and recommendations, the studies were sufficiently congruent to form the basis for new technology promotion policies supported by all three major parties. The most significant of these was the creation and funding by the government of a number of "technology integration centres" (TICs) to bridge the gap between academic research and commercialization.

In 2009, Lord Mandelson, Secretary of State for the Department for Business Innovation and Skills, commissioned a study by an eminent Austrian-born UK entrepreneur and scientist, Dr. Hermann Hauser, to examine business-focused "Technology and Innovation Centres" and their existing and potential role in enabling knowledge transfer in the United Kingdom.[84] Hauser studied TICs in a number of countries and the existing knowledge transfer institutions in the U.K. Drawing particularly on Germany's Fraunhofer-Gesellschaft as a model, Hauser set forth a blueprint of recommendations for the establishment of a national network of TICs in the U.K. supported in part by long-term, stable government funding.[85] The Labour government stepped down in May 2010 before Hauser's recommendations could be implemented. However, the Conservative Party, the biggest party within the coalition which emerged after the election, had just released a report by James Dyson, an eminent British industrial designer and entrepreneur, examining how Britain could become Europe's leading generator of new technology. He offered wide-ranging recommendations with respect to educational reform, knowledge exploitation, support for high-tech start-ups and established companies, and promoting increased esteem for science and engineering in the United Kingdom. Like Hauser, Dyson addressed the challenge of translating university research into commercial products, and recommended that –

> *"New university/industry research institutions capable of becoming centres of excellence in a particular research field should be given government sponsorship. These institutions should provide space for interactions, promote staff moving between business and academia and allow sharing of expensive resources."*[86]

---

[84]*The Manufacturer*. 2012. "What is the High Value Manufacturing TIC?" January 6.
[85]Hauser, Hermann. 2010. *The Current and Future Role of Technology and Innovation Centres in the U.K.* March. The Guardian commented that "The German model tends to become fashionable in Westminster every 20 years or so, like flared trousers." *The Guardian*. 2012. "British Politicians Are Praising Germany: But Is It Just a Passing Fad?" September 18.
[86]Dyson, James. 2010. *Ingenious Britain: Making the UK the Leading High Tech Exporter in Europe*. p. 38.

The 2010 Comprehensive Spending Review

The Coalition government of the Conservative and Liberal Democratic parties which took power in May 2010 was committed to eliminate the country's deficit and implement steep cuts in government spending.[87] In October 2010, the Chancellor of the Exchequer for the coalition government adopted a Comprehensive Spending Review, characterized by deep cuts in research budgets. The amount allocated to the Research Councils for equipment and facilities was cut in half, forcing a consolidation of existing research infrastructure. Austerity forced hard choices on the government, which decided to maintain spending on priorities at the expense of other programs, stop support for public engagement, and simplify the technology-transfer landscape by rationalizing the clutter of multiple tech-transfer organizations.[88] The Spending Review committed £200 million annually through 2014-15 to support manufacturing and business development, emphasizing the commercialization of technology and support for potential high growth companies. The government indicated that these funds would be allocated to support the Manufacturing Advisory Services to fund "growth hubs," to support high growth companies, and to create a network of Technology and Innovation Centres (TICs). £200 million was earmarked over five years for "an elite network of research and development Technology and Innovation Centres."[89]

The TIC Concept

In the fall of 2010, the Government announced that it would invest £200 million to create a network of Technology and Innovation Centres over a 4-year period. The Technology Strategy Board would be charged with creating and supervising the TICs. In the 2010-11 Parliamentary enquiry into Technical Innovation Centres, there was considerable support for a government project that would build upon existing research organizations rather than creating entirely new entities that would displace or compete with established ones. Professor Ric Parker, Director of Research and Technology for the Rolls-Royce

[87]"The [current] UK government's raison d'être, its overarching narrative since it was elected in 2010, has been to eliminate the UK's structural deficit over the lifetime of this Parliament." Trades Union Congress. 2011. *German Lessons: Developing Industrial Policy in the UK.* December 8. p. 11.

[88]Interview with Professor Nigel Titchener-Hooker, Professor Chris Mason , University College London and Professor Gary Lye, Engineering and Physical Sciences Research Council, London June 11 2012.

[89]HM Treasury. 2012. *Spending Review 2010.* Cm 7942. Pp. 52. October. Mike Oldham, head of the TIC program (now designated "Catapult") observes that "we got through the October 2010 savage expense review…we survived intact, one of the few programs that did." Interview, London, June 12, 2012.

Group, said that "we can't afford greenfield sites if we are going to make any real impact on this."[90]

TSB Chief Executive Iain Gray told Parliament that the TSB had surveyed "nearly 100 centres" around the country and concluded that "a shortlist of a couple of dozen" were "operating at what I would call a reasonably good regional maybe even national level, but what we really want to do is to identify a small number of world-leading centres—six to eight—that can operate on a world stage."[91] In January 2011, the TSB published a prospectus for the TICs establishing basic principles, criteria for selecting their areas of specialization, and guidelines for how they would be developed and run. The prospectus clarified a number of basic points about the nature of the TICs:

- **Physical sites:** The TICs would not be virtual but would operate from one or a small number of linked physical sites. They would provide access for businesses to "the best technical expertise, infrastructure, skills, and equipment that would otherwise be outside the reach of individual companies."[92]
- **Funding model:** The TICs will be funded through a combination of core funding provided by the TSB and public and private contract research revenues. The core funding from the TSB is expected to enable sustainable investment over the course of five years with the expectation of continuing for another five years.[93]
- **Scale:** The TICs are envisioned as deploying a technical staffs of roughly 150 highly qualified individuals to achieve annual turnover of £20 to £30 million. Each TIC will need to attract £10 to £15 million per year from the private sector to be viable.
- **Governance:** Each TIC will be run by an "autonomous, business led management board." The TSB will supervise the TICs utilizing an

---

[90]House of Commons, Science, and Technology Committee, Technical and Innovation Centres enquiry. Oral evidence. December 15, 2010. Q4 Ev 7. Dr. Tim Bradshaw, Head of Enterprise and Innovation for the Confederation of British Industry, testified that "There are already existing centres in the commercial space, operating commercially, that we do not need to replicate or tread on the toes of…If we are trying to get an impact on growth and commercial return quickly, then it is probably better to start, by and large, with things we have already and build from there."

[91]House of Commons, Science, and Technology Committee, Technical and Innovation Centres enquiry. Oral evidence. December 15, 2010. Q84 Ev 24.

[92]Interview with Mike Oldham, Head of Catapult Centres Programme. London. June 12, 2012.

[93]In addition to core funding, David Willets, the Minister for Science and Universities, suggested in 2011 that the TICs might acquire existing assets from other public entities. "There are some interesting questions about what is happening to some of the assets of the [Regional Development Authorities] which have not been fully resolved. You may find that some things that currently belong to someone else such as a university or an RDA, could become part of this new [TIC] legal entity. I don't think that's a barrier." House of Commons, Science, and Technology Committee, Technical and Innovation Centres enquiry. Oral evidence. January 12 2011. Q123 Ev 31.

internal advisory oversight committee supported by a small staff or "programme team."[94]

- **Timing:** All seven TICs were expected to be fully operational in 2013.[95]
- **Metrics:** The ultimate measure of success of the TICs will be wealth creation in the United Kingdom, but given the subjectivity and long time frame entailed in such an assessment, intermediate measures will be used, including the value of work won competitively, the number of new and repeat customers per year, number of new projects, intellectual property created and new businesses created.[96]

Criteria for Selecting Technology Areas

The Technology Strategy Board indicated that it would select the target technology areas in which the TICs would specialize according to five criteria: 1) Global markets for technology worth billions of pounds per year; 2) world-leading UK research capabilities in the technology area; 3) existence of a commercial base in the United Kingdom capable of exploiting the technology to capture a "significant share of the value chain;" 4) an opportunity to "anchor" knowledge-intensive activities and wealth creation in the United Kingdom; and 5) an alignment with strategic national priorities. The TSB indicated that its first priority was establishment of a TIC for high-value manufacturing.[97]

Branding

One aspect of the Fraunhofer admired by British policymakers was the strength of the Fraunhofer brand, which is admired around the world and is closely associated with the excellence of German products and engineering. TSB Chief Executive Gray emphasized that British TICs should be given a collective brand name "that we can all buy into and use" as a "very powerful mechanism for the UK."[98] In the end, it was decided to call the TICs "Catapult Centres" based on the use of "catapult" as a verb meaning "to thrust forward or move quickly."[99]

---

[94]House of Commons, Science, and Technology Committee, Technical and Innovation Centres enquiry. Oral evidence. January 12 2011. Q130 Ev 32.
[95]*Professional Engineering.* 2012. "Catapults to Fame and Fortune." July 31.
[96]TSB. 2011. *Technology and Innovation Centres: A Prospectus.* January.
[97]*Professional Engineering.* 2011. "Innovation to Boost Economy. February 7.
[98]House of Commons, Science, and Technology Committee, Technical and Innovation Centres enquiry. Oral evidence. December 15, 2010. Q104 Ev 27.
[99]*Mondaq.com.* 2012. "Innovation and Research Strategy: A Summary (December 2011)." January 18. *InnovateUK.* "Technology Strategy Board Reveals Name for Technology and Innovation Centres." December 8.

Intellectual Property

The House of Commons Science and Technology Committee said in its report on the TIC enquiry that "the management of intellectual property rights will be crucial to an effective working relationship between TICs, academia, and business. We ask the TSB to set out principles for IP management including an outline of current best practice in its TIC implementation plan but we reject the need for prescription."[100] The TSB has established guidelines for the handling of intellectual property by the Catapult Centres which provide that each Catapult will agree with the TSB as part of its grant funding agreement an IP policy establishing the approach to be taken and the processes which will be adopted. The IPR arrangements are expected to facilitate the acceleration of the commercialization of new technologies and the growth of high-technology industries in the United Kingdom. Ownership of the IPR will vary according to funding source:

- If research is conducted exclusively under core funding, the Catapult itself is expected to own the IPR and license it to business users.
- If R&D is performed pursuant to a contract with a company, the company is expected to receive exploitation rights for the IPR but the IPR arrangements must not inhibit future use of the Catapult's background IPR.
- If collaborative R&D is jointly funded by the Catapult and business(es), it is expected that all partners will agree to "appropriate arrangements to share the rights to exploit IP created." The TSB will not allow Catapults to use core funding for collaborative R&D projects.[101]

Relationships with Universities:

In a number of Catapults built on existing research organizations, ongoing relationships between the intermediate research organizations and universities already exist. However, the way the Catapults will interact with universities and other basic research institutes is not yet clear.[102] A key question will be the IP arrangements, which will be negotiated by the Catapults, and both

---

[100]House of Commons, Science and Technology Committee. *Technology and Innovation Centres.* Second Report of Session 2010-11. pp. 37.
[101]Technology Strategy Board. 2012. *Catapult Centres—Guidelines for Management of Intellectual Property.*
[102]Interview with David Way, Director of Knowledge Exchange and Special Projects, Technology Strategy Board, and Mike Oldham, Head of Catapult Centres Programme, London, June 12, 2012. David Way of the TSB indicates the UK will not follow the Fraunhofer model in this respect, faulting it because "all [Fraunhofer] centers are run by academics." Mike Oldham, who heads the Catapult program, say that "how [for the Catapults] to reach into the research base" for technologies sought by industries "is one of the really difficult questions that we have."

universities and industry are reportedly saying, "give us some IP models we can use."[103]

Building on Existing Organizations:

The TICs are being built out of existing applied research and technology transfer entities, a process which is currently under way. TICs will not incorporate the whole landscape of applied research organizations but pick and choose a few of the best ones. A partial taxonomy of the types of entities involved is set forth below.

- **Research and Technology Organizations (RTO):** RTOs are independent "special knowledge organizations" which develop and transfer technology to industry. Many of them trace their origins to the Research Associations (RAs) which were set up between the 1920s and the 1940s to access government funding from what was then the Department of Scientific and Industrial Research as well as funding from industry. They have evolved into a heterogeneous array of commercial and not-for-profit entities which concentrate on laboratory and technical consultancy services. Most of them have a specific sector focus.[104]
- **Knowledge Transfer Partnerships:** Knowledge Transfer Partnerships (KTPs) involve three actors, a UK-based company, a knowledge base partner (in most cases a university) and a university graduate, a "KTP Associate," who is supervised by a mentor with appropriate technical skills ("Academic Supervisor"). The company designates a project that is strategic in character, capable of conveying economic benefit to the firm, and must involve an effort the company could not have undertaken without the assistance of the knowledge base partner. The KTP Associate works at the company to execute the project with the

[103]Interview with David Way, Director of Knowledge Exchange and Special Projects, Technology Strategy Board, and Mike Oldham, Head of Catapult Centres Programme, London, June 12, 2012.
[104]House of Commons Science and Technology Committee. Technology and Innovation Centres Enquiry. Written Evidence Submitted by Rolls-Royce, TIC 82. Ev. 68-69. Some of the most impressive RTOs are backed by Rolls-Royce, which has developed a network of seven Advanced Manufacturing Research Centers (AxRCs), five of which are located in the United Kingdom. In these organizations, Rolls-Royce, other industrial partners, and universities conduct research on manufacturing process technologies. The UK-based magazine *Machinery* observed in 2008 that one of the AxRCs, based in Rotherham, Yorkshire, in partnership with Boeing, had "established itself as probably the highest profile manufacturing process development facility in the UK over its short life." *Machinery*. 2008. "The Final Piece." September. The AxRCs "operate in the traditionally difficult space for manufacturing where university research is applied and developed in conjunction with full-scale capital equipment, overcoming the 'Valley of Death' between technology validation and productionisation."

frequent presence and contribution of the Academic Supervisor. The company may or may not ultimately offer employment to the KTP Associate but many actually do. Roughly 1000 KTPs are operating at any given point.[105] KTPs are funded by grants, drawn from contributions from over 15 public organizations and are supervised by the Technology Strategy Board.

- **Knowledge Transfer Offices:** Knowledge Transfer Offices (KTOs) are offices established within universities to disseminate information about current university-based research to business, nonprofits, and government agencies. Most of them are funded through the Higher Education Innovation Fund, allocated on the basis of faculty size and knowledge-transfer income. Some KTOs provide back office services for collaborative research agreements with companies, train business employees and coordinate industry networks. The 2010 Dyson report noted that businesses have complained that KTOs display "unrealistic expectations in IP negotiations.[106]
- **Knowledge Transfer Networks:** Knowledge Transfer Networks (KTNs) are networks designed to raise nationwide awareness of specific technology areas, facilitate research, communication and information sharing, and connecting companies in need of technology with entities capable of providing it. Funded by the Technology Strategy Board, the UK's KTNs are now hosted on __connect, an online knowledge sharing platform.

High Value Manufacturing Catapult

The first TIC opened by the Technology Strategy Board was the High Value Manufacturing TIC, launched in October 2011 and subsequently dubbed the HVM Catapult[107]. The TSB put the establishment of the HVM Catapult on a fast track, reflecting, the government's emphasis on revival of manufacturing.[108]The HVM Catapult's activities will embrace all forms of manufacturing utilizing metals and composites, as well as process manufacturing technologies and bio-processing. It will support a broad range of UK industrial sectors, including pharmaceuticals and biotechnology, electronics, food and beverage processing, chemicals, automotive, and energy.[109]

---

[105]The KTPs were launched in 2003, superseding the Teaching Company Scheme, which had been established in 1975.

[106]Dyson. 2010. *Ingenious Britain.* Op Cit p. 13.

[107]*The Manufacturer.* 2011. "First National Technology and Innovation Centre Opens." October 14.

[108]*Professional Engineering.* 2011. "Innovation to Boost Economy." February 7.

[109]UK Trade and Investment. 2012. *UK Advanced Engineering.* Pp. 29. The scale of the HVMC exceeds that of most Fraunhofer institutes. The HVMC is expected to staff up to a level of 750-1000 people, compared with a typical Fraunhofer staffing of 400 people per institute. The HVMC

The HVM Catapult aggregates seven UK research centers that are already operating. These organizations are very differentiated, with the Advanced Manufacturing Research Centre (AMRC) heavily oriented towards aerospace and EMG (Warwick Manufacturing Group) oriented towards automotive technology. "A big part of Catapult was to get them all to agree to a common management structure and a common strategy. The big question is centrifugal force…[some of them] almost saw each other as enemies."[110] The seven manufacturing research centers were receiving government funding through the regional development agencies. With the abolition of the RDAs in 2011 the centers confronted the loss of government funding and were thus "incentivized" to cooperate with Catapult.[111]

One of the organizations being folded into the HVM Catapult, the Advanced Manufacturing Research Centre, is a collaboration between the University of Sheffield and roughly 60 private companies which include BAE, Boeing, and Rolls Royce, which enjoys a world-class reputation for high quality collaborative research. The AMRC was established in 2004 with an eye to building on Sheffield's centuries-old expertise in metallurgy and metalworking to develop new manufacturing techniques involving metals as well as new materials.[112] The AMRC has already had a dramatic impact on Sheffield's economy. Drawn by the prospect of the proximity of a "world-class facility," companies such as Rolls-Royce have opened new factories in the Sheffield area, drawing with them smaller supply-chain companies. Richard Wright, Executive Director of the Sheffield Chamber of Commerce, observed of the AMRC that—

> *It was a long-term investment at a time when manufacturing was not really the thing in the UK, in 2002-3 when there was a different impression of manufacturing. If you take a long-term view of things you can build something special. Because we have invested in that area we are now seeing the benefits in terms of a buoyant advanced manufacturing center in the region. Rolls Royce has decided to open two factories here. It has allowed the region to up itself by more than one gear.*[113]

Another important component of the HVM Catapult, The Manufacturing Technology Centre (MTC) in Coventry, opened in 2011 and is already booming, having outgrown its original 12,000 square meter facility. From a start with 8 employees and 3 projects in April 2011 it had grown to 80 people working on

---

currently has 30 industrial members with a target of 50. Interview with Will Barton. Interim CEO, High Value Manufacturing Catapult. London. June 11, 2012.

[110]Ibid.

[111]Ibid.

[112]*The Observer*. 2012. "New Hope Rises from the Fields where Miners Fought and Lost." July 8.

[113]Ibid.

160 projects by June 2012. MTC is looking to double its current facility size and will open a 4,000 square meter academy which will train 2.000 apprentices over the next 10-15 years. Clive Hickman, MTC's CEO, identifies target opportunities for innovative manufacturing to support technology areas such as ultra-large diameter wind turbines, flexible plastic televisions, low-cost batteries for electric vehicles, and partial and fully composite road vehicles.[114]

The HVM Catapult will make a concerted effort to obtain funding from the European Union and to assist small and medium enterprises in securing EU funds.[115] Professor Nigel Perry, CEO of one of the HVM Catapult's member organizations, told Parliament in 2010 that TICs would play a "crucial role" in the pursuit of EU money and that "we will be developing a dedicated team inside CPI whose sole target will be to identify and target those [EU] framework programmes." Asked by an MP whether the United Kingdom had "not been extraordinarily successful thus far," he responded "you can say that, yes."[116]

Other Catapults. A number of other planned Catapult centers have been announced by the TSB.

Fraunhofer Comes to the United Kingdom

In addition to serving as a partial model for the Catapult program, in 2012 the Fraunhofer-Gesellschaft indicated that it would open a Fraunhofer institute in Glasgow focusing on photonics, the first Fraunhofer to be established in the United Kingdom. Photonics had been considered and rejected by the TSB as a possible technology area for the establishment of a Catapult.[117] The Fraunhofer Centre of Applied Photonics will be affiliated with the University of Strathclyde, which has an excellent reputation for photonics research and commercialization.[118] The head of one of the UK's Catapults commented that "they've picked an area we're not investing in."[119] The establishment of the

---

[114]*Professional Engineering*. 2012. "Fast Growth at Technology Centre Bodes Well for Manufacturing." June 14.

[115]Mike Oldham, the head of the Catapult network, comments that "frankly, small companies should avoid EU programs unless they know what they are doing. The primary value of the EU programs is not money, but building links to become part of a supply chain." Interview, London, July 12, 2012.

[116]House of Commons, Science, and Technology Committee, Technical and Innovation Centres enquiry. Oral evidence. December 15, 2010. Q76 Ev 21. Professor Richard Brook, President of the Association of Independent Research and Technology Organizations, who testified with Perry, commented that "the process of applying for [EU] framework projects is quite painful in many respects and puts industry off. SMEs find it particularly hard to bear the risk that is involved. The TICs can be the champions of helping them into Europe." Ibid.

[117]The TSB indicated that the sheer diversity of photonics application made it difficult to pin down focused ideas in a coherent manner. *Optics.org*. 2012. "Photonics Misses out on UK 'Catapult' Center." March 28.

[118]*Optics.org*. 2012. "Fraunhofer Sets up UK Applied Photonics Unit." March 30; *The Scotsman*. 2012. "Germans' GBP9m Scots Laser Research Centre." May 30.

[119]Interview. London. June 11, 2012.

TABLE APP-A4-1 Planned Catapult Centers

Theme	Planned Research Focus
Connected digital economy	Cloud computing, identity management, blending of digital and physical worlds
Offshore renewable energy	Offshore wind, wave and tidal power
Cell therapy	Cell therapies in pre-clinical stages
Satellite applications	Satellite communications, broadcasting, positioning, observation
Future cities	Urban systems and infrastructure

SOURCES: *Catapult—Connected Digital Economy—Priorities and Capabilities—Conclusions from the Consultation Exercise, April to June 2012.* July 15 2012. *Renewable Energy Focus.* 2012. "UK Announces £50m Offshore Renewable Energy Catapult." February 9. Interview with David Way, Director of Knowledge Exchange and Special Projects, Technology Strategy Board, London, June 12, 2012. *BBC News.* 2012. "Plan to 'Catapult' UK Space Tech." January 4. Future Cities Interest Group. "Future Cities Catapult FAQs." *<https://connect.innovateuk.org/web/future-cities-special-interest-group/faq>*. Technology Strategy Board. 2012. "Next Catapult Centres to Focus on Future Cities and Transport Systems." March 21.

Fraunhofer in Scotland was supported by German companies operating in the United Kingdom. The Fraunhofer reportedly told the Scottish government that "we won't invest unless we get government core funding."[120]

OUTLOOK

While it is too soon to assess the prospects of the Catapult initiative, a few observations are possible. On the positive side, the effort has seen Britain's political parties work together on an initiative to attack the country's relative weakness in the area of applied and translational research. A substantial effort has been made to avoid past mistakes such as the failure to provide government core funding to the Faraday Centres. The existing research institutes which are being incorporated into the Catapults are in a number of cases already world-class organizations, and the provision of additional funding and strategic direction is likely to have a positive effect. The TSB has drawn up its vision for Catapult on the basis of extensive consultations with stakeholders and is implementing it with impressive speed.

On the negative side, a number of observers have expressed concern that core funding of £200 million over five years is inadequate. Professor Nigel

[120]Interview with Mike Oldham, Head of Catapult Centres Programme. London. June 12, 2012.

Perry, CEO of the Centre for Process Information, told Parliament in 2010 that £200 million was "a great start, but I'm not sure it's' enough."[121] The Fraunhofers currently receive more than eleven times more core funding from governments than the Catapults will get.[122]And the misimpression that doomed the Faraday Centres—that the Fraunhofer model involved primarily private funding—may still surround the Catapult effort.[123] It is not clear that public funding of the Fraunhofer proportion, which involves two thirds or more of the operating budget and additional funds for equipment and facilities, will be available to the Catapults in an era of stringent fiscal austerity. The 60 Fraunhofer institutes tend to concentrate on narrower and more focused technology areas than the Catapults will do, recalling comments to the effect that past British efforts at applied research spread too little money over too wide a technological landscape. And, as was frequently observed in the Parliamentary enquiry on Technology and Innovation Centres, many of the features of the German innovation system that facilitate the operation of the Fraunhofer are simply not present in Britain.

At the same time, the Catapult initiative enjoys support from across the political spectrum in Britain, a dynamic that has not always characterized similar initiatives in the past. It is being implemented at a time when the traditionally distant relationship between the British research community and industry is improving. The Catapults are being formed by incorporating some of the best existing applied research institutes in the world, such as the Advanced Manufacturing Research Centre (AMRC) in Sheffield. While the Catapult project is modest in scale compared with the Fraunhofer, it is reasonable to expect that it will improve the competitiveness of British industry across a relatively broad front and may enable the United Kingdom to retain a competitive edge in a few areas in which it already leads.

---

[121]Perry commented that "we've done a lot of benchmarking across Europe with CPI, and what we see is that the average institute across Europe has about €25 million or £25 million, and about 200 people. You've been to see the Fraunhofers. There are 59 there and that is the average size of those 59. If you look at the £200 million, we mustn't forget that in the one third, one third, one third model, that should, with time, grow to have the economic impact of around £600 million effectively,. On that scale, I think it's a great start, but it has some way to go yet." House of Commons, Science, and Technology Committee, Technical and Innovation Centres enquiry. Oral evidence. December 20, 2010. Q69 Ev. 19

[122]The Fraunhofer s received the equivalent of £447 million in core government funds in 2010, the most recent year for which figures are available, whereas the Catapult program will receive 40 million per year in comparable funding.

[123]In October 2011, the British engineering publication *Professional Engineering* made the statement "In Germany, about 70 percent of the research work carried out by the Fraunhofers is funded by business." *Professional Engineering.* 2011. "Technology and Innovation Centres will Stimulate Rebalancing of the Economy." October 12.

Appendix A5

The Carnot Initiative in France

In 2006, France launched the Carnot initiative, inviting its public applied research organizations to apply for designation as "Carnot Institutes," a newly-created seal of excellence named in honor of Nicolas Leonard Sadi Carnot, a 19th century French scientist.[1] Carnot institutes benefit not only from the honor but from government funds correlated to the amount of revenue which each institute derives from contract research for French industry. Sometimes referred to as "Fraunhofer Lite," the Carnot initiative is designed to incorporate some of the best features of Germany's Fraunhofer-Gesellschaft, Germany's successful applied research organization, without attempting to replicate that institution in its entirety—a recognition that it has taken sixty years to build the Fraunhofer system and that some aspects of the German innovation system that contribute to Fraunhofer's effectiveness are not present, at least to the same degree, in France.

The 34 institutes that currently bear the Carnot designation account for about 15 percent of the research personnel in France, with 19,000 research professionals and 7,700 Ph.D. students. The consolidated annual budget is about 2 billion Euros ($2.54 billion at 1:1.27). The institutes generate 17,000 A-grade publications per year. They engage in 7800 direct annual research contracts with companies generating total revenues of about 350 million Euros. Roughly half of the institutes' research is financed by companies.[2]

---

[1]Carnot exemplified "applied research that [yielded] scientific results at the highest level." Tasked with improving steam engines, he established the second law of thermodynamics, which applies to many machines in use today, such as motor vehicles and household electrical appliances. Institutes Carnot, 2007.

[2]Jean-Michel Le Roux, *Carnot Program* (2012).

THE EVOLUTION OF FRENCH INNOVATION POLICY

Postwar French policy measures in the innovation sphere have tended to create new measures alongside existing ones; instead of evaluating and culling out unsuccessful programs "the idea is often to create a new measure with more or less the same objectives but with a larger budget and/or a larger coverage."[3] The Carnot initiative was part of a broader effort that began in the early 2000s to reform the French innovation system and improve the relationship between the national research base and private industry, particularly small and medium enterprises (SMEs). Since 1983, when the government of Francois Mitterrand began dismantling the institutions of France's postwar *dirigiste* industrial policy, French science and innovation policy has progressively moved away from centralized, government-directed "top down" measures based heavily on large enterprises toward decentralized, industry-initiated "bottom up" measures to encourage and incentivize entrepreneurialism by small and medium-sized firms. During the *trente glorieuses* (thirty glorious years of postwar economic expansion), the government held a large ownership stake in the economy, provided financial backing for "national champion" companies, and tasked public applied research organizations with executing *grandes programmes*, large-scale R&D programs to support the development of strategic industrial sectors. The *grandes programmes* gave certain French industries a lasting technological advantage (rail, atomic energy, aviation, telecommunications), but "by absorbing most of the R&D funds, they deprived other sectors of even the most basic support for technology innovation."[4] By the beginning of the 1980s, *dirigisme* was becoming fiscally unsustainable and less relevant to the economic challenges arising out of the emerging information technologies and the small-firm driven innovation observable in the United States.

The slowing of global economic growth that began in the early 1970s drew French policymakers' attention to the potential role that could be played by innovation in stimulating economic growth. It was recognized that the country exemplified "the so-called European paradox of tending to be strong in basic research and weak in applied research."[5] France's basic research was dominated

---

[3]Eparvier, Patrick. 2007. *Country Review France.* United Nations University and University of Maastricht. March. p. 31. In 2006, France created a "High Council for Science and Technology" to identify research and technology priorities. The new High Council took its place alongside an already established "High Council for Research and Technology" responsible for evaluating research projects and providing advice on R&D spending. The latter High Council acknowledged the creation of the former and issued a statement to the effect that the distribution of roles between the two Councils needed clarification. *Ibid.* citing High Council of Research and Technology. 2005. "Opinion of the High Council of Research and Technology on the Pact for Research."

[4]Blanka Vavakova, "Reconceptualizing Innovation Policy: The Case of France," *Technovation* 26 (2006) p. 453; OECD, *Technology and Industry Outlook 2012*, p. 292.

[5]A 1993 study of the French innovation system reported that of 2040 firms and research centers conducting R&D, only 7 percent had an R&D staff with more than 50 scientists and engineers. A

by the National Center for Scientific Research (CNRS), the largest basic research institution in Europe. CNRS scientists accounted for numerous Nobel prizes but were relatively uninterested in applied science or technology development.[6] The lack of cooperation between public research organizations and French industry had long been lamented by policymakers.[7] Historically, most corporations have demonstrated a low level of interest in R&D.[8] The French educational system has been chronically unable to produce sufficient numbers of qualified graduates with the skills needed by French industry and the research base.[9] Most of the best students at the *Grandes Ecoles*, the elite universities that train the country's science, business, and government leaders, showed little interest in doing research or obtaining doctorates.[10]

President Francois Mitterand's 1983 "*tournant de la rigeur*" (roughly, turn to austerity or U-turn) laid the foundations for a wave of privatizations and the systematic dismantling of the institutions of the *dirigiste* state. Referring to these reforms, one recent observer commented that "looking across the wealthy democracies, one would be hard-pressed to find any country that shifted so far away from its postwar economic strategy as the France…of Francois Mitterand and Jacques Chirac."[11] Throughout the mid and latter 1980s and 1990s, French policymakers moved away from the *grandes programmes,* targeting small and medium-sized enterprises and public-private partnerships.[12] At the same time, as a 2009 EU assessment observed, rather than disappearing, "the [large] state-owned or previously state-owned companies were able to adapt to the changing

---

group of about 150 firms accounted for 75 percent of the industrial R&D in France. Chenais, Francois. 1993. "The French National System of Innovation." Op Cit. p. 209.

[6]National Institutes of Science and Technology Policy, 2009. *Analysis of Recent Trends in Science, Technology and Innovation Policies in Selected Countries*, March, p. 31. CNRS Nobel Laureates include physicist Alfred Kastler, who discovered the optical pump; Pierre-Gilles de Gennes, who conducted research on molecular states in liquid crystals; biologist Jean Dausset and chemist Jean-Marie Lehn. "France's Research Center," *Die Zeit* (July 22, 1994) JPRS-EST-94-023.

[7]"It must be said that the concern of public authorities about the lack of cooperation between national enterprises and publicly funded research institutes and their component institutes' laboratories is almost congenital to French national research policy." Vavakova, Blanka. 2006. "Reconceptualizing Innovation Policy: The Case of France." *Technovation* 26. p. 445.

[8]The comparatively low R&D outlays by private sector reflect the structure of French industry, in which low and medium technology sectors account for a substantial proportion of total employment and value-added. 84 percent of business expenditure for R&D is undertaken by the manufacturing sector. Of the French companies with the largest R&D outlays, only 6 percent are high technology companies, the remainder being medium or low tech companies. European Commission. 2011. *Erawatch Country Reports 2011: France.* p. 13.

[9]"Lack of Engineers, Trade Deficit Hamper Electronics Industry," *Electronique Actualites* (April 1, 1983); *Report of the INRIA Visiting Committee* (December 18-20, 2008), p. 4.

[10]National Institue of Science and Technology Policy. 2009. *Analysis of Recent Trends in Science, Technology, and Innovation Policies in Selected Countries.* March. p. 31.

[11]Jonah D. Levy, "The Return of the State? French Economic Policy Under Nicolas Sarkozy," Paper presented to the 12th Biennial Meeting of the European Union Studies Association, Boston, MA, March 3-5, 2011, p. 6.

[12]European Commission, Enterprise Directorate-General, *INNO-Policy Trench Chart—Policy Trends and Appraisal Report: France* (2008) p. 7.

world and to compete at the international level," and today France remains highly competitive in aerospace, telecommunications, rail transport, and atomic energy.[13]

Mitterand presided over the 1982 enactment of the *Law on Programming of Research and Technological Development*, which had the goal of encouraging innovation by small and medium enterprises, improving cooperation between public research institutes and industry, and encouraging R&D by companies. Pursuant to the law the Agence Nationale de Valorization de la Recherche (ANVAR) was tasked with administering a special fund to promote innovation, particularly by SMEs.[14] In the same year, France introduced research tax credits, which reimbursed companies for one-half of the increase in their R&D expenditure, a measure patterned on a comparable credit in the United States.[15] By 1987, 3500 firms were applying for the credit annually, a figure which more than doubled in the subsequent decade, with recipient firms benefitting from tax reductions or reimbursements totaling 3 billion francs per year.[16] The credit was modified in 2004, 2006, and 2008 in a manner that transformed it into "France's most powerful innovation support tool." In 2009, the R&D tax credit was worth 4.7 billion Euros to French industry, equaling 60 percent of all government funding to industry.[17]

The 1982 law also provided the framework for "an explosion of links" between private and public sector research that was so pervasive that two decades later it was "no longer possible to speak of a separation between private

---

[13]European Commission, Directorate-General for Research, *ERAWATCH Country Report 2008: An Assessment of Research System and Policies—France*, EUR 23766 EN/2Y, 2009, p. 35. Despite this assessment, the French government has intervened to bail out large companies that have fallen into crisis. "France Defends Auto Bailout Amid Protectionism Row," *France 24* (December 2, 2009). "EU Gives Alstom Bailout Backing," *BBC News* (July 7, 2004).

[14]ANVAR was established within CNRS in 1967 with the mission of transferring research results from public laboratories to companies. Over time it became a principal institution for supporting innovation by SMEs. Vavakova, "Reconceptualizing Innovation Policy" (2006) op. cit. pp. 445-447. By 1991, ANVAR had provided assistance to over 15,000 SMEs in France. One example was Prosyst, a company created in Valenciennes in 1986. Initially ANVAR granted aid to Valenciennes University to develop an error diagnostics card for programmable robots. Several laboratory members decided to form Prosyst to commercialize the system. ANVAR provided funds to help them develop a business plan, and assisted Prosyst in the late-state development and commercial launch of the new product. Today Prosyst still exists as a developer of industrial simulation platforms. "ANVAR's Role in Innovation Discussed," *Courrier Anvar*, (April 1991) JPRS-EST-91-015.

[15]*L'Usine Nouvelle. 1987.* "France to Start Experimental R&D Tax Credit in Fiscal 1988." JPRS-ELS-88-006. February 23, 1988.

[16]Mustar and Laredo, "Innovation and Research Policy in France (1980-200) or the Disappearance of the Colbertist State." *Research Policy* 31 (2002).

[17]In 2004, the R&D credit was renewed indefinitely, removing the uncertainty companies would otherwise experience in planning research investments not knowing whether the credit would be available. After 2004, "on the basis of a simple declaration companies [could] benefit from a tax reduction for a large range of research spending such as R&D personnel expenses, R&D personnel expenses, R&D subcontracting or patenting costs." European Commission, 2011. *Erawatch Country Reports 2011: France.* p. 14.

and public sector research."[18] The core of the 1982 Research Law consisted of policy measures to foster cooperation between public research organizations and companies. The mission of the public bodies was modified by addition of an explicit mandate to contribute "to the application and valorization of research results." Procedures were established require each research institute to account for all relationships with companies. The law provided stipulations for the temporary movement of individuals in both directions between institutes and companies. New legal structures were established to facilitate public-private collaborations. A 2006 analysis of the impact of the 1982 Research Law by a CNRS scholar commented that the law—

> *...by its discourse, changes in the juridical framework, institutional reforms, and multitude of new policy measures and structures has appeared to many observers and actors as a powerful push to the transfer of technology from academia to industry.*[19]

Despite the reforms initiated in and after 1982, for the next two decades a French productivity gap in research and innovation was widely noted. Comparatively high levels of public R&D expenditure relative to the UK and other countries did not yield increased research results. Private sector R&D spending lagged and exploitation of industrial research results was lackluster.[20] It was frequently observed that France did not have enough middle-size companies (50 to 500 employees) or high-growth small companies with above-average levels of job creation, R&D and expert performance.[21] Although some

---

[18]Between 1983 and 1996, the number of research contracts between CNRS and industry increased tenfold, from a few hundred projects to over 3000. During the same period the number of industrial contracts with public research organizations increased 8-fold from 500 million francs in 1983 to 4.1 billion francs in 1997. Mustar and Laredo, "Disappearance of the Colbertist State," (2002) op. cit. p. 64.

[19]Vavakova, "Reconceptualizing Innovation Policy" (2006) op. cit. p. 446. A 1994 study of the French national innovation system observed that "relations between public research and the enterprise sector...is one area where changes were deepest over the last year, thus rendering the critique of academic research as unrelated to the enterprise sector no longer pertinent today." Phillipe Mustar, "La Politique d'innovation en Frances: le Colbertisme Entame" in F. Sachwald (ed.), *Les Defis de la Mondialisation: Innovation at Concurrence*, cited in Vavakova, pp. 446-47.

[20]"Financing—The Budget of Research," *Zero un Informatique* (March 17, 1986) JPRS-EST-86-008 (June 18, 1986) "Industrial R&D: Increased Support and New Programs," *Electronique International Hebdo* (November 7, 1991) JPRS-EST-92-006 (February 24, 1992).

[21]Bussilet, Sophie, Patrick Eparvier and Elisabeth Zaparucha. 2007. *Case Study France*. European Commission: DG Research. December. p. 9. The shortage of mid-sized firms partially reflects the country's tax and regulatory regime. France has a very large number of firms with 49 employees. Many regulations take effect when firm headcount reaches 50. According to a 2008 report on economic growth chaired by Jacques Attali, formerly an advisor to Mitterand, 34 laws and regulations begin to apply when a company reaches 50 employees. Efforts to reform this system have not resulted in much progress. "A Lack of Enterprise: France Needs More Start-ups and Mittelstand Firms," *The Economist* (November 17, 2012).

technology-intensive French sectors were performing well in international competition, as a mid-1980s report to then Industry Minister Laurent Fabius pointed out, "all our successes are occurring in the fields associated with governmental contracts, whereas American industry's performances are very far from being limited to the captive markets of the state."[22]

> *The blame for such inefficiencies has been consistently placed on a vast and ossified public science and research system, with recurrent and non-competitive research funding, jobs for life, the absence of a culture of exploitation, and a lack of synergy between public research and industry.*[23]

Lionel Jospin, a Socialist who served as Prime Minister between 1997 and 2002, sought to promote innovation and entrepreneurialism by the country's small and medium businesses as a means of reducing chronically high unemployment levels. His Economics Minister, Dominique Strauss-Kahn, made numerous visits to Silicon Valley and met with U.S. high-tech executives like Steve Jobs, Bill Gates, John Chambers, and Michael Dell.[24] An extensive effort was made to reach out to entrepreneurs and industrial leaders but also to social interest groups affected by new institutional arrangements "that promoted risky high-technology start-ups without undermining the institutional foundations of French capitalism."[25] A 2009 assessment of the French innovation system by the European Commission's Directorate-General for Research gave the country high marks for its public funding of R&D but noted that "domains of world level scientists and technological excellence exist, but are often specialized in stable/mature research fields." The Directorate cited "poor knowledge circulation between academic research (universities/CNRS) and business and noted "low demand for research outcomes from potential new companies."[26]

In 1999, France enacted the *Innovation and Research Act,* which prioritized the creation of new innovative firms and the transfer of learning from

---

[22]"French companies in general and small/medium companies in particular do not devote sufficient resources to R&D." European Commission, Directorate-General for Research. *ERAWATCH Country Report 2008: An Assessment of Research System and Policies—France* (2009) p. 19.

[23]Emmanuel Muller, Andrea Zenker and Jean-Marie Heraud, "France: Innovation System and Innovation Policy" in Fraunhofer ISI, GIGA and Georgia Tech STIP. *New Challenges for Germany in the Innovation Competition* (August 2008) p. 144.

[24]In a 1992 interview when he was serving as Minister of Industry, Strauss-Kahn emphasized the key role of application of research. "Of what use is innovation if we do not have the techniques to industrialize and commercialize the products that result from this innovation? It is unthinkable for an advanced industrial society, such as France, to content itself with simply being a research center. We need to master the technologies that allow us to produce the advanced products which we conceptualize and design." "Let's Create Conditions for Good Growth," *Electronique International Hebdo* (November 5, 1992) JPRS-EST-92-039 (December 29, 1992).

[25]Gunnar Trumbull, *Silicon and the State: French Innovation Policy in the Internet Age* (Washington, D.C.: Brookings Institution Press, 2004) p. 13.

[26]Ibid, pp. 3-4.

public research organizations to French industry. The law established fiscal incentives for innovating firms, created a new and simplified legal framework for innovating companies, authorized government researchers to participate in innovative start-ups without losing their civil service status for 6 years, and established new mechanisms to foster public-private research collaboration.[27]

In 2003, the Ministry of Industry and the Ministry of Research and Technology jointly drew up the *Innovation Plan,* a series of recommended measures to support innovation in France. This plan was augmented in 2005 by the research ministry's *Pact for Research,* which set forth proposals that provided the basis for the *Law for Research,* enacted in 2006. The purpose of the *Law for Research* was to support greater cooperation between various actors in the research ecosystem, to network public and private research activities, to provide improved conditions for scientific careers, and to encourage the integration of the French research system into the European Research Area (ERA).[28] Concurrently, the French research bureaucracy was overhauled and new entities established for innovation:

- **The High Council for Science and Technology (HCST)** was established in September 2006 to advise the President and the government on policies affecting scientific research, innovation, and technology transfer. It is comprised of 12-21 eminent individuals from the fields of science and technology who serve 4-year terms.
- **The National Agency for Research (ANR),** established in 2005, was tasked with financing research projects and managing innovation programs, such as the Carnot Institutes and the research projects that are part of the competitiveness clusters. The ANR operates on the basis on annual calls for research project proposals, and selects projects based on scientific and technical excellence criteria based on peer review evaluation.[29]
- **The OSEO Group,** formed by the merger of a number of government organizations, emphasizes the promotion of innovation by small and medium enterprises (SMEs). It includes the former Agence Nationale de Valorisation de la Rechesche (ANVAR), which was established in 1974 to promote application of public research discoveries, with an

[27]Muller, Emmanuel, Andrea Senker and Jean-Alain Heraud. 2009. *France: Innovation System and Innovation Policy.* Karlsruhe: Fraunhofer ISI Discussion Paper. April. p. 3.
[28]Fraunhofer ISI, German Institute of Global Area Studies, and Georgia Tech. 2008. *New Challenges for Germany in International Competition.* August. p. 128.
[29]The ANR makes "thematic" and "white" (non-thematic) calls for proposals. The former consist of priorities identified by the government and the latter can relate to any topic, with the most promising and original proposals selected. Chong, Stephanie Fen and Emilie-Pauline Gallie. 2007. *Linking Two Instruments for a Better Innovation Policy-Mix: The Case of the National Research Agency and the Competitiveness Clusters.* Working Pwper IMRI July. pp. 16, 18.

emphasis on SMEs. It operates under the designation "OSEO Innovation."

- **The Research Evaluation Agency (AERES)** was created by the Law on Research in 2006, and is charged with objectively and systematically evaluating research institutes, scientists, and programs. It is comprised of 25 French and foreign individuals drawn from the research base.

Competitiveness Clusters

In 2004 the French government launched an ambitious "Competitiveness Clusters" (*pôles de compétitivité*) initiative seeking to promote interaction between companies, educational institutions, and public research organizations within limited geographical areas. This program became operational in 2005 for an initial three-year period and was extended to 2009-12. The program is based on the award of subsidies to innovative projects involving collaborations between companies, research organizations, and universities. An evaluation of the second phase of the program (2009-12) jointly commissioned by the Directorate General for Competitiveness, Industry and Services (DGCIS) and the Directorate for Territorial Cohesion and Regional Competitiveness (DATAR) concluded in 2012 that the program should be extended through 2020.[30]

Regional Initiatives

After 1980, the 26 French regions began to increase their role in the national innovation system. Although the regions do not have legislative authority, they receive substantial allotments of the national tax income that they can direct to their developmental priorities. These are negotiated with the central government and embodied in *Contrats de Projet-Etats-Regon* (state-region contracts), seven-year region-based innovation initiatives. The contracts reflect the central government's intervention at the regional level in collaboration with regional authorities (OSEO project, competitiveness clusters) and the regional and other local authorities' own initiatives, which include research promotion, creation of research networks, support for young researchers, support for start-ups, mobilization of risk capital, and promotion of technology transfer mechanisms.[31]

---

[30]*TCI-Network*. 2012. "French Competitiveness Clusters' Evaluation Published." June 26.

[31]European Commission, Enterprise Directorate-General, *INNO-Policy Trend Chart—Policy Trends and Appraisal Report: France* (2008) p. 18. Although the regions have steadily increased their spending on research, innovation, and technology transfer, regional spending on R&D represented only 5 percent of French public expenditures on R&D for 2008. European Commission. 2011. *Erawatch Country Reports 2011: France*. p. 23.

Research and Higher Education Clusters (PRES)

The Law for Research provided for the establishment of Research and Higher Education Clusters (PRES) to encourage increased collaborations between research actors and universities. PRES was intended to end the fragmentation of university research and to facilitate sharing of research resources, equipment, and tasks within a given geographic region.

National Research and Innovation Strategy (SNRI)

Beginning in 2009, France has adopted 5-year National Research and Innovation Strategies (SNRI) setting national priorities that will guide policymaking and resource allocation. The first 5-year strategy was developed by nine working groups drawn from French business, research and civil society communities. In the first five year plan, three priorities have been identified: (1) healthcare, nutrition and biotechnology; (2) environmental and eco-technology; and (3) information, communications, and nanotechnology.

Advanced Thematic Research Cluster (RTRA)

Based on legislation passed in April 2006 on Thematic Networks for Research, this program supports "research and higher education actors who want to launch together a specific scientific project of high-quality and international visibility giving them a global scope."[32]

THE CARNOT PROGRAM

The Carnot initiative arose out of the *Pact for Research*, which sought to reinforce the activities of existing public research institutes that were already involved in research partnerships with private and/or public entities. The program designates what are regarded as the best of those institutes as "Carnot Institutes," and awards them government funding through the ANR, with the level of funding linked to the volume of each Institute's contract research revenue from industry. The Carnot designation is intended to be a "seal of excellence." The Carnot program was inspired by the example of Germany's Fraunhofer-Gesellschaft.[33]

Carnot institutes are selected and monitored by ANR. Institutes are periodically invited via a call for proposals to seek Carnot designation by demonstrating their strong affinity with companies. The ANR provides support funding to Carnot institutes based on an incentive formula that takes into

---

[32]Accessed from <*http://erawatch.jrc.ec.europa.eu/erawatch/*>.

[33]European Commission, Enterprise Directorate-General. 2008. *INNO-Policy TrendChart—Policy Trends and Appraisal Report: France*. p. 15.

account revenues from contract research for public and private entities, income that flows from the ownership of intellectual property, and income from SMEs. ANR funding increases proportionally as each of the other revenue streams grows. Eligible income for ANR matching funding includes income from research partnership contracts financed by public or private entities (excluding states, the EU, international organizations, and national agencies) which are either end users of the results or entities acting as intermediaries for user companies.

Although ANR selects Carnot institutes, the Carnot designation is bestowed by the Ministry of Higher Education and Research. The criteria for Carnot designation require a candidate institution to—

- Clearly define its research strategy;
- Maintain or create a sound internal organization;
- Retain downstream research in-house to enrich more applied research; and
- Be substantially engaged in contract research with other socio-economic actors.

The purpose of the Carnot award is to foster stronger links and partnerships between public and private research in France. The most important priority of the program is knowledge transfer from public research organizations to other entities, particularly private companies, through contract research, licenses, and IPR creation in public, academic, and non-profit institutes. The first Carnot designations, in 2006 and 2007, were for a renewable 4-year period. Subsequent designations run for 5 years. By 2011 34 Carnot institutes had been designated, distributed widely across France and involving about 25,000 researchers. The consolidated research budget for the institutes in 2011 was 1.9 billion Euros. Because the Carnot institutes are public organizations, over half of their funding is derived from various government sources in the form of core funding or research contract revenue.[34] However in 2011, the institutes were also generating about 350 million Euros from partnerships with industry, of which 60 million Euros was from SMEs.

Partnerships between Carnot institutes and companies take a number of forms:

- Direct partnership research contracts with companies;

[34]The UK's 2010 Hauser Report, which surveyed a number of applied research organizations, indicated that in 2008, the Carnot institutes received 59 percent of its income from core and other public funding. Hauser, Hermann. 2010. *The Current and Future Role of Technology and Innovation Centres in the UK.* p. 11.

TABLE APP-A5-1 Overview: Carnot Institutes

Institute	Core Business	Key Applications	Annual budget (million €)	Personnel
3BCAR	Chemistry	Renewables	46.2	836
ARTS	Materials, mechanics, processes	Energy, ICT	82.0	1,260
BRGM	Energy, environment, earth sciences	Minerals, environment	108.0	989
CALYM	Life sciences	Treating lymphoma	14.6	301
CEA LETI	Microelectronics, nanotechnology	ICT	225.0	1,539
CEA LIST	ICT	Telecom, robotics, metrology	58.5	681
CED2	Chemistry, energy, environment	"Green chemistry"	19.1	303
Cetim	Materials, mechanics, processes	Manufacturing, energy transport, logistics	39.0	900
CIRIMAT	Materials, mechanics, processes	Aerospace, energy, health	11.0	213
CSTB	Building and territory management	Construction	34.6	351
Curie Cancer	Life sciences, health	Cancer treatment	21.8	247
Energies de futur	Energy, environment	Energy, renewables	130.0	1,555
ESP	Energy, environment	Energy and propulsion systems	25.3	310

I@L	Building, materials, mechanics, processes	Construction, health, manufacturing, transportation	61.5	1,328
ICEEL	Energy, environment, earth sciences, materials, mechanics, processes	Sustainable use of natural resources, green industrial processes	73.0	1,435
ICM	Life sciences, health	Treatment of brain, neurological diseases	29.6	397
ICSA	Life sciences	Animal health	66.1	954
IFPEN Transports Energie	Transportation, fuel propulsion, engines	Transportation, energy, environment	41.5	363
Ifremer EDROME	Earth sciences	Development of marine resources	13.0	246
INRIA	ICT	Telecom, aeronautics, software, energy, health care	255.0	3035
ISIFoR	Energy, environment	Sustainable engineering of fossil resources	30.5	474
LAAS CNRS	ICT, microelectronics, nanotechnology	ICT, robotics, Microsystems	32.3	522
LISA	Chemistry, life sciences, health	Exploitation of lipids	8.4	166
LSI	ICT, life sciences	Software	16.5	520
M.I.N.E.S	ICT, energy, materials, mechanics, processes, earth sciences	Clean energy, green building, materials engineering, intelligent systems	80.5	1820

TABLE APP-A5-1 *Continued*

Institute	Core Business	Key Applications	Annual budget (million €)	Personnel
MICA	Chemistry, energy, micro- and nanotechnology, materials, mechanics, processes	Functional materials	64.9	879
ONERA ISA	Transport, engines, fuel, materials, mechanics, processes	Aerospace	14.7	860
PASTEUR MI	Life sciences, health	Infectious diseases	71.5	772
PolyNat	Chemistry, materials, mechanics, processes	Biosourced materials	15.9	273
Qualiment	Life sciences, health	Food products	42.9	516
STAR	Chemistry, energy, environment, materials, mechanics, processes, ICT, microelectronics	Aeronautics, energy, health, ICT, transport	29.4	650
Telecom & Societe Numerique	ICT	ICT, ICT applications	71.7	2,200
Voir et Entendre	Life sciences, health	Vision and hearing	13.8	236

SOURCE: Association Instituts Carnot, <*http://www.instituts-carnot.edu/en/instituts-carnot*>.
NOTE: Personnel includes permanent staff and students.

- Collaborative contracts in response to requests for proposals from ANR, the French Interministerial Single Fund, and the EU Framework Programme;

- Joint research teams and laboratories;
- Supervision of PhD students financed by companies

The Carnot institutes are a heterogeneous group of publicly-owned research organizations ranging in headcount from 213 professional employees (CIRIMAT) to 3,034 (INRIA). A number of them are public research organizations which were founded during the *dirigiste* era and at one time executed *grandes programmes,* and which have reoriented their mission and approach in recent decades.[35] In contrast to Taiwan's ITRI and Germany's Fraunhofer-Gesellschaft, some of the Carnot institutes engage in basic as well as applied research. Parent organizations of the various institutes include universities and CNRS, and other governmental or quasi-governmental entities. Many Carnot institutes operate through multiple research centers in various parts of France.[36]

The goal of the Carnot Institutes is the improvement of society through renewable energy, personal health care, improved transportation and mobility, civil safety, homeland security, and information and communications technology (ICT). Accordingly, the Carnot Institutes concentrate on a number of thematic areas:

- Life sciences and health technology.
- Materials, mechanics, and processing.
- Earth sciences and natural resources.
- ICT—micro- and nano- technologies.
- Building, civil engineering, and land use planning.
- Environment and energy, propulsion, and chemistry.[37]

Contracts with Industry

The nature of the research collaborations between Carnot institutes and companies vary considerably from institute to institute. In general, collaborations may involve very specific research sought by a company with respect to which the company pays for the full cost of the project; R&D

---

[35]CEA LETI, for example, is the microelectronics research organization of CEA, the French Atomic Energy Commission. Founded in 1965 in Grenoble, LETI worked closely with Thomson Semiconductor on the development of submicron CMOS technology and participated in JESSI and other major European Community joint research projects. "Integrated Circuits: Looking Beyond the Borders," *Industries et Techniques* (June 1, 1988) JPRS-EST-88-005 (July 27, 1988).
[36]INRIA, a Carnot institute specializing in computer science, operates research centers in Paris, Saclay, Grenoble, Bordeaux, Nancy, Rennes, Sophia Antipolis, and Lille. Carnot ARTS, which specializes in engineering, has research centers in Angers, Bordeaux, Paris, Lille, Aix-en-Provence, Cluny, Metz, Dijon, and Valduc.
[37]Joachim Rams, President, Instituts Carnot, "France: Carnot Institutes," International Energy Agency, *Transforming Innovation into Realistic Market Implementation Programmes*, Workshop Summary, April 27-28, 2010, p. 84.

collaborations, consortia, and joint laboratories involving cost sharing; and special projects for small and medium enterprises. Many contracts involve personnel from both the Carnot institute and the industrial partner. CEA LETI (microelectronics and nanotechnology) administers PEPITE, a program that provides short term (6-12 months) project engineering for small companies seeking to use mature technologies held by CEA LETI. Carnot IFPEN Transports Energie offers Joint Industry Projects (JIPs), consortia in which the institute does all of the research itself, shares the results with participating companies, and retains industrial ownership. IFPEN also performs "research demonstrators which perform the last step in the validation of a technology prior to industrialization."[38]

AiCarnot

The Association des Instituts Carnot (AiCarnot) acts as a coordinator and network developer of the various individual Carnot institutes. It secures public financial support for each institute, works to define the medium-term objectives, and defines and manages intellectual property rights policies.[39] AiCarnot has developed long term relationships with regional and thematic entities providing technological support to companies, including OSEO, the French competitive clusters, research clusters, professional organizations, and the French Chamber of Commerce.

Evaluations

French public research organizations, including Carnot institutes, are periodically subject to external evaluations supervised by the government Agency for the Evaluation of Research and Higher Education (AERES). The evaluations are conducted by independent experts (Visiting Committees) who not only have no connection to the institution under examination and in a majority of the cases are not even French. Some of the AERES reports are published.

Intellectual Property

AiCarnot has promulgated a code of best practices with respect to IP and knowledge and technology transfers (KTT) for the Carnot institutes. The code provides with respect to research partnerships between the Carnot institutes and "socio-economic actors" (in most cases companies), each party will have proprietary rights with respect to the R&D results they develop alone during the collaboration. Results that the parties develop together are jointly owned, with

[38] *<http://www.ifpenergiesnouvelles.com/development-industriel/de-la-recherche-a-l-industrie>*.
[39] *Ibid*, p. 8.

conditions regarding the exercise of IP rights "defined according to specified and negotiated terms, for example, in proportion with their contributions in terms of inventing and funding." A free right of use of the partnership's research results is held by the Carnot institute solely for the purpose of subsequent research. Transfer of IPR by Carnot institutes is to be considered on a case by case basis with an appropriate compensation. The rights to prior knowledge which the parties bring to the collaboration cannot be modified by the collaboration unless specifically negotiated. At the same time, the partners grant free access to their prior knowledge for the sole purposes of the joint research. Licenses for the technology developed by the partnership "will be limited to a definite period and to specific fields and territories."[40]

Support for Manufacturing

Twenty Carnot institutes offer competencies in "materials, mechanics, and processes," which support French manufacturing companies and industries. A number of institutes offer companies access to on-site pilot manufacturing facilities on which they can prove processes and equipment, and/or tools and platforms through which factory environments can be simulated. CEA LETI, for example, which specializes in microelectronics, features 8,000 square meters of CMOS compatible clean rooms equipped with 200 and 300mm fabrication tools. The STAR Carnot Institute (Science and Technology for Research Applications) operate a multi-purpose production platform for advanced materials, tools for materials deposition, and clean rooms.[41]

Some institutes, such as Carnot CETIM, are virtually entirely dedicated to industrial automation technologies and systems, simulation of industrial processes, metrology, and other themes directly relevant to manufacturing.[42] CETIM's website details 120 recent projects that have boosted competitiveness of French industry, generally through incremental improvements in materials and industrial production processes. Some examples:

- **AEML:** France's AEML Company (Ateliers Electriques et Metallurgiques due Loiret) manufactures equipment for use in construction and color preparation for motor vehicle bodies. AEML wanted to repatriate part of its Chinese production of metal clamps to

---

[40]AiCarnot, "The Carnot Institutes' Code of Best Practices for Intellectual Property and Knowledge and Technology Transfers," (2009).

[41]Carnot CEA LIST (smart digital systems) is equipped with robotics and robotics platforms, CAD stations for integrated circuit design, reactors for diamond synthesis, ionizing radiation metrology platforms, immersive rooms to create virtual environments, and experimental network diagnostic tool platforms. *<http://www.instituts-carnots.eu/en/node/250/>.*

[42]Founded in 1965 at the behest of French machinery companies, CETIM has been involved in factory automation and industrial materials and processes for many years. "Very High Speed Machining Gaining Ground: 3-Year Research Program Launched," *L'Usine Nouvelle* (November 14, 1991) JPRS-EST-92-008 (March 12, 1992).

France, and to do so it sought to develop a new model clamp that could be patented and manufactured at costs that would be competitive with Chinese costs. AEML asked Carnot CETIM to identify a variety of electrogalvanized steel that was less expensive and which could be stamped in a more productive manner than the existing steels in use. CETIM used digital simulation tools to identify a type of steel, DC03, that was cheaper and which provided a superior performance. As a result of this project, AEML's factory in Meung-sur-Loire, France, employs an "operating sequence [which] allows it to compete with Chinese manufacturing costs."[43]

- **Inoforges:** Inoforges, a small French company engaged in the manufacturing of brass, copper, and aluminum parts, asked CETIM to review its plans for manufacturing connecting aluminum parts for welded frames intended for rail transport. CETIM conducted simulations with its software "Forge 200S" which revealed that the company's planned method would result in large amounts of power consumption with a risk of poor material homogeneity and high stresses on the tooling. CETIM and the company's engineers simulated a redesigned part. The new part and associated processes generated material savings of nearly 15 percent and reduced process costs by about 10 percent.[44]
- **Rivard:** Rivard SA is a manufacturer of cleaning systems based on high water pressure. At the beginning of 2005 it needed to carry out an emission control plan for its painting and coating operations to comply with new regulations concerning volatile organic compounds (VOC). To do this and to achieve greater efficiency in production, the company turned to CETIM. CETIM engineers proposed that the company set up an automated and continuous process involving the preparation of paint, in-line feeding of electrostatic pulverizing guns, and a control system measuring consumption using flowmeters and volumeters. The new process cut VOC emissions by 50 percent and yielded a 30 percent reduction in paint consumption during the production process.[45]

Networks

The Carnot institutes comprise a major network covering many of the competencies necessary to sustain a high technology economy, including information and communications technology, microelectronics, construction, factory automation, energy, medicine, chemicals, and advanced materials. To

---

[43]CETIM, "Repatriating the Manufacturing of Clamps from China." *<http://www.cetim.fr/cetim/en/Services/References/AEML>*.

[44]*<http:// www.cetim.fr/cetim/en/Services/References/Inoforges>*.

[45]*<http://www.cetim.fr/cetim/en/Services/References/Rivard>*.

leverage these disciplines, the institutes have formed "Carnot alliances" in a number of market sectors in order to bring to bear the skill sets of the entire network on particular technological challenges. Companies can access these networks regardless of the point of entry. The sectors covered are environment, chemistry, transport, ICT/micro and nano technology, healthcare, sustainable construction and mechanics, materials, and processing. In some cases, Carnot institutes have formed joint laboratories among themselves to pursue themes of mutual interest.[46]

The Carnot institutes are also heavily networked with other French research organizations and educational institutions. The Carnot Institute ARTS (Actions on Research for Technology and Society), for example, which specializes in engineering, is led by the engineering school Arts & Metiers Paris Tech and has research partnerships with five universities and seven public research organizations including CNRS and CNAM (Conservatoire National des Arts et Metiers). Some Carnot institutes are involved in the governance of French Competitiveness Clusters.[47] Finally, the Carnot institutes enjoy extensive European and international networks.[48]

Training

A significant proportion of the Carnot institutes' professional staff is comprised of PhD students. At the largest Carnot institute, INRIA (computer science), 1270 of the institute's 3035 professional staff are PhD students. "The contributions of young doctoral and post-doctoral researchers from a variety of fields are a noteworthy benefit of the Carnot initiative."[49] IFPEN Transports Energie operates IFP Schoul on its premises that offers graduate-level training to 600 young engineers. IFPEN and IFP Schoul jointly operate IFP Training, an international training organization for managers, engineers, and technicians from

---

[46]Carnots CETIM and Cirimat, both of which are involved in materials, mechanics, and processes, have formed Cetimat, a joint laboratory specializing in surface engineering and the in-service behavior of materials. *<http://www.cetim.fr/cetim/en/News/Headliner/Headlines/A-joint-laboratory-for-functionalising-surfaces>*.

[47]The Institute Carnot BRGM, specializing in earth sciences, participates in the governance of seven competitiveness clusters.

[48]CEA LETI Carnot (microelectronics) belongs to the Heterogeneous Technology Alliance (HTA) that is comprised of LETI, Germany's Fraunhofer Gesellschaft, the VTT in Finland, the CSEM in Switzerland, and the French Institute for New Energy Technologies (LITEN). The partners have agreed to be interdependent, according each other access to their technologies and coordinating their investments. HTA concentrates on micro and nano technologies and its mission is to "bridge the valley of death" between academia and industry. "Heterogeneous Technology Alliance: Vision—Mission—Strategy." Heterogeneous Technology Alliance. October 17, 2011. *<http://www.hta-online.eu/fileadmin/MEDIA/Press_and_Media/HTA___White_Paper_111017.pdf>*. CEA LETI also has research partnerships with Caltech's Kavli Nanoscience Institute, IBM and Japan's Micromachine Center.

[49]LETI. *LETI, a Carnot Institute. <http://www.let.fr/en/discover-leti/partnerships/let-a-carnot-institute>*.

the oil and gas, chemical, and automotive industries.[50] Carnot institutes' research contracts with companies sometimes stipulate that the institutes will train employees of the companies in the technologies that are the subject of research.[51]

EU Engagement

The Carnot institutes are major players in the European Union's Seventh Framework Programme (FP7), which provides major EU funding for R&D. Carnot CEA LIST (smart digital technology) alone is involved in about one hundred FP7 projects, including 12 in which it is the project leader.[52]

Spin-off

A number of institutes receiving the Carnot designation had already established a long history of spin-offs, a practice that continues under the Carnot initiative. INRIA, an information technology institute that is the largest Carnot by budget and headcount, has created 106 spin-off companies.[53] These include Esterel Technologies, an innovative software SME which certifies and verifies "critical codes;" Trusted Logic, a developer of smart cards; and the high-performance computing firms Activeon, Kerlabs, and Sysfera.[54] A 2008 assessment of INRIA by a Visiting Committee commented that "INRIA has continued its successful policy aimed at creating new spin-off companies. Along with educating students, this is one of the most effective mechanisms for technology transfer. The number of new companies created roughly followed the overall growth of the institute."[55] The Carnot Institute Telecom & Societe Numerique averages 25 spin-offs per year.[56]

Carnot spin-offs benefit from a number of measures implemented by the French government during the past decade to encourage start-ups. The Ministry for Higher Education and Research has funded the establishment of incubators in close proximity to public research organizations.[57] The *Jeune Entreprise Innovante* status was authorized in the framework of the Finance

---

50 *<http://www.ifpenergiesnouvelles.com/formation/mission-formation>.*

51 LETI. *Collaborating with LETI. <http://www.let.fr/en/how-to-collaborate/collaborating-with-leti>.*

52 *<http://www.instituts-carnot.eu/en/Node/250>.*

53 Institute Carnot Fiches en anglais, "Institute Carnot INRIA." INRIA provides financial and technical support for start-ups through INRIA-Transfert, which administers a number of start-up funds.

54 *Report of the INRIA Visiting Committee* (December 18-20, 2008) p. 3.

55 *Ibid*, p. 3.

56 Carnot Institute TELECOM-EURECOM, *Strategy and Technologies Transfer* (December 2008).

57 In the period 2000-2005, 1,415 projects of innovative companies were hosted in these incubators, nearly 50 percent of which were based on research results of the public research organizations. Muller, Zenker and Hearud, *France: Innovation System and Public Policy* (2009) *op. cit.* p. 19.

Law in 2004, making small innovative firms exempt from numerous taxes and social security payments.[58] Through the Investments for the Future Programme (PIA), a stimulus program, the government funded the establishment of two special venture funds to assist start-ups, the National Fund for Digital Society (2010) and the National Seed Fund (2011).[59] PIA is also underwriting the establishment in 2012 of thirteen projects of "Mutualized for Innovation Platforms" (PFMI), which are intended to provide SMEs with pooled resources (people, equipment, associated services) to support innovative projects and R&D.[60]

INRIA

INRIA, a Carnot institute specializing in computer science, has the largest budget and professional staff in the Carnot network, and is recognized as one of the best institutions of its kind in the world. Founded in the era of *dirigisme* as part of a *grande programme*, it achieved the status of a public scientific and technological establishment in 1985.[61] INRIA's first chairman, Jacques-Louis Lions, was one of the greatest applied mathematicians of the Twentieth Century. He built INRIA into an institution that applied mathematics, computer science, and automated controls in a deep and integrated manner. He welcomed interaction with industry as a source of proximity to new research challenges, and placed an emphasis on training, particularly, at the doctoral level, working closely with French universities. INRIA introduced its motto in 1994, "scientific excellence and technology transfer," reflecting its emphasis on applications and relationships with industry. INRIA contributed particularly to the French telecommunications sector, where it forged close relationships with Alcatel and France Telecom. In 1998 it set up a subsidiary, INRIA-Transfert, to function as an incubator for start-ups.[62]

INRIA closely interacts with the French academic research infrastructure. A 2008 report by a Visiting Committee indicated that of the 160 projects then underway at INRIA, over 75 percent involved academic or other research organizations, and that of the 1200 Ph.D. students then working at the institute, only 25 percent were fully paid by INRIA. The Committee commented that

---

[58] *Ibid.*

[59] OECD, *OECD Science and Industry Outlook 2012*, p. 292.

[60] "Thanks to this mutualization, SMEs and medium-sized companies can have access to high technology equipment that they would have been unable to purchase or use on their own." Groupe Caisse des Depots, "Results of the Call for Projects 'Mutualized for Innovation Platforms'," (August 3, 2012).

[61] INRIA, originally known as IRIA, was created in 1967 as part of the French Computer Development Plan. <*http://www.slideshare.net/fullscreen/INRIA/planstrat2008-2012en/6*>.

[62] *Ibid.*

> *The original and distinctive INRIA "model" based on team-projects, each with a clear research focus and vision, has an extremely fruitful impact on the academic system, fostering the emergence of regional poles of excellence with international visibility, as well as interdisciplinary, and individual collaborations. This flexible and dynamic team-project model shapes the bottom-up creativity of researchers in a coherent national strategy.*[63]

Collaboration with Fraunhofer

In 2007, ANR and Germany's research ministry, BMFB, launched an initiative to foster collaboration between the Carnot institutes and Germany's Fraunhofer-Gesellschaft in the area of applied research. This effort was funded by the two governments at 10 million Euros over a 3-year period. Eleven bilateral research groups were established, including the following:

- The Fraunhofer institutes for Cell Therapy and Immunology IZI and Reliability and Microintegration IZM joined with the Institut Carnot FEMTO to develop "lab on a chip" technology to enable fast and affordable blood sample analysis at the doctor's office, eliminating the need for a trip to the major laboratory.
- The Fraunhofer Institute for Interfacial Engineering and Biotechnology IGB and the Institut Carnot CIRIMAT launched a joint project to develop bone tissue using innovative biomaterials for use in surgery.
- The Fraunhofer Institutes for Physical Measurement Techniques IPM and the Institut Carnot IEMN undertook a joint project to develop antennas for terahertz radiation for security controls on substances such as drugs and explosives.[64]

CONCLUSION

The Carnot initiative is one facet of a broader effort by France to improve its innovation system in a manner that will improve the competitiveness of French industry and create jobs. The purpose of the Carnot effort is to "make the transition from fundamental research to industrial research more fluid."[65] The Carnot model incentivizes each institute to ascertain what types of technology French industry is seeking and to secure contracts to develop those technologies, and the government rewards success in obtaining contracts with

[63] *Report of the INRIA Visiting Committee* (December 18-20, 2008) p. 4.
[64] Fraunhofer-Gesellschaft. 2009. "Research News: German French Camaraderie." Press Release. November 16.
[65] Jean-Michel Le Roux, *Carnot Program* (October 2012).

matching funds. The sheer number of annual contacts with companies (7,800) is impressive. The spinoff efforts by some Carnot institutes are very robust and the track record of the French institutes in this area appears to eclipse that of the Fraunhofer.

At the same time, even if highly successful, the Carnot initiative, standing along, cannot remedy some of the more intractable weaknesses in the French innovation system, such as manpower shortages, the acknowledged shortcomings of the educational system, chronic underinvestment in research by industry, and the comparative lack of interest by young people in careers in engineering and science.[66] In 2011, five years after the inception of the Carnot initiative, a study by two French academics concluded that research cooperation between public and private sectors in France contributes less to companies' innovation capacity than is the case in Germany, based on an econometric study of the share of innovative products in total turnover. The European Commission attributed these findings to the difficulty encountered by companies in cooperating effectively with public research organizations, the complexity of the knowledge-transfer system, and the difficulty private companies experience in finding the right research partners.[67] These findings underscore the fact that addressing the challenges facing France in innovation will take a major effort spanning many years.

---

[66]In 2011, the European Commission commented on "the low level of interest shown by [French] companies for innovation. This is due to the weak culture of innovation characterizing French companies." European Commission. 2011. *Erawatch Country Reports 2011: France.* p. 14.

[67]Robin, Stephanie and Tosben Schmidt. 2011. "Partenaniats Public/Prive et Innovation dans les Enterprises" in MINEFI. *L'Innovation dans les Entreprises: Moteurs, Moyens et Enjeaux* cited in European Commission. 2011. *Erawatch Country Reports 2011: France.* p. 14.

Appendix B

An Evaluation of the MEP: A Cross Study Analysis

Jan Youtie
Georgia Institute of Technology

The Manufacturing Extension Partnership (MEP) has used a diversity of evaluation methods and metrics over its more than two decade history. Many of these studies are sponsored by the program itself. State economic development agencies and federal oversight bodies also engage evaluation efforts. These evaluations have been carried out by various private consultants, university researchers, policy foundations, government examiners, and MEP itself. The aim of this paper is to look across these evaluation reports to identify the methods used, focus of these evaluations in terms of geography and unit of analysis, and the findings. Not all of these reports have recommendations, but for those that do, this paper will present these recommendations so that similarities and trends may be discerned. This work updates a previous effort to catalog and analyze MEP evaluations conducted by the author along with Philip Shapira (Youtie and Shapira 1998; Shapira 2003). The results will show several ongoing themes in MEP assessments including efforts to present and validate its mission, balance multiple program objectives, and adapt program direction to the need for innovative technologies and products.

1. OVERVIEW OF THE MEP EVALUATION SYSTEM

Discussions of MEP evaluation system typically focus on three primary program objectives: (1) delivering service to a broad range of manufacturing enterprises (market penetration), (2) maintaining optimal revenue levels for program operation (revenue generation), and (3) having an impact on clients that are served (client impact).[1] These objectives are not always in alignment, which can lead to conflicts (Figure APP-A-1.). For example the need for market

[1]See for example, Oldsman (2004), Sears and Blackerby (1998), Voytek et al., (2004).

penetration emphasizes widespread service through training or short-term engagements, which can be associated with less revenue generation and lower-levels of impact on clients. Efforts to raise revenue have the potential to be associated with fewer clients served. Efforts to raise impact can also mean larger projects with a smaller set of clients. Maximizing service impact can even effect revenue generation as some types of projects with the potential to yield high impact are stressed over other types of projects that have short-term cost-cutting appeal to potential customers.

Given these potentially diverging objectives, the MEP uses multiple evaluation methods to address them. The core performance measurement activity at MEP is the client survey. This survey, which has been administered since 1996, is conducted approximately one year after an engagement with a client company. The survey draws on a standard nationwide reporting system. Centers populate this system on a quarterly basis with reporting information about center, client, and project and event attributes. The survey is administered by a third party company once a quarter primarily using telephone and web-based methods. The questionnaire has included various questions over the years but the core questions ask clients to report changes in quantitative outcomes such as sales, employment, cost savings, and capital investment. In addition, the current survey asks about factors in using the MEP and strategic challenges the company faces. Prior questionnaires included items about customer satisfaction, but the MEP has reduced these questions, because of lack of variability in overwhelmingly positive responses. The current questionnaire asks clients about the likelihood of recommending the MEP to other companies.

Survey results concerning fiscal year 2010 indicated that the MEP served 34,299 companies. Of the 7,786 participating in the survey (out of 9,654 qualified to do so given their receipt of more intensive services), these clients' aggregate results attributed to MEP assistance were $8.4 billion in sales, $1.3 billion in cost savings, $1.9 billion investments, and 72,075 jobs created or retained. An economic impact model applied to the results indicated that for every $1 of federal investment, $32 of economic growth is returned (MEP 2011; see also MEP 1994, 1997, 1998). Client responses associated with high impact numbers are authenticated and, if found valid, are profiled as success stories using a mini-logic-based case study method.

The survey is designed to fulfill Government Performance and Assessment Act (GPRA) requirements for metrics on program impacts. Program metrics are reported relative to measurable goals which the program sets as part of GPRA. The survey also forms the basis for tracking center-level performance, since 2003, through a system of Minimally Acceptable Impact Measures (MAIM). [2] MAIM is comprised of five measures: (1) Bottom-line

---

[2]NIST MEP Reporting Guidelines, Center Operations, version 6.1. Effective Quarter 2, 2011.

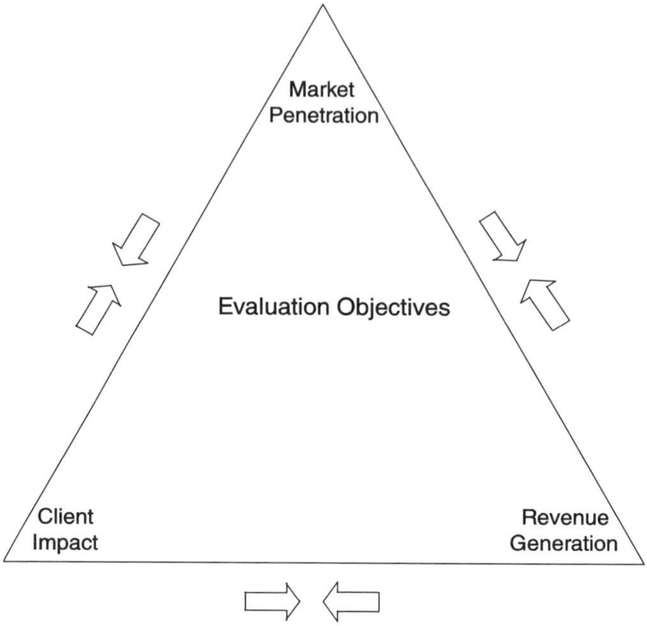

FIGURE APP-B-1 Evaluation objectives.

Client Impact, (2) Cost Per Impacted Client, (3) Investment Leverage Ratio, (4) Percent Quantified Impact, and (5) Survey Response Rate. In early 2012, the MEP instituted a new center performance methodology using a balanced scorecard approach, termed MEP sCOREcard. Roughly half of the score comes from seven center metrics (new sales, retained sales, jobs, investment, cost savings, clients served, and new clients served) and the other half from diagnostics of how the center performs on six dimensions: innovation practice, next generation strategy, market understanding, business model, partnerships, and financial viability.

MEP's enabling legislation calls for external panel reviews of centers. These reviews are typically conducted every other year for most centers, with panel composition including several directors from other MEP centers. Reviews focus on center operating plans, annual reports, and progress toward goals and follow established criteria based on the Baldridge quality framework. This model enables learning across centers in addition to the primary goal of providing feedback for the center under review. If the reporting and review processes continue to indicate poor performance, panel reviews are conducted more frequently and the MEP may re-compete center contracts.

MEP also has invested in other types of special evaluation studies. MEP has used logic-based case studies to examine clients with exceptionally high impacts from MEP services (Cosmos Corporation 1997, 1998, 1999) and to understand how new services are being implemented (SRI and Georgia Tech

(2009a). The logic models portray the functioning of MEP services, showing inputs, work processes, intermediate outcomes, and impacts on technology adoption, business performance, and broader impacts on the economy. Comparison group studies of clients and non-clients have been widely used to control for selection bias in that MEP is likely to attract clients already on the path to higher productivity (Jarmin 1997, 1997; Nexus Associates 1996, Oldsman 1996, Nexus Associates 1999, Youtie et al., 2008, 2010) and to account for observable and unobservable factors (Cheney et al., 2009). The 1990s saw the role of evaluations of several service delivery pilots including inter-firm networking (Welch et al., 1997) and SBDC-MEP partnership programs (Yin et al., 1998). Experimentation in evaluation approaches was evidenced in several workshops and evaluation working group sessions in the 1990s; this experimentation period has yielded to a more standardized evaluation process at the national level in the 2000s.

In addition to evaluations of the national program, several states have conducted evaluations of their particular state centers or system of centers. New York invested in independent evaluations of their centers in the 1990s (Oldsman et al., 1996, Nexus Associates, 1996). Pennsylvania conducted two highly regarded assessments of the centers in the Industrial Resource Center system (Nexus Associates 1999, Deloitte 2004). The state of Ohio included MEP program analysis as part of their assessment of the Third Frontier and other technology-based economic development programs (SRI and Georgia Tech 2009). Many states' requirements of centers focus on activity reporting metrics, but the presence of these large scale evaluations bring forth the possibility of differences between national program and state-level evaluation and reporting requirements.

Most centers do not have additional evaluation programs beyond what is required of the national or state sponsor. However, a few centers have developed and maintained special capabilities to support their evaluation efforts. The Michigan Manufacturing Technology Center created a longitudinal data set of survey-based metrics for full-scale comparison, the Performance Benchmarking Service (Luria 1997, Luria and Wiarda 1996). The Georgia MEP has administered a survey of non-clients and clients which has been used for evaluation as well as for conducting studies to address particular needs of the center from time-to-time (Shapira and Youtie 1998; Youtie et al, 2008, 2010).

Trade associations have been responsible for several evaluations of aspects of the program. The National Association of Manufacturers included a question about service use as part of its survey of members' technology adoption practices (Swamidass 1994). The Modernization Forum (a former association of MEP centers) sponsored several studies to provide information in support of the program's rationale as well as to aide in system set-up in the early years of the MEP. In the 2000s, the Advanced Small Manufacturers Coalition or AMSC (the current association of MEP centers), along with other non-profit organizations, has turned the emphasis of MEP evaluations towards strategic redirection of the program (NAPA 2003, 2004; AMSC, 2009; Stone & Associates and CREC 2010; MEP Advisory Board, Yakimov and Woolsey 2010) and international benchmarking efforts (Ezell and Atkinson 2011).

This evaluation system has resulted in a substantial body of evaluation studies related to the MEP program. The table in the Appendix represents 39 evaluative studies in the 1990s and another 26 in the 2000s. Thirty percent of these studies are published in academic journals or books, including in two journal special issues published in the 1990s (*Research Policy* 1996 issue and *Journal of Technology Transfer* 1998 issue). More than 10 percent are official federal government publications, with the remainder comprised of state government publications, conference proceedings, "gray" literature reports and white papers, dissertations, and informal memoranda. The most common method used in these evaluations is the customer survey, which was utilized by roughly one-third of the studies represented in this paper. Six of the works in this table used case study methodology (although a few others had case examples within primarily quantitative papers) while five linked client data to administrative databases at the state or national level. Sixteen of the studies utilized comparison groups, which signify the sophisticated nature of the evaluations in terms of controlling for factors besides extension services that could affect client outcomes. Seven of the evaluations involve benefit-cost and/or fiscal impact analysis to represent public and private returns from the program. This paper will show that the characteristics of the evaluations reflect the nature of the program's evaluation system (which in turn reflects the nature of the program itself). The 1990s was a period of system build-up and exploration in both the program and the evaluation system, whereas greater standardization occurred in the 2000s. Hence, this paper divides the literature into these two groupings to represent trends in evaluation methods and results over the 20-year period.

2. EVALUATION STUDY RESULTS: 1990s

Evaluation studies in the 1990s used diverse and sometimes novel methods to understand program processes and effects. This mix of studies was heavily influenced by the MEP's setting up of an evaluation working group in the 1993-1999 time period, producing a formal evaluation framework, and sponsoring four workshops on the evaluation of industrial modernization from 1993-1997.[3] Feller et al. (1996), Shapira et al. (1996), and Sears and Blackerby (1998) address this early evaluation system as a whole, discussing issues in performance measurement and program improvement amidst conflicting goals such as addressing important program goals while avoiding over-burdening of

---

[3]Atlanta Workshops on the Evaluation of Industrial Modernization, Aberdeen Woods, GA, 1993, 1994, 1996, 1997.

the client. As a result, a range of studies was produced with attention given to measurement of penetration, intermediate effects and short and longer-term outcomes on clients and the broader economy.

Swamidass (1994) conducted a survey of members of the National Association of Manufacturers (NAM) to assess their use of modern technologies and techniques. The survey found that only 1 percent of manufacturers say government is an important source of assistance in technology investment decisions; however many MEP centers are known through their university or center name rather than as a source of government assistance. This diversity makes efforts to measure penetration of the program, outside of program counts of manufacturers served, difficult.

How MEP fits in with other service providers has been an important dimension of evaluation efforts related to the market penetration objective of the program. The high cost of private firm service to small manufacturers has long been considered a major barrier to these operations' productivity. Whether MEP competes or complements private sector consulting was the subject of a major study sponsored by the Modernization Forum and carried out by Nexus Associates through surveys of MEP clients, a comparison group of manufacturers, and private consultants (Modernization Forum and Nexus Associates 1997; Oldsman 1997). Seventy-three percent of manufacturer responses suggested that MEP complemented consultants' work while only 7 percent of MEP clients reported that the MEP offered the same services as private consultants. Moreover, MEP clients were more likely to experience substantial benefits, in that the probability of a typical MEP customer improving its performance was 5.4 times higher than a manufacturer that acquired consulting services on its own. In another study of manufacturers in the Appalachian region, these enterprises were found to be tradition-bound in their ratings of various types of information sources—preferring their internal staff, customers, and suppliers. This study suggested that MEP service efforts would need time to build credibility and trust in their clientele base (Glasmeier et al, 1998). The MEP engaged in several formal efforts to collaborate with other service providing organizations. Shapira and Youtie (1997) found that MEP sponsorship led to greater service coordination than individual center efforts alone or state government demands would have provided, which in turn, generally improved the services to MEP clients, albeit at a significant expenditure of resources for validating and coordinating with these providers. On the other hand, Yin et al (1998) found that SBDC centers in a special MEP-SBDC pilot program did not have substantially better service delivery than a comparison group of MEP centers with their own SBDC collaboration initiatives.

Partnership with state governments has been an important element to MEP's funding formula. This formula has assumed an equal contribution of state and federal funding, with remaining revenue coming from client fees or other sources. A simulation of the federal-state relationship concluded that two-thirds of the states would not provide state funds if federal funding was discontinued. (MEP 1998)

Center-to-center comparisons were the subject of a few evaluation studies. Chapman (1998) conducted a distinctive data envelope analysis of MEP centers. This work showed that different centers were at the frontier of different service areas, with no one center consistently in the lead. Wilkins (1998) also performed center comparisons involving 14 centers, similarly finding that no one center excelled on all measures.

The mix of services and delivery methods was the subject of various evaluations in the 1990s. GAO's seminal 21-year-old manufacturing center review found there was a misalignment between the legislation establishing the centers—which emphasized technology transfer from the federal laboratories—and the needs of small manufacturers for assistance with proven technologies (GAO 1991). A 1993 National Academies study reiterated that although the program's enabling legislation focuses on technology upgrading, center specialists emphasize that a broader range of management and training, as well as technology, services are required. (National Academy of Sciences, 1993). Youtie and Shapira (1997) observed that the type of outcome is associated with service mix; marketing and product development services were 60 percent more likely to lead to sales outcomes, energy products more likely to lead to cost savings, plant layout and environmental projects more likely to capital expenditure avoidance, and quality projects not strongly associated with any type of outcome. Oldsman and Heye (1998) performed a simulation which showed that services which enable a manufacturer to raise piece prices generate more profit than services which enable reduction in scrap rate. Luria (1997) maintained that the program's service mix attracts cost-minimization/lean companies that are not on the path to increasing productivity. Cosmos Corporation led case studies of high impact projects with 25 manufacturing clients (1997) six manufacturing clients (1998) and seven different highly transformed manufacturers (1999). The results indicated the importance of integration of services and making discontinuous changes across multiple systems with leadership by top management.

Although most MEP technical assistance services are delivered on a one-on-one basis to a single manufacturing client at a time, MEP invested in a networking service delivery pilot from 1996 to 1998. This pilot had an extensive evaluation component, capped by a survey 99 members of 13 separate business networks. The results indicated that the median net benefit of network participation to the firm was $10,000, while some members experienced significantly higher benefits, raising the mean to $224,000 (Welch et al., 1997). Kingsley and Klein (1998) further found, in a meta-analysis of 123 case studies of networks, that networks with private sector leadership and funding were more likely to be associated with new business outcomes.

Intermediate outcomes were a major source of examination in 1990s-era evaluations. Several client survey-based studies qualitatively indicated that a higher percentage of companies engage in implementation following MEP assistance. Two-thirds of Georgia MEP customers took action on center recommendations (Youtie and Shapira 1997). Nearly 30 percent of Massachusetts center customers would not have carried out changes without MEP assistance (Ellis 1998). Many client surveys also suggested positive views of performance improvement, with the GAO (1995) finding that 73 percent of manufacturing respondents across the nation had better business performance and Ellis (1998) indicating that 71 percent of Massachusetts manufacturers improved their competitiveness as a result of center assistance.

Technology adoption was an important focus of several evaluation studies. Shapira and Rephann (1996) observed that manufacturing technology assistance program participants in West Virginia were more likely to adopt individual technologies and be amenable to technological investments than non-participants, but did not have significantly higher aggregate adoption across a range of technologies. The Luria and Wiarda (1996) Performance Benchmarking database analysis indicated that MEP customers adopted most technologies (with the exception of information technologies) more quickly than non-MEP customers. Evidence from case studies of centers in Northern Pennsylvania, Michigan, Minnesota conducted by Kelly (1997) led to the conclusion that the use of one-on-one services militates against widespread diffusion of knowledge and skills important for advanced technology usage.

Most of the evaluations try to get at business outcomes such as productivity as measured by value added. Several challenges in doing so are observed in these studies. The lion's share of impacts was found to come from a small number of manufacturing clients, with many reporting small or no impacts (Oldsman 1996). Most manufacturers had difficulty calculating impacts, and the timing of measurement was found to be an issue in that customers overestimate benefits, especially sales impacts, and underestimate costs close to point of survey, except for small number of high impact projects (Youtie and Shapira 1997).

Customer surveys tended to present positive outcomes. Quantitative business outcomes tended to present a more moderate picture, however, particularly when comparison groups were used to control for other factors and explanations besides program assistance. Some comparison group studies surveyed all manufacturers in a particular region (as in Youtie et al. 1998) or in a national sample (as in Luria 1997 and Luria and Wiarda 1996). Others linked MEP customer information to administrative datasets at the Census Bureau or Department of Labor and selected enterprises from these datasets to match client profiles (Jarmin 1997, 1999; Oldsman 1996; Nexus Associates 1996). Most of these studies focused on productivity as measured by value-added per employee, although other outcomes metrics were used as well. Jarmin (1997, 1999), Shapira and Youtie (1998), and Nexus Associates (1999) found clients to have higher growth in value-added per employee than non-clients. These analyses tended to focus on a few centers/network of centers (Georgia in the case of Shapira and Youtie, and Pennsylvania in the case of Nexus Associates).

Jarmin's analysis of eight MEP centers from 1987 to 1992 found productivity increases in clients over non-clients ranging from 3.4 to 16 percent. Nexus Associate's analysis of Pennsylvania centers reported higher labor productivity of 3.6-5 percent in clients as compared with non-clients. The average Georgia client had $366,000 to $440,000 higher in value-added than non-clients. Other comparison group-based evaluations found fewer differences between served and un-served manufacturers. Analysis of the Performance Benchmarking dataset showed that MEP clients do better than non-clients in sales growth, job growth, and adoption of some process improvements, but clients are not significantly better than non-clients in growth in wages, profits, and productivity (Luria 1997). Evaluation of the New York program indicated that participating manufacturers added 5.7 percent fewer workers than similar, non-participating companies (Oldsman 1996). Because the MEP seeks to enhance productivity, implementation of efficiency measures may result in a diminishment of some factory worker positions. This reduction is not automatically a drawback as the program's aim to promote long-term manufacturing competitiveness can lead to some declines along other dimensions.

The costs and benefits of manufacturing extension beyond those of the individual clients served was the subject of another set of studies. The results of these studies were reasonably positive. Cost-benefit analyses by Shapira and Youtie (1995), Nexus Associates (1996), and Michigan Manufacturing Technology Center (1996) demonstrate net public and private benefits of MEP assistance outpaces costs by a ratio in the 1:1 to 3:1 range. A Pennsylvania study (Nexus 1999) reported much more positive net returns to the state investment of 22:1. Thompson (1998) found the taxpayer payback to Wisconsin varied from slightly below break-even to positive. Several of these studies put forth methods to address various issues in cost-benefit analysis such as accounting for the full range of private and public costs and benefits, addressing returns and investments over time, and giving consideration to zero-sum re-distribution of benefits and value-added capture through downward adjustment of sales impacts for export sales and value-added (Shapira and Youtie 1995).

3. EVALUATION STUDY RESULTS: 2000s

The studies conducted in the 2000s reflected a different climate than was seen in the previous decade. Whereas the 1990s was a period of program expansion and experimentation, the 2000s saw substantial fluctuations in the program's budget, a systematizing of services, and consolidation of the number of centers as certain centers were combined into statewide or regional programs. The MEP evaluation system itself became more standardized as the evaluation working group of the 1990s was ended, center-level personnel became reporting rather than evaluation specialists, and metrics decisions were raised to the level of the center director rather than to center-level evaluators. This systematization is reflected in the MEP evaluation plan and metrics published by Voytek and colleagues in 2004 (Voytek et al., 2004). Evaluations in the 2000s presented in this paper were distinctive in their greater use of international comparisons and program assessment using expert panelists and document review. The table in the Appendix shows that six of the nineteen assessments published in the 2000s involved expert panelists or document review. Five used survey methods and three used comparisons with services or manufacturers in other countries.

Market penetration was addressed in several of these studies. Stone & Associates and CREC (2010) found penetration to be a concern in that the MEP only serves 10 percent of manufacturers, 2 percent with in-depth assistance. Although this level of service could be argued a reflection of cherry picking of clients, Deloitte (2004) reported that Pennsylvania manufacturing extension centers did not engage in "creaming;" a comparison of the credit rating of Pennsylvania manufacturing clients and a matched group indicated that the differences were not statistically significant. On the other hand, GAO (2011) examined the relationship between fees charged and penetration, finding that 80 percent of MEP centers were very or somewhat likely to prioritize revenue generation projects with larger clients.

Concerns about mission in terms of the programs' relationship with the private sector were raised to higher policy levels in the 2000s. Four governmental assessments—OMB (2002), National Commission on Fiscal Responsibility and Reform (2010), Schact (2011) and GAO (2011)—were devoted to this issue. OMB's Program Assessment Rating Tool (PART) evaluated the MEP program purpose and design, strategic planning, management, and results and accountability. Rated moderately effective, the assessment determined that "It is not evident that similar services could not be provided by private entities." The National Commission on Fiscal Responsibility and Reform concluded that MEP provides services that exist in the private sector. GAO's review of the cost share match requirements for centers reported that rural areas, which often are too costly for private consultants to serve, also were harder for the MEP to serve as centers increasingly found it necessary to develop cost share. Schact's Congressional Research Service report also addressed the issue of the appropriate level of federal investment for the program. To address the concerns raised in the OMB assessment, the National Academy of Public Administration (NAPA 2003) used a panel and document review and interview process to conclude that barriers to productivity improvement continue for small and medium-sized manufacturers and these firms are underserved by the private market.

Management of the program is another major substantive area of study in 2000-era evaluations. The 2003 NAPA review concluded that "MEP is effective in its core mission of helping small manufacturers reduce the barriers to productivity improvement" (p. 44). The GAO (2002) PART review of the program also gave MEP top ratings along many program management dimensions including program purpose, need, program and performance goals, strategic planning, collaboration, quality evaluations, budgetary goal alignment, financial management, proposal and grantee oversight, and achievement of performance goals and cost efficiencies and effectiveness. Shrank and Whitford (2009) found the program to advance experimentation, diversity, and access to local knowledge. Although center-to-center variability continued to be observed as it was in the Chapman and Wilkins studies, center-level evaluations also indicated the existence of stronger and weaker performing centers. An analysis of groups of large and small center MAIM scores in 2001, 2003, and 2005 observed that there were no consistent top performing centers from period to period, although a few centers landed near the top in many of the periods under analysis(Youtie 2005). NAPA (2004) found strong performance differences between centers; a substantial association was evidenced between high performing centers and number of clients served, years in operation, number of full time equivalent (FTE) employees for the center per million dollars of federal investment, and ratio of state dollars. A study of the manufacturing extension center in Arkansas found that it and its partners complied with MEP's implementation resource criteria and program goals (Russell 2011).

Program impacts continued to be a focal point of a set of evaluations. MEP metrics became systematized and focused primarily on the results of the client survey. As Luria noted, centers are accustomed to the survey; however it is marked by issues such as large numbers of clients that cannot monetize the effects of program assistance, the significant role of outliers, attribution concerns, and the importance of focusing on value-added.[4] MEP has sought to respond to some of these issues, for example, by applying value-added adjustments to sales results from the MEP survey in its bottom-line client impact MAIM metric (Voytek et al, 2004).

Several states conducted evaluations of their centers' outcomes and assessments of individual centers or regions of centers were also published during this period. The Deloitte assessment of Pennsylvania extension service impacts concluded that productivity and fiscal impact results from the Nexus Associates 1999 evaluation persisted into the 2000-2003 time period based on findings that the client mix in the more recent period was the same as it was in the earlier study, customer dissatisfaction had not increased, and MEP customer survey results showed the Pennsylvania centers to be high performers in terms of impacts. In contrast, Davila's evaluation of the Chicago center found that in an earlier period, clients were more likely than non-clients to have adopted new machinery and equipment, but by the next year, clients were similar to the general population in this regard (Davila 2004). Survey-based analyses of Georgia clients and non-clients in 2008 and 2010 maintained prior findings of higher increases in value-added per employee for clients. The Georgia results also differed sharply from those in comparable regions in the UK and Spain in that manufacturing extension customers in Georgia were more apt to engage in product and process innovation than similar non-customers (Roper et al, 2010).

---

[4]See Luria (2011).

Georgia survey results were also used to address certain issues, for example, rural-urban differences; the Georgia survey found that although rural and urban manufacturers were similar in their adoption of hard technology, rural manufacturers were less likely to use soft technologies and supply chain integration than their urban counterparts (Youtie and Shapira 2005). SRI and Georgia Tech (2008) found that Ohio MEP customers were more likely to have retained jobs than non-customers, controlling for industry, size, and other factors.

Updating the original Jarmin study into the period in which the MEP achieved full national coverage became a priority for the program. SRI and Georgia Tech released a study in 2009 which aimed to provide this update (Cheney et al., 2009). The study was not without problems, in that measures of selection bias used in the earlier study—such as proximity to an MEP office or location in a metropolitan area—did not work in the 2009 study when MEP offices were more widely accessible. The study used two different models to assess productivity impacts of clients and non-clients from 1997 to 2002 and the results of these two models diverged sharply, with one indicating positive effects and the other, negative effects. However, both models found positive and significant productivity growth was associated with MEP assistance for smaller establishments, certain types of MEP services, and a short delay (one or two years) after receipt of service.

The most significant focus of 2000s-era evaluations was the need for a strategic shift in the program's orientation towards product development, marketing, and technology rather than cost reduction. The Next Generation Manufacturing (NGM) survey, led by ASMC, indicated that one-quarter of small manufacturers did not engage in world-class practices in six strategic areas. NAPA (2004) found that the MEP cost reduction service orientation was not sufficient to maintain small and medium-sized enterprise (SME) competitiveness. Helper and Wial (2010) suggested that client-level assistance decisions may not pay sufficient attention to national manufacturing policy goals. Ezell and Atkinson (2011) reported, in an international comparison of 16 countries' manufacturing programs, that the MEP has not made the transition from continuous improvement to growth as many of the other programs had. However, Shapira et al (2011) found that these programs are heavily embedded in their countries' national innovation system, which limits the ability for facile adoption across country borders. MEP added a set of growth services, Eureka! Winning Ways (E!WW), in 2006 as a pilot and 2007 as a full-service launch to address, in part, these concerns. SRI and Georgia Tech (2008) found that customers of the E!WW pilot program had a distinctive profile in terms of using and experiencing benefits from this service. This profile included attributes such as being in industries with job losses, private family-owned firms, concentrated structure for product development, and history with technological implementation. The observation suggests that there are demand considerations involved in implementation of this type of service area.

4. RECOMMENDATIONS FROM EVALUATION STUDIES

Thirteen of the evaluation studies profiled in this paper included recommendations. Half of these studies explicitly referred to the need for MEP to transition to services that involve new product and market offerings. The integration of advanced technology into products has also been referenced. In addition, there are three, relatively diverging recommendations concerning funding—two for substantial increase federal funding (Stone & Associates and CREC 2010; Helper and Wial 2010), a second for greater flexibility in the funding formula (NAPA 2004), and a third for sun-setting federal funding (National Commission on Fiscal Responsibility and Reform 2010). The NAPA (2004) and Ezell and Atkinson (2011) studies promoted the need for more attention to exporting and global supply chains. MEP Advisory Board (2010) advocated greater green service offerings and (as did Deloitte 2004) workforce retention and development related assistance.

For some 15 years MEP has received recommendations to transition to product development, marketing, and technology-intensive services. The program has certainly integrated aspects of this counsel into its Next Generation MEP strategic plan[5]—which promotes an enterprise wide and market position orientation to manufacturers rather than a problem-based one —and into certain service offerings, for example E!WW growth services. Given that these recommendations persist in studies published through 2011, however, greater examination of the factors that constrain these types of offerings and approaches, such as those noted in the E!WW pilot evaluation (SRI and Georgia Tech 2008), may merit further attention. The NAPA (2004) study also pointed to the funding mix and business model as constraints on widespread use of these types of services. The need for a different funding mix would allow centers to devote more resources to these types of services rather than to revenue generation. Product development services and greater integration of advanced technology also requires different partnership arrangements, extensive industry familiarity, and cross-geographic boundary operations.

5. DISCUSSION

This collection of manufacturing extension evaluations illustrates the usefulness and merit of understanding who the customer is, what types of assistance are given, what investments and resources are used, and what kinds of outcomes are produced. MEP's system, which shifted from exploratory to standardized, has shown the usefulness of maintaining center, customer, and

---

[5]Manufacturing Extension Partnership, Next Generation Strategy, Gaithersburg, MD <*http://www.nist.gov/mep/ngs.cfm*>. (Accessed January 30, 2012).

project information and post-service customer surveying to understanding the value that clients place on the service. While surveys of customers (and non-customers) are the most commonly used method, a broad mix of methods has been relied upon such as cost-benefit and economic and fiscal impact analyses, economic models which draw on administrative databases, performance benchmarking information sets, and logic-based case studies. A range of outcomes have been examined in these studies, including capital investments, cost savings, technology adoption, sales, and employment. Still productivity as measured by value-added per employee remains the "gold standard" for assessing program effects. One of the challenges facing these evaluations is the potential effect of observed and unobserved factors in the economy, industry, and service provision to name a few of these factors. Many evaluations have used comparison groups to distinguish MEP's intervention from these other factors. Despite the use of comparison group studies, challenges remain in efforts to distinguish the effect of MEP assistance, which is often small in scale, from that of other extraneous factors.

External reviews of the MEP have also been widely used. These reviews typically have relied on interviews, document review, site visits, and analysis of existing program and customer data. These types of reviews have been useful in providing a perspective that is less tied to day-to-day service issues and more strategic and forward-thinking. While the largest share of, particularly quantitatively-oriented evaluations, are useful for program justifications, these external assessments in particular play a central role in enabling program learning about un-served market demand, opportunities not yet fully addressed, and relationships within the manufacturing policy and service system, including the private sector system.

This is not to say that learning from quantitative studies is rare. Although many evaluations emphasize quantitative impacts, these studies have implications for learning. In terms of resources, studies evidenced that the more resources devoted to a client, the more likely the client is to achieve impacts associated with the assistance. In addition, product development and marketing projects have been more likely than other substance areas to yield sales impacts. At the same time, important soft impacts continue to be de-emphasized in favor of quantitative outcomes that help with program justification. Similarly evaluations tend to be oriented around projects, with other forms of assistance such as training and supplier linkage still not receiving much attention. The potential downside is that what the program measures can be what the delivery system emphasizes.

Most consistent across the 20 years has been the recommendation that MEP shift from an emphasis on cost reduction toward product and technologically oriented services. To be sure, MEP has taken considerable steps

TABLE APP-B-1 Summary of Recommendations in Evaluation Studies

Author/Year	Recommendations
Luria (1997)	Nurture distinctive manufacturers with proprietary or design-intensive products and encourage other manufacturers to follow this strategy.
Modernization Forum and Nexus Associates (1997), Oldsman (1997)	Enhance work with consultants through using resources to identify them, application materials, project proposals, selecting consultants, developing standard contracts for working with them
Oldsman and Heye (1998)	Help companies become more distinctive as well as more efficient.
NAPA (2004)	Focus more on technology diffusion, product development, supply chain integration, and integrating the national network with other state/university providers and private sector firms. Provide greater flexibility in funding.
Davila (2004)	New evaluation metrics and approaches should be developed to measure adaptive learning, worker benefits, public benefits and costs, spillover effects, and how to allocate scarce MEP resources.
Deloitte (2004)	Lower barriers to access; offer more business strategy, product innovation, worker retention services; continue process innovation; support and grow advocacy and research for SMEs
Youtie (2005)	Examine consistently high performing centers' service mix and survey practices
SRI and Georgia Tech (2008)	Focus marketing on target customer characteristics; think strategically about participant mix; select and train growth coaches; use team-based approaches and fewer idea limitations in service design; evaluate after one year
Helper and Wial (2010)	Enhance funding for MEP to provide more product and market development services and greater coordination between MEP and other federal programs for manufacturers.
Stone & Associates and CREC (2010)	Expand program scale from current levels of 7,000-8500 firms served to 30,000 with substantially greater funding
MEP Advisory Board, Yakimov, Woolsey (2010)	Streamline innovation and growth services, target green services, emphasize exporting, develop talent

National Commission on Fiscal Responsibility and Reform (2010)	Eliminate federal funding for the MEP
Ezell and Atkinson (2011)	Offer technology acceleration programs and practices, exporting, energy efficiency, quality, standards, and design

in this direction through its strategic planning and growth services offerings. This is an important area for a fresh round of evaluation studies to catalog and describe interventions, pilot new measurement approaches and datasets (such as patents or SBIR awards) and their relationship to existing metrics, and include recommendations that go into greater depth concerning service areas and outcomes that have yet to be addressed.

6. CONCLUSION

Does MEP work? Would the country be substantially worse off without the program? Despite this collection of evaluations, one finds limitations in the ability to fully address these questions. MEP is a modestly funded program so it is difficult to separate out the effects of the program from other larger influences. The program's small size makes it hard to judge it against the broader economic, industrial, and other forces that affect the manufacturing sector in the United States if not globally. In addition, the zero-sum issue (helping one U.S. manufacturer could hurt another) is at play, notwithstanding certain social benefits of having stronger, more internationally competitive manufacturers in the U.S. economy rather than weaker ones. While some national agency reviews and audits of the program call attention to weaknesses in the larger manufacturer sector as an indication of the value of the MEP, one might question the usefulness of this yardstick to judge the orders-of-magnitude smaller MEP program.

Another set of limitations concern the inability of MEP clients to fully report quantitative impacts. The majority of MEP clients do not report quantitative impacts in response to the independent client survey. Many SME clients do not have the capabilities and systems to calculate outcomes in the way that is requested. Shapira et al (2004) found that just under half of the companies interviewed were unable to provide data on total production costs. White et al. (2012) observed that 27 percent of manufacturer responses in the Census of Manufacturers in 2002 and 2007 had imputed values for missing data about the value of shipments (i.e., sales); this same figure for total cost of materials was 42 percent and imputed figures were even higher in certain industries. Moreover, the Cosmos Corporation case studies conducted of exemplary MEP engagements—which had evaluation specialists visiting client companies, interviewing respondents, and reviewing company records—typically found additional quantitative impacts that were not reported to MEP through surveys.

Spillover effects also are difficult to assess without detailed surveys of MEP clients' customers, notwithstanding the application of econometric models to proxy these indirect effects.

Within these limitations, this paper addresses the question of the MEP's effectiveness by observing that the lion's share of evaluations of the MEP found the program to be effective. To examine the question of what would happen in the absence of the MEP, the most helpful evaluations are the 15 studies with a comparison group. Some of these comparison group-based evaluations use administrative databases, which have limited information but greater representativeness of the population, while others use surveys of served and un-served firms, which have some issues in representing the un-served firm population but tend to include more variables of interest for controlling extraneous effects. In addition, most of these are at the state or local level, with only three being nationally representative studies. Within these strictures, all 15 of the studies found some positive and significant effects in firms served by the program relative to un-served firms after controlling for size, industry, and other factors. These effects included improvement in sales growth, productivity (value added per employee) growth, job retention (and in one case, job growth), and technology and process improvement adoption. Six of these also found negative results in non-client gains over client parameters in wage, profitability, productivity, job growth, technology and networking adoption. Thus, while there are some mixed results, most studies find the presence of the MEP to be beneficial.

To understand the effects of the program, it can be useful to trace the chain of assistance to outcomes in the case study volumes published by Cosmos Corporation. Although these case studies are not randomly selected, they do illustrate how program assistance can lead to firm outcomes. A few exemplars follow. One case study company re-vamped its product line, turning profitable for the first time in a few years, by applying cost-based marketing practices, which it learned about from an extension center training program and subsequent engagement. A technological problem was preventing another company from achieving millions of dollars of new orders until the local MEP connected two research laboratories to help the company address this problem. Another company's mindset and confidence were changed after going through an MEP center-led growth services ideation session, leading to new business lines and sales. The new president of a family-owned business responded to supply chain pressures by transforming the firm, with the help of the extension center, through strategic planning, quality and cellular manufacturing initiatives, adoption of systems integrating manufacturing and accounting, customer-profitability analyses, and human resource training and apprenticeship opportunities; the results included substantial sales gains and payroll increases. Manufacturing clients directly attribute these outcomes to the MEP assistance. These studies illustrate the diverse approaches to serving and benefitting manufacturers.

REFERENCES

American Small Manufacturers Coalition (2009). Next Generation Manufacturing Study: Overview and Findings. Washington DC, June.

Chapman, R. (1998). Using Data Envelopment Analysis to Assess Performance of Manufacturing Extension Centers, NISTIR 6198, Gaithersburg, MD: National Institute of Standards and Technology, July.

Cheney, D, Ordowich. C, Youtie, J, Fernández-Ribas, A , Shapira, P (2009). Evaluating the Impact of MEP Services on Establishment Performance: A Preliminary Empirical Investigation. Technical Report. Arlington, VA: November.

Cosmos Corporation (1997).MEP Successes: A Case Study Approach. NIST Special Publication 916. Washington DC: Government Printing Office.

Cosmos Corporation (1998). MEP Successes: Case Study Series Exemplary Projects Case Studies. Washington DC: Government Printing Office, November 1998.

Cosmos Corporation (1999). MEP Successes: Case Study Series Transformed Firms Case Studies. Washington DC: Government Printing Office, April 1999.

Davila, N. (2004). Evaluating Manufacturing Extension: A Multidimensional Approach. Economic Development Quarterly. Vol. 18, No. 3, pp. 286-302.

Deloitte (2004), Manufacturing Pennsylvania's Future: Regional Strategies that Build from Current Strengths and Address Competitive Challenges. Submitted to Industrial Resource Centers, PA Department of Community and Economic Development and Team PA Foundation.

Ellis, S. (1998). Evaluation of Massachusetts Manufacturing Partnership: Selected Findings. In P. Shapira and J. Youtie, editors. Manufacturing Modernization: Implications of Evaluation Results for Program Improvement and Policy Development: Proceedings of Fourth Workshop on the Evaluation of Industrial Modernization Programs, Atlanta, Georgia: Georgia Institute of Technology.

Ezell, S., Atkinson, R. (2011). International Benchmarking of Countries' Policies and Programs Supporting SME Manufacturers. Information Technology and Innovation Foundation: Washington DC, September.

Feller, I. Glasmeier, A., Mark, M. (1996). Issues and perspectives on evaluating manufacturing modernization programs. Research Policy Vol. 25. No. 2, pp. 309-319.

Frazier, R.M. (2011). The Imperatives of Successful Policy Implementation: An Evaluation of the Hollings National Institute of Standards and Technology-Manufacturing Extension Partnership (NIST-MEP) Program's Implementation in Arkansas (Doctoral dissertation). University of Arkansas.

General Accountability Office (2011). NIST Manufacturing Extension Partnership Program Cost Share, GAO-11-437R. Washington DC: Government Printing Office.

General Accounting Office (1995). Manufacturing Extension Programs: Manufacturers' Views of Services, GAO/GGD-950210216BR. Washington, DC.

General Accounting Office (1991). Technology Transfer: Federal Efforts to Enhance the Competitiveness of Small Manufacturers, GAO/RCED-92-30. Washington, DC.

Glasmeier, A., Fuellhart, K. Feller, I., Mark, M. (1998). The Relevance of Firm-Learning Theories to the Design and Evaluation of Manufacturing Modernization Programs. Economic Development Quarterly, Vol. 12, No. 2, pp. 107- 124, May.

Helper, S., Wial, H. (2010). Strengthening American Manufacturing: A New Federal Approach. Washington DC: Brookings.

Hollings Manufacturing Extension Partnership Advisory Board, with Yakimov, G., Woolsey, L (2010). Innovation and Product Development in the 21st Century. Gaithersburg, MD, February.

Jarmin, R. S. (1998). Manufacturing Extension and Productivity Dynamics: Preliminary Evidence. In P. Shapira and J. Youtie, editors. Manufacturing Modernization: Implications of Evaluation Results for Program Improvement and Policy Development, Proceedings of Fourth Workshop on the Evaluation of Industrial Modernization Programs, Atlanta, Georgia: Georgia Institute of Technology.

Jarmin, R. S. (1999) "Evaluating the Impact of Manufacturing Extension on Productivity Growth" Journal of Policy Analysis and Management, 18 (1): 99-119.

Kelly, M. (1997). From Mission to Commercial Orientation: Perils and Possibilities for Federal Industrial Technology Policy, Economic Development Quarterly, November, Vol. 11, No. 4, pp. 313-328.

Kingsley, G., Klein, H. (1998). Interfirm Collaboration as a Modernization Strategy: A Survey of Case Studies, The Journal of Technology Transfer, Vol. 23, No. 1, Spring, pp.65-74.

Luria, D. (1997). Toward Lean or Rich? What Performance Benchmarking Tells Us About SME Performance, and Some Implications for Extension Center Services and Mission. In P. Shapira and J. Youtie, editors. Manufacturing Modernization: Learning from Evaluation Practices and Results: Evaluation of Regionally-Based S&T Programs Proceedings of Third Workshop on the Evaluation of Industrial Modernization Programs, Atlanta, Georgia: Georgia Institute of Technology, 6-29.

Luria, D. (2011). Evaluating the MEP Evaluation. Presentation at the workshop on Strengthening American Manufacturing: The Role of the Manufacturing Extension Partnership. Washington DC, November 14, 2011.

Luria, D., Wiarda, E. (1996). Performance benchmarking and measuring program impacts on customers: lessons from the Midwest Manufacturing Technology Center, Research Policy, Vol. 25, No. 2, pp. 233-246.

Manufacturing Extension Partnership (1994). Making a Difference for America's Manufacturers. Gaithersburg, MD: National Institute of Standards and Technology.

Manufacturing Extension Partnership (1997). MEP National Data Highlights. Gaithersburg, MD: National Institute of Standards and Technology, December.

Manufacturing Extension Partnership (1998). Review of Mission and Operations of Regional Centers of the Manufacturing Extension Partnership, Gaithersburg, MD: National Institute of Standards and Technology, February.

Manufacturing Extension Partnership (1998). Review of Mission and Operations of Regional Centers of the Manufacturing Extension Partnership, Gaithersburg, MD: National Institute of Standards and Technology, February.

Manufacturing Extension Partnership (2011). The Manufacturing Extension Partnership: Partnering for Manufacturing Innovation and Growth, June 2011, *<http://www.nist.gov/mep/upload/MEP-PARTNERING-IMPACTS-FEB2011.pdf>*.

Michigan Manufacturing Technology Center (1996). Fifth Year Review. Ann Arbor, MI: Michigan Manufacturing Technology Center.

Modernization Forum and Nexus Associates (1997). Competition or Collaboration: The Role of Manufacturing Extension Centers in the Private Consulting Market. Working with Consultants: A Tool Kit for Manufacturing Extension Centers, Volume 3. Washington, DC: NIST MEP; Oldsman, E. 1997, May. Manufacturing extension centers and private consultants: collaboration or competition? Technovation. 17 (5), 237-244.

National Academy of Public Administration (2003). The National Institute of Standards and Technology's Manufacturing Extension Partnership Program Report 1: Re-examining the Core Premise of the MEP Program. Washington DC, September.

National Academy of Public Administration (2004). The National Institute of Standards and Technology's Manufacturing Extension Partnership Report 2: Alternative Business Models, Washington DC, May.

National Academy of Sciences (1993). Learning to Change: Opportunities to Improve the Performance of Small Manufacturers. Washington DC: National Academy Press.

National Commission on Fiscal Responsibility and Reform (2010). $200 Billion in Illustrative Savings, a supplement to a draft proposal by the co-chairs. Washington, DC.

Nexus Associates, Inc. (1996). Evaluation of the New York Manufacturing Extension Program: Final Report, prepared for the New York Science and

Technology Foundation/ Empire State Development. Belmont, MA, March 18.

Nexus Associates, Inc. (1999). The Pennsylvania Industrial Resource Centers: Assessing the Record and Charting the Future. Report prepared the Ben Franklin/IRC Partnership Board. Belmont, MA, October

Office of Management and Budget (2002). Program Assessment Rating Tool (PART), Detailed Information on the Manufacturing Extension Partnership Assessment, Washington, DC: Government Printing Office.

Oldsman, E., Heye, C. (1998). Waste not, want not—A look at the impact of manufacturing extension centers . Journal of Technology Transfer, Vol. 23, No. 1, pp. 37-41.

Oldsman, E (1996). Does manufacturing extension matter? An evaluation of the Industrial Technology Extension Service in New York. Research Policy, Vol. 25, No. 2, pp. 215-232.

Oldsman, E., (2004). Evaluating SME Programs: Learning from the NIST Manufacturing Extension Partnership. Paper presented at the SME Evaluation Workshop, Mexico City, September 23, 2004.

Roper, S, Youtie, J, Fernandez-Ribas, A. (2010). Knowledge, Capabilities and Manufacturing Innovation: A USA–Europe Comparison. Regional Studies Vol. 44, No. 3, pp. 253-279

Schact, W. (2011). Manufacturing Extension Partnership Program: An Overview, 7-5700. Washington DC: Congressional Research Service.

Schrank, A., Whitford, J. (2009). Industrial Policy in the United States: A Neo-Polanyian Interpretation. Politics and Policy Vol. 37, No. 4, pp. 521-553.

Sears, D, Blackerby, P. (1998). The MEP Evaluation Approach: Past, Present and Future In P. Shapira and J. Youtie, editors. Manufacturing Modernization: Implications of Evaluation Results for Program Improvement and Policy Development, Proceedings of Fourth Workshop on the Evaluation of Industrial Modernization Programs, Atlanta, Georgia: Georgia Institute of Technology, 161-170.

Shapira, P. (2003). Evaluating manufacturing extension services in the United States: experiences and insights in : P. Shapira and S. Kuhlmann (editors), Learning from Science and Technology Policy Evaluation: Experiences from the United States and Europe, Cheltenham, UK and Northampton, MA: Edward Elgar, 2003.

Shapira, P, Youtie, J, Kay, L (2011) "Building Capabilities for Innovation in SMEs: A Cross-Country Comparison of Technology Extension Policies and Programs," International Journal of Innovation and Regional Development Vol. 3-4, pp. 254-272.

Shapira, P., Youtie, J. (1995). Assessing GMEA's Economic Impacts: Towards a Benefit-Cost Methodology. Atlanta, Georgia: Georgia Institute of Technology, March (Revised October, 1996).

Shapira, P., Youtie, J., Roessner, J.D., (1996). "Current Practices in the Evaluation of U.S. Industrial Modernization Programs," Research Policy, Vol. 25, No. 2, 1996, pp. 185-214.

Shapira, P., Youtie, J. (1997). Coordinating Manufacturing Extension Services, The Journal of Technology Transfer, Vol. 22, No. 1, Spring, pp. 5-10.

Shapira, P., Youtie, J. (1998). Evaluation Industrial Modernization: Methods, Results, and Insights from the Georgia Manufacturing Extension Alliance" The Journal of Technology Transfer, Vol. 23, No. 1, Spring, pp. 17-28.

Shapira, P., Rephann, T. (1996). The adoption of new technology in West Virginia: Implications for manufacturing modernization policies, Environment and Planning C: Government and Policy, Vol. 14, pp. 431-450.

Shapira, P., Youtie, J., Wang, J., Hegde, D., Cheney, D., Franco, Q., Mohapatra, S. (2004). Re-assessing the Value of Information and its Impact on Productivity in Small and Midsize Manufacturers. Atlanta Georgia and Arlington, VA: Georgia Tech Policy Project on Industrial Modernization and SRI International.

SRI and Georgia Tech (2008) Eureka! Winning Ways: Analysis of Early Client Experiences. Arlington, VA, August.

SRI and Georgia Tech (2009) Making an Impact: Assessing the Benefits of Ohio's Investment in Technology-based Economic Development Programs. Arlington, VA, September.

Stone & Associates and the Center for Regional. Economic Competitiveness (2010). Re-examining the Manufacturing Extension Partnership Business Model: Alternatives for Increasing the Program's Impact on US Manufacturing Sector Performance" Gaithersburg, MD, October.

Swamidass, P. (1994). Technology on the Factory Floor II: Benchmarking Manufacturing Technology Use in the United States, Washington DC: The Manufacturing Institute.

Thompson, C. (1998). Local Politics, National Policy, and the Taxpayer-Payback of Manufacturing Extension, The Journal of Technology Transfer, Vol. 23, No. 1, Spring, pp.37-42.

Voytek, K., Lellock, K., Schmit, M. (2004). Extension Partnership Program Developing Performance Metrics for Science and Technology Programs: The Case of the Manufacturing. Economic Development Quarterly, Vol. 18, No. 2, May, pp. 174-185.

Welch, D., Oldsman, E. Shapira, P., Youtie, J., Lee, J. (1997). Net benefits: An assessment of manufacturing business networks and their impacts on member companies. Chapel Hill, NC: USNet and Regional Technology Strategies.

White, T.K., Reiter, J., Petrin, A. (2012). Plant-level Productivity and Imputation of Missing Data in U.S. Census Manufacturing Data. National Bureau of Economic Research Working Paper No. 17816.

Wilkins, T. (1998). Benchmarking Manufacturing Extension Centers. In P. Shapira and J. Youtie, editors. Manufacturing Modernization: Implications of Evaluation Results for Program Improvement and Policy Development: Proceedings of Fourth Workshop on the Evaluation of Industrial

Modernization Programs, Atlanta, Georgia: Georgia Institute of Technology.

Yin, R., Merchlinsky, S., Adams-Kennedy, K. (1998). Evaluation of MEP SBDC Partnerships. Bethesda, MD: COSMOS Corporation.

Youtie J., Shapira, P. (1997). Tracking Customer Progress: A Follow-up Study of Customers of the Georgia Manufacturing Extension Alliance, Journal of Technology Transfer, Vol. 22, No. 2, Summer, pp. 43-52.

Youtie, J., Shapira, P. (1997). GMEA 1997: Review of Results. Evaluation Working Paper E9701. Atlanta, Georgia, March.

Youtie, J., Shapira, P. (1998), Summary of Manufacturing Extension Impact Studies, Georgia Tech Policy Project on Industrial Modernization, Atlanta, GA: Georgia Institute of Technology.

Youtie, J., Shapira, P. (2005). The Challenge of Manufacturing Innovation: Industry, Rurality, and Competitiveness in the State of Georgia. Evaluation Working Paper E2501. Atlanta, Georgia, June.

Youtie, J., Shapira, P., Kay, L., Rivera, A., Lynch, B., Cutler, G., Fernandez-Ribas, A. (2008). Innovation in Manufacturing: Needs, Practices, and Performance in Georgia 2008-2010. Atlanta Georgia: Georgia Tech Program in Science, Technology, and Innovation Policy.

Youtie, J., Shapira P., Kay, L, Dodonova, D., Sabbarese, D., Morales, C. (2010). Innovation in Manufacturing: Needs, Practices, and Performance in Georgia 2010-2012. Atlanta Georgia: Georgia Tech Program in Science, Technology, and Innovation Policy.

SUMMARY OF MANUFACTURING EVALUATION STUDIES

Author/Year	**Method**	**Focus**	**Main Findings**
GAO (1991)	Interviews, legislative history, on-site visits	Four original Manufacturing Technology Centers	Legislation establishing these centers set transferring laboratory technologies to small businesses, but the centers find that most manufacturers need off-the-shelf technologies.
National Academy of Sciences (1993)	Expert panel, workshops	Seven Manufacturing Technology Centers	Despite emphasis of enabling legislation on technology upgrading of small manufacturers, a broader range of services is required; sunset provisions encourage large company focus.
MEP (1994)	Center surveys of customer impacts	MEP customers	Benefits per company anticipated by 610 firms responding to MEP center surveys in 1994 included 5.5 jobs added or saved, $43,000 savings in labor and material costs, and an increase of almost $370,000 in sales. Benefits exceeded federal costs by 8:1 ratio.
Swamidass (1994)	Member survey	National Association of Manufacturers members	Only 1 percent of manufacturers say government is an important source of assistance in technology investment decisions, suggesting that market penetration of modernization services is low.

Author/Year	Method	Focus	Main Findings
GAO (1995)	Survey of MEP manufacturing customers	Nationwide	73 percent of 389 respondents indicated that their overall business performance had been improved.
Shapira and Youtie, (1995)	Benefit-cost study	Georgia, MEP customers	Combined net public and private economic benefits exceed costs by a ratio of 1.2:1 to 2.7:1
Feller, Glasmeier, and Mark (1996)	Review of existing evaluations	Various federal and state programs assisting SMEs	Methods do not address key areas of significant program-level need.
Luria and Wiarda (1996)	Benchmarking survey, comparison group	Michigan MTC customers; nationwide manufacturers	MEP customers improve faster than comparable firms in a comparison group. However, assisted firms had smaller increases in computer-based technologies. 17 key technology and business performance metrics used; ITI Performance Benchmarking Service dataset.
Michigan Manufacturing Technology Center (1996)	Benefit-cost study	Michigan, MTC customers	Combined net public and private economic benefits exceed costs by a ratio of 1.45:1
Nexus Associates (1996)	Survey of NYMEP customers, comparison group, benefit-cost study	NYMEP customers	NYMEP generated $30 million to $110 million of value-added income; 510 to 1920 jobs. Benefit cost ratio of 0.14:1.0 to 0.51:1.0. Cobb-Douglas Production Function; A priori prediction of high impact oversampling; ITI

Author/Year	Method	Focus	Main Findings
			Performance Benchmarking Service dataset is control group.
Oldsman (1996)	Customer Survey, comparison group	New York Industrial Technology Extension Service customers	Total annual cost savings for the 1,300 companies participating in the program between July 1990 and March 1993 is $30 million. Majority companies said their ability to compete was improved as a result of the program. The average customer added 5.7 percent fewer workers than similar, non-participating companies.
Shapira and Rephann (1996)	Survey with comparison group, multivariate regression	West Virginia, manufacturing extension customers and non-customers.	Participation in a manufacturing technology assistance program is not yet associated with higher levels of aggregate new technology use, but it is found to associated with adoption of specific technologies and receptivity to new technology investment. The study's results also confirm the value of training and suggest that a strategy of targeting smaller and medium-sized plans with services focused on multiple clustered locations may be effective in stimulating new technology use among these manufacturers.

Author/Year	Method	Focus	Main Findings
Shapira, Youtie, and Roessner (1996)	Review of existing evaluations	Various federal and state programs assisting SMEs	Most evaluations emphasize activity reporting and customer valuation to avoid burdening clients, but important issues require comparative approaches and control groups.
Cosmos Corporation, NIST MEP, 1997	Case studies	25 MEP engagements in 13 states	Structured case studies of MEP projects show that program services help smaller manufacturers to modernize their operations, improve quality, and increase profitability through such means as reducing waste, redesigning plant layouts, and improved inventory control and employee training.
Jarmin (1997)	Longitudinal study, comparison group	Longitudinal Research Databases, 1987-1992, MEP customer data from 8 centers	Manufacturing extension clients had 3.4-16 percent higher growth in value-added per worker than non-clients . Standard value-added production function; Controls for self-selection.
Kelly (1997)	Case studies of 3 centers	Northern Pennsylvania, Michigan, Minnesota	MEP's focus on one-on-one assistance fails to address problems that limit the diffusion of knowledge and skills in using more advanced technologies.
Luria (1997)	Performance Benchmarking Service dataset, comparison	Michigan MTC customers	Customers improved to a greater extent than non-customers in sales growth, employment

Author/Year	Method	Focus	Main Findings
	group		growth, and adoption of certain process improvements and technologies. However, center customer growth in wage rates, profitability, and labor productivity were not significantly different from that of non-customers. The author attributes the results to the center's service mix, which attracts companies that are not on a rising productivity path, combined with intense customer price pressures.
MEP (1997)	Telephone survey of MEP customers by U.S. Census Bureau	Nationwide, MEP customers	MEP customers report $110 million increased sales, $16 million from reduced inventory levels, $14 million in labor and material savings, 1,576 net jobs created, 1,639 total jobs retained as a direct result of MEP services. Information provided 9-10 months after project close.
Modernization Forum and Nexus Associates (1997), Oldsman (1997)	Survey, comparison group	750 MEP clients, 800 comparison companies, 202 private consultants nationwide	Only 7 percent of MEP clients report that the MEP offers the same services as private consultants. The probability of a typical MEP customer improving its performance is 5.4 times greater than a manufacturer that secured consulting services on its own.

Author/Year	Method	Focus	Main Findings
Shapira and Youtie (1997)	Case studies and analysis of reporting data	6 MEP centers and their partnerships	MEP sponsorship has led to increased service coordination not readily obtained through individual center efforts alone or through demands of state governmental funders. Increased service coordination, in turn, has mostly improved the assistance delivered to firms, though significant expenditure of resources was required to achieve these benefits.
Welch, Oldsman, Shapira, Youtie, and Lee (1997)	Survey of manufacturing network customers	99 members of 13 separate business networks	The median net benefit of network participation to the firm is $10,000 (the average was $224,000).
Youtie and Shapira, (1997)	Customer survey - longitudinal tracking study	Georgia, MEP customers	68 percent assisted firms took action, with more than 40 percent percent reporting reduced costs, 32 percent improved quality, 28 percent capital investment . Customers overestimate benefits and underestimate costs close to point of survey, except for small number of high impact projects.
Youtie and Shapira (1997)	Customer survey; project-impact analysis	Georgia, MEP customers	Product development, marketing projects are 60 percent more likely to lead to sales increases; energy projects are most likely to lead to cost savings;

Author/Year	Method	Focus	Main Findings
			plant layout, environmental projects help companies avoid capital spending. Quality projects do not rate highly anywhere, although they require the largest MEP customer time commitment.
Chapman (1998)	Data envelopment analysis of MEP reporting data.	Compares 51 MEP centers using second half of 1996 data.	Centers excel in different areas. (Specifically, MEPs on the frontier in one area may move out of/not be on the frontier in another area).
Cosmos Corporation (1998)	Logic-based case studies	6 case studies from 6 centers	Integration of certain interventions lead to substantial outcomes.
Ellis (1998)	Surveys of MEP customers	Massachusetts MEP customers	29 percent MMP customers may not have undertaken changes without MMP assistance. 71 percent of MMP customers reported some improvement in competitiveness.
Glasmeier, Fuellhart, Feller, and Mark (1998)	Survey of 51 manufacturers	Information requirements of plastics industries the Appalachian Regional Commission's counties in Ohio, Pennsylvania, and West Virginia.	Firms most often use traditional information sources because of their credibility and reliability, so MTCs need time to establish a history to demonstrate their effectiveness to firms.

Author/Year	Method	Focus	Main Findings
Kingsley and Klein (1998)	Meta-analysis of 123 case studies	Cases of industrial networks in Europe, North America, and Asia	Network membership can be built with the sponsorship of parent organizations and with public funding, but the networks that generate new business are associated with private sector leadership and at least some private sector funding.
MEP (1998)	Telephone survey of MEP customers by U.S. Census Bureau	Nationwide, MEP customers	MEP customers report increased sales of nearly $214 million, $31 million in inventory savings, $27 million in labor and material savings, and a $156 million increase in capital investment as a direct result of MEP services. Information provided 9-10 months after project close.
MEP (1998) (with Nexus Associates)	Simulation model	MEP centers nationally	2/3 of states would end state funding if federal funding were ended; 60-70 percent of centers would not be able to maintain a focus on affordable, balanced service.
Oldsman and Heye (1998)	Simulation	Hypothetical metal fabrication firm	Reducing scrap by 2 percent raises profit margins by 1.2 percent, but increasing piece price by 2 percent adds $200,000 a year.
Sears and Blackerby (1998)	MEP evaluation plan and metrics	National system	MEP's evaluation is designed to contribute to center-level and system-level performance through attention to customer

Author/Year	Method	Focus	Main Findings
			information, data quality, data analysis, and producing results.
Shapira and Youtie (1998)	Survey of manufacturers, comparison group	Georgia manufacturers with 10+ employees	The average client plant had a value-added increase of $366k-$440k over non-clients. Cobb-Douglas Production function; Controls include use of other public and private sector service providers.
Thompson (1998)	Benefit-cost study, simulation	Wisconsin taxpayers	Taxpayer payback ratios of 0.9:1.0 to 3.5:1 from the point of view of the state taxpayer who receives a federal subsidy. However, there is considerable variation in payback ratios by industry and by service type. Increasing *sales* shows the greatest taxpayer-payback.
Wilkins (1998)	Center management benchmarking	14 MEP centers	No single measure designates a high or low performing center. Costing rate of $200-$400 per hour resulted. Field staff tend to develop more projects than they close. 75 percent of centers have moved from subsidizing services to generating positive cash flow.
Yin, Merchlinsky, Adams-Kennedy (1998)	Survey and case studies, comparison group	7 pilot centers (receiving $750,000 over 3 years to	Pilot and comparison centers did not differ markedly either in the nature of their partner

Author/Year	Method	Focus	Main Findings
		establish a manufacturing SBDC) and 7 comparison centers with SBDC relationships but no special funding	relationships with SBDC or in the seamlessness of their service delivery.
Cosmos Corporation (1999)	Logic-based case studies	7 case studies from 5 MEP centers	A transformed firm has made significant changes in four of five systems; many paths to transformation were observed.
Nexus Associates (1999)	Quasi-experimental comparison group study, fiscal impact analysis	SME client cohorts from 1988/1989-1998/1999, longitudinal research dataset	On an annualized basis, IRC clients increased labor productivity by 3.6-5.0 percentage points and output by 1.9-4.0 percentage points more than they would have done without assistance. Productivity gains resulted in an inflation-adjusted $1.9 billion increase in gross state product between 1988-1997. A benefit-cost analysis finds returns to state investment of 22:1.
Jarmin (1999)	Panel, longitudinal study	Longitudinal Research Database, Annual Survey of Manufacturers 1987-1993, MEP customer data from 9 centers	The timing of observed productivity improvements at client plants is consistent with a positive impact of manufacturing extension.

Author/Year	Method	Focus	Main Findings
OMB (2002)	Program assessment	National MEP program	MEP is rated "moderately effective" with concerns raised that the program serves only 7 percent of manufacturers, that competition with the private sector exists, efforts to pursue self-sufficiency without federal funds have not be pursued.
NAPA (2003)	Panel review	National MEP program	SME barriers to productivity improvement persist and these firms are underserved by the market.
NAPA (2004)	Panel review	National MEP program	MEP has performed well but its basic service model, funding formula, and role of the national office need to evolve.
Davila (2004)	Chicago Manufacturing Center clients and Performance Benchmarking Service firms	Chicago area	Clients were more likely than non-clients to have adopted machinery and equipment but in the next year, this difference was insignificant.
Deloitte (2004)	Detailed review of NAICS growth, company credit scores, comparison group	SME clients of Industrial Resource Centers (IRCs) in Pennsylvania	IRCs are not creaming; mean credit rating of 2.92 for clients v. 2.81 for non-clients not significantly different in multinomial regressions. Impacts reported in the Nexus 1999 evaluation continue to hold true.

Author/Year	Method	Focus	Main Findings
Voytek, Lellock, and Schmit (2004)	MEP evaluation plan and metrics	National system	Center-level metrics can be used to diagnose performance.
Youtie (2005)	MEP system metrics	MEP centers in the ntional system	No consistent "top center" exists in terms of key Minimally Acceptable Impact Measures (MAIM) metrics.
Youtie et al. (2008)	Survey of clients and non-clients	Georgia MEP	Georgia Tech clients experienced a 12 percent increase in value-added per employee over non-clients.
SRI and Georgia Tech (2008)	Logic-based case studies	8 case studies from 4 MEP centers	Companies in industries with job losses, private family-owned firms, concentrated structure for product development, history with technological implementation were more apt to experience benefits from E!WW. 3 of 8 case study firms gained tangible benefits and 4 of 8 took steps toward growth.
ASMC (2009)	Survey	Manufacturers in 18 states	One-quarter of manufacturers do not engage in world-class practices in six strategic areas.
SRI and Georgia Tech (2009)	Longitudinal study, comparison group	Ohio MEP served and unserved firms in Ohio unemployment security data	Comparing 443 MEP clients to 14,062 non-client manufacturers, the average manufacturing establishment shrank by 6.28 employees

Author/Year	Method	Focus	Main Findings
			between 1998 and 2002. The average firm that received treatment during this period shrank by just 0.53 employees. This difference is statistically significant at the 0.1 level.
Cheney et al., (2009)	Longitudinal study, comparison group	Census of Manufacturers, MEP customer data, all 59 centers	Manufacturing extension clients had growth in value-added per worker vs. non-clients ranging from -5.8 percent to 1.7 percent for 713,330 observations depending on whether a difference in differences or lagged dependent variable model is estimated.
Schrank and Whitford (2009)	Review of SME policy	US policy	MEP promotes experimentation, diversity, and access to local knowledge.
Helper and Wial (2010)	Policy review	Great lakes region manufacturing	MEP serves individual manufacturers willing to pay for assistance rather than what may be needed to implement national policy goals.
Roper et al. (2010)	International survey models	Manufacturers in Georgia USA and three regions in Europe	University intervention is significant in Georgia product and process innovation but complementaries are not well captured.
Stone & Associates and CREC (2010)	Interviews, document review	National, MEP clients	MEP serves 10 percent of manufacturers, 2 percent with indepth assistance.

Author/Year	Method	Focus	Main Findings
MEP Advisory Board, Yakimov, Woolsey (2010)	Advisory board review	National MEP program	Streamline innovation and growth services, target green services, emphasize exporting, develop talent.
National Commission on Fiscal Responsibility and Reform (2010)	Budget analysis, working groups	National MEP program	MEP provides services that exist in the private sector and helps firms that should close.
Youtie et al. (2010)	Survey of clients and non-clients	Georgia MEP	Georgia Tech clients had $11,000 higher value-added per employee than non-clients.
Ezell and Atkinson (2011)	International comparison	MEP and similar programs in 10 other countries	Countries' manufacturing support programs have made a transition from continuous improvement to growth but the US program is under-funded.
GAO (2011)	Program assessment, surveys	National MEP program	Clients fees comprise more than half of non-federal funds for 26 centers; 80 percent of MEP very or somewhat likely to prioritize revenue generation projects with larger clients.
MEP (2011)	Client and fiscal impact	Nationwide, MEP customers	$1 of federal investment in MEP generates $32 of return in economic growth, $3.6 billion in new sales nationally.
Russell (2011)	Historical documents,	Little Rock, Arkansas	The center and its partners comply with

Author/Year	Method	Focus	Main Findings
	interviews, survey		MEP's implementation resource criteria and program goals.
Schact (2011)	Legislative history review	National MEP program	Issues around federal funding for the program and the match level of this funding remain.
Shapira et al (2011)	International comparison	7 technology extension services in 6 countries	Countries' technology extension services reflect their distinctive national innovation systems.

Appendix C1

MEP Center Data (FY2010-2011)

SOURCE: Data provided by MEP staff by email, February 9, 2011.

SIZE AND FEDERAL CONTRIBUTION (FY2010-2011)

Center Name	State	Total Income	Federal Amount	Program Income	Federal share (%)
California Manufacturing Technology Consulting (CMTC)	CA	$28,764,606	$9,981,903	$8,800,800	34.7
Ohio MEP	OH	$22,379,692	$4,545,417	$12,950,000	20.3
Texas Manufacturing Assistance Center (TMAC)	TX	$16,773,856	$5,727,871	$6,788,889	34.1
New York MEP	NY	$21,149,691	$5,468,093	$6,931,598	25.9
New Jersey MEP	NJ	$8,511,104	$1,679,580	$4,000,000	19.7
Michigan Manufacturing Technology Center (MMTC)	MI	$9,504,662	$2,268,003	$3,232,143	23.9
Corporation for Manufacturing Excellence (MANEX)	CA	$7,112,710	$1,577,000	$3,810,710	22.2

Center Name	State	Total Income	Federal Amount	Program Income	Federal share (%)
Wisconsin MEP	WI	$8,518,303	$2,591,886	$2,353,000	30.4
Enterprise Minnesota	MN	$9,042,992	$2,106,003	$2,612,200	23.3
South Carolina MEP	SC	$6,804,009	$2,268,003	$2,200,000	33.3
Missouri Enterprise	MO	$7,063,343	$2,109,748	$2,289,448	29.9
Illinois Manufacturing Extension Center	IL	$6,144,226	$2,108,686	$2,105,492	34.3
Impact Washington	WA	$6,275,095	$1,863,215	$2,300,000	29.7
North Carolina MEP	NC	$5,404,996	$2,216,244	$1,768,341	41.0
Mid-America Manufacturing Technical Center (MAMTC)	KS	$5,476,200	$1,864,950	$2,000,000	34.1
Massachusetts MEP	MA	$5,325,083	$2,364,771	$1,496,125	44.4
Illinois Chicagoland (new)	IL2	$6,806,588	$2,500,000	$1,250,000	36.7
(DVIRC)	PA	$7,113,960	$1,344,000	$2,369,960	18.9
Georgia MEP	GA	$4,024,356	$2,552,258	$1,100,000	63.4
Connecticut State Technology Extension Program (CONNSTEP)	CT	$3,447,989	$1,027,489	$2,403,000	29.8

Center Name	State	Total Income	Federal Amount	Program Income	Federal share (%)
Iowa Center for Industrial Research and Service (CIRAS)	IA	$5,374,030	$1,859,206	$1,514,824	34.6
Alabama Technology Network	AL	$4,752,583	$1,780,800	$1,500,000	37.5
A.L. Philpott MEP	VA	$3,597,487	$1,276,799	$1,720,000	35.5
MEP Utah	UT	$9,143,913	$908,832	$2,016,241	9.9
Florida MEP	FL	$3,870,666	$1,268,438	$1,350,000	32.8
Tennessee MEP	TN	$3,711,662	$1,252,228	$1,252,228	33.7
New Mexico MEP	NM	$3,110,313	$1,360,802	$980,000	43.8
MANTEC, Inc.	PA	$3,886,270	$674,823	$1,600,000	17.4
Catalyst Connection	PA	$2,548,504	$1,092,004	$956,500	42.8
Northwest Pennsylvania Industrial Resource Center (NWPIRC)	PA	$3,590,377	$702,745	$1,156,000	19.6
Colorado Association for Manufacturing and Technology (CAMT)	CO	$3,132,704	$655,201	$1,200,000	20.9
Indiana MEP-Purdue Technical Assistance Program	IN	$2,539,409	$671,287	$1,092,574	26.4
Arizona MEP (new)	AZ	$5,229,722	$1,000,000	$750,000	19.1

Center Name	State	Total Income	Federal Amount	Program Income	Federal share (%)
MRC	PA	$3,206,326	$588,003	$1,151,922	18.3
North Dakota Manufacturing Extension Partnership	ND	$2,370,782	$471,682	$1,163,100	19.9
Oregon MEP	OR	$2,372,554	$778,620	$821,240	32.8
Arkansas Manufacturing Solutions	AR	$2,315,981	$941,110	$605,000	40.6
Nevada Industry Excellence	NV	$1,876,690	$756,001	$739,308	40.3
Maine MEP	ME	$2,375,322	$863,520	$540,000	36.4
Mississippi Technology	MS	$1,650,071	$1,003,782	$396,289	60.8
MEP of Louisiana	LA	$2,150,301	$588,870	$782,191	27.4
Kentucky Manufacturing Assistance Center (KMAC)	KY	$1,898,619	$529,004	$828,392	27.9
Nebraska MEP	NE	$1,724,979	$599,430	$730,000	34.7
Idaho TechHelp	ID	$2,113,359	$508,801	$808,458	24.1
Oklahoma Manufacturing Alliance	OK	$1,485,897	$981,121	$275,000	66.0
Northeastern Pennsylvania Industrial Resource Center (NEPIRC)	PA	$1,959,218	$504,001	$702,500	25.7
Vermont MEP	VT	$1,714,483	$396,483	$798,000	23.1

Center Name	State	Total Income	Federal Amount	Program Income	Federal share (%)
Industrial Modernization Center IMC (new)	PA	$1,432,558	$713,414	$460,000	49.8
IMC-PA	PA	$2,095,214	$363,414	$718,000	17.3
Delaware MEP	DE	$1,602,400	$302,400	$750,000	18.9
PRIMEX	PR	$1,394,310	$510,718	$483,592	36.6
Alaska MEP	AK	$1,960,534	$806,401	$146,401	41.1
RIMES	RI	$1,308,301	$450,238	$500,476	34.4
New Hampshire MEP	NH	$1,547,503	$421,303	$529,000	27.2
South Dakota MEP	SD	$1,695,453	$250,000	$643,511	14.7
Manufacturing Works	WY	$1,193,002	$372,049	$450,000	31.2
Montana Manufacturing Extension Center (MMEC)	MT	$1,397,136	$512,000	$276,283	36.6
NW Wisconsin Manufacturing Outreach Center (NWMOC)	WI	$1,816,099	$403,201	$324,202	22.2
UM MAP - MD	MD	$1,140,374	$400,971	$240,000	35.2
High Technology Development Corporation (HTDC)	HI	$966,266	$437,066	$166,400	45.2
West Virginia MEP	WV	$776,217	$328,540	$157,677	42.3

Center Name	State	Total Income	Federal Amount	Program Income	Federal share (%)
Total		$299,671,050	$88,450,428	$106,037,015	
Max		$299,671,050	$88,450,428	$106,037,015	66.0
Min		$776,217	$250,000	$146,401	9.9
25th percentile		$1,747,759	$511,039	$614,628	23.2
75th percentile		$5,397,255	$1,755,495	$1,756,256	36.7
median		$2,829,409	$886,176	$1,036,287	32.0
average		$8,802,862	$2,622,772	$3,028,118	32.1

SME POPULATIONS, CONTACTS, AND CLIENTS

State	Center Name	SMEs contacted	Impacted Clients	#SME in service area
CA	California Manufacturing Technology Consulting	3,563	571	31,596
OH	Ohio Manufacturing Extension Partnership	3,273	434	15,978
NY	New York Manufacturing Extension Partnership	1,817	404	18,617
FL	Florida Manufacturing Extension Partnership	1,633	212	14,192
TX	Texas Manufacturing Assistance Center	1,422	282	20,823
MI	Michigan Manufacturing Technology Center	1,043	225	13,472
MA	Massachusetts Manufacturing Extension Partnership	1,005	155	7,596
KS	Mid-America Manufacturing Technology Center	992	85	3,143
WI	Wisconsin Manufacturing Extension Partnership	928	170	7,488
NH	New Hampshire Manufacturing Extension Partnership	867	41	2,058
SC	South Carolina Manufacturing Extension Partnership	854	158	4,260
GA	Georgia Manufacturing Extension Partnership	852	152	8,575
TN	Tennessee Manufacturing Extension Partnership	823	130	6,616
PR	Puerto Rico Manufacturing Extension Inc	796	33	2,075

State	Center Name	SMEs contacted	Impacted Clients	#SME in service area
NC	North Carolina Manufacturing Extension Partnership	720	167	9,951
IA	Iowa Center for Industrial Research and Service	710	152	3,718
PA	Catalyst Connection	635	139	3,282
NJ	New Jersey Manufacturing Extension Program	633	133	9,170
UT	Utah Manufacturing Extension Partnership	614	87	3,312
OK	Oklahoma Manufacturing Alliance	598	134	3,925
MN	Enterprise Minnesota	538	166	7,850
MI	Missouri Enterprise	516	103	6,783
IL	Illinois Manufacturing Extension Center - Downstate	513	126	4,711
CA	Corporation for Manufacturing Excellence (Manex)	506	89	12,395
AL	Alabama Technology Network	491	124	4,822
HI	Hawaii HTDC- MEP	467	28	970
WA	Impact Washington	435	86	7,582
AK	Alaska Manufacturing Extension Partnership	397	43	501
DE	Delaware Valley Industrial Resource Center	374	113	4,130
IN	Indiana MEP Purdue Technical Assistance Program	355	68	8,790
CT	Connecticut State Technology Extension Program	349	136	4,889

State	Center Name	SMEs contacted	Impacted Clients	#SME in service area
VA	GENEDGE ALLIANCE	342	112	5,668
ME	Maine Manufacturing Extension Partnership	309	62	1,798
NM	New Mexico Manufacturing Extension Partnership	306	73	1,535
AR	Arkansas Manufacturing Solutions	299	83	3,007
PA	IMC-PA	292	70	996
OR	Oregon Manufacturing Extension Partnership	285	61	5,658
MS	Mississippi Technology Alliance	258	59	2,558
PA	Manufacturers Resource Center	256	102	1,537
CO	Colorado Association for Manufacturing and Technology	253	56	5,240
RI	Rhode Island Manufacturing Extension Services	234	41	1,800
NV	Nevada Industry Excellence	230	37	2,015
PA	Northwest Pennsylvania Industrial Resource Center	223	151	1,633
PA	MANTEC	222	111	2,519
PA	Northeastern Pennsylvania Industrial Resource Center	219	117	1,131
AZ	Arizona Manufacturing Extension Partnership	211	59	5,005
KY	Kentucky Manufacturing Assistance Center	174	32	4,064
WI	Northwest Wisconsin Manufacturing Outreach Center	170	42	2,017

State	Center Name	SMEs contacted	Impacted Clients	#SME in service area
KY	Manufacturing Extension Partnership of Louisiana	159	68	3,353
MT	Montana Manufacturing Extension Center	140	44	1,300
ID	Idaho TechHelp	131	54	1,948
NE	Nebraska Manufacturing Extension Partnership	130	43	1,948
MD	University of Maryland Manufacturing Assistance Program	124	51	3,601
WY	Manufacturing-Works	124	41	568
VT	Vermont Manufacturing Extension Center	116	42	1,079
ND	North Dakota Manufacturing Extension Partnership	112	22	752
WV	West Virginia Manufacturing Extension Partnership	84	27	1,384
DE	Delaware Manufacturing Extension Partnership	82	47	634
SD	South Dakota Manufacturing Extension Partnership	79	28	1,045

Bottom Line Impacts

State	Center Name	Dollars per impacted client	Dollars per SME in area
MS	Mississippi Technology Alliance	$3,628,316	$83,687
AR	Arkansas Manufacturing Solutions	$1,752,289	$48,367
PA	Northeastern Pennsylvania Industrial Resource Center	$406,158	$42,016
VA	GENEDGE ALLIANCE	$1,777,412	$35,122
MT	Montana Manufacturing Extension Center	$903,744	$30,588
AK	Alaska Manufacturing Extension Partnership	$258,127	$22,155
UT	Utah Manufacturing Extension Partnership	$787,255	$20,680
PA	Catalyst Connection	$449,238	$19,026
ME	Maine Manufacturing Extension Partnership	$501,969	$17,309
PA	IMC-PA	$227,337	$15,977
PA	MANTEC	$350,835	$15,460
IA	Iowa Center for Industrial Research and Service	$357,066	$14,598
WY	Manufacturing-Works	$189,488	$13,678
ND	North Dakota Manufacturing Extension Partnership	$441,145	$12,906
PA	Manufacturers Resource Center	$190,727	$12,657
VT	Vermont Manufacturing Extension Center	$317,441	$12,356

State	Center Name	Dollars per impacted client	Dollars per SME in area
IL	Illinois Manufacturing Extension Center - Downstate	$442,084	$11,824
NM	New Mexico Manufacturing Extension Partnership	$245,324	$11,667
SD	South Dakota Manufacturing Extension Partnership	$410,214	$10,991
DE	Delaware Manufacturing Extension Partnership	$143,990	$10,674
MA	Massachusetts Manufacturing Extension Partnership	$511,091	$10,429
TN	Tennessee Manufacturing Extension Partnership	$512,985	$10,080
SC	South Carolina Manufacturing Extension Partnership	$266,418	$9,881
PA	Northwest Pennsylvania Industrial Resource Center	$101,261	$9,363
GA	Georgia Manufacturing Extension Partnership	$520,977	$9,235
PR	Puerto Rico Manufacturing Extension Inc	$573,356	$9,118
AL	Alabama Technology Network	$324,107	$8,335
AZ	Arizona Manufacturing Extension Partnership	$692,556	$8,164
CT	Connecticut State Technology Extension Program	$283,121	$7,876
WV	West Virginia Manufacturing Extension Partnership	$399,595	$7,796
KS	Mid-America Manufacturing Technology Center	$282,617	$7,643

State	Center Name	Dollars per impacted client	Dollars per SME in area
NJ	New Jersey Manufacturing Extension Program	$505,089	$7,326
KY	Manufacturing Extension Partnership of Louisiana	$356,425	$7,228
KY	Kentucky Manufacturing Assistance Center	$875,661	$6,895
NC	North Carolina Manufacturing Extension Partnership	$409,470	$6,872
HI	Hawaii HTDC- MEP	$237,034	$6,842
NY	New York Manufacturing Extension Partnership	$304,926	$6,617
TX	Texas Manufacturing Assistance Center	$472,872	$6,404
OH	Ohio Manufacturing Extension Partnership	$233,556	$6,344
OK	Oklahoma Manufacturing Alliance	$181,349	$6,191
NH	New Hampshire Manufacturing Extension Partnership	$289,945	$5,776
MI	Missouri Enterprise	$372,441	$5,656
ID	Idaho TechHelp	$202,963	$5,626
WI	Wisconsin Manufacturing Extension Partnership	$219,172	$4,976
WA	Impact Washington	$415,484	$4,713
NV	Nevada Industry Excellence	$245,103	$4,501
IN	Indiana MEP Purdue Technical Assistance Program	$572,077	$4,426
MN	Enterprise Minnesota	$205,336	$4,342

State	Center Name	Dollars per impacted client	Dollars per SME in area
CA	California Manufacturing Technology Consulting	$236,000	$4,265
DE	Delaware Valley Industrial Resource Center	$153,771	$4,207
MI	Michigan Manufacturing Technology Center	$249,156	$4,161
WI	Northwest Wisconsin Manufacturing Outreach Center	$172,018	$3,582
OR	Oregon Manufacturing Extension Partnership	$295,514	$3,186
MD	University of Maryland Manufacturing Assistance Program	$224,834	$3,184
NE	Nebraska Manufacturing Extension Partnership	$142,483	$3,145
RI	Rhode Island Manufacturing Extension Services	$128,873	$2,935
FL	Florida Manufacturing Extension Partnership	$181,849	$2,716
CA	Corporation for Manufacturing Excellence (Manex)	$316,356	$2,272
CO	Colorado Association for Manufacturing and Technology	$199,414	$2,131

JOB IMPACTS

State	Center Name	Sum of Job Impacts	Jobs per client	Jobs per SME
OH	Ohio Manufacturing Extension Partnership	4,163	101.5	7.33
NY	New York Manufacturing Extension Partnership	4,871	41.6	4.31
FL	Florida Manufacturing Extension Partnership	2,022	47.0	4.04
TN	Tennessee Manufacturing Extension Partnership	2,205	31.5	2.21
IL	Illinois Manufacturing Extension Center - Downstate	3,160	51.0	1.76
MA	Massachusetts Manufacturing Extension Partnership	1,642	58.6	1.57
AR	Arkansas Manufacturing Solutions	4,624	55.7	1.54
TX	Texas Manufacturing Assistance Center	4,751	34.2	1.45
PA	Catalyst Connection	1,071	48.7	1.42
OK	Oklahoma Manufacturing Alliance	823	17.5	1.30
MS	Mississippi Technology Alliance	2,851	48.3	1.11
CT	Connecticut State Technology Extension Program	1,390	31.6	1.07
GA	Georgia Manufacturing Extension Partnership	1,592	21.8	1.04
VA	GENEDGE ALLIANCE	1,431	53.0	1.03
UT	Utah Manufacturing Extension Partnership	1,057	25.2	0.98
PR	Puerto Rico Manufacturing Extension Inc	1,491	9.9	0.91

State	Center Name	Sum of Job Impacts	Jobs per client	Jobs per SME
CA	California Manufacturing Technology Consulting	2,822	32.4	0.85
SC	South Carolina Manufacturing Extension Partnership	1,269	12.4	0.83
WA	Impact Washington	699	25.0	0.72
PA	Northwest Pennsylvania Industrial Resource Center	1,087	32.9	0.52
NJ	New Jersey Manufacturing Extension Program	2,620	23.4	0.46
MN	Enterprise Minnesota	1,159	10.4	0.46
AL	Alabama Technology Network	1,752	14.1	0.36
NC	North Carolina Manufacturing Extension Partnership	1,753	12.9	0.36
PA	Northeastern Pennsylvania Industrial Resource Center	1,519	9.6	0.36
OR	Oregon Manufacturing Extension Partnership	1,112	13.1	0.35
MI	Michigan Manufacturing Technology Center	1,731	29.3	0.35
PA	Manufacturers Resource Center	630	15.4	0.31
IN	Indiana MEP Purdue Technical Assistance Program	2,083	13.4	0.27
MI	Missouri Enterprise	985	6.5	0.26
IA	Iowa Center for Industrial Research and Service	1,409	10.8	0.21
KY	Manufacturing Extension Partnership of Louisiana	380	7.0	0.20
KS	Mid-America Manufacturing Technology Center	321	7.6	0.16

State	Center Name	Sum of Job Impacts	Jobs per client	Jobs per SME
MT	Montana Manufacturing Extension Center	316	8.5	0.16
WI	Wisconsin Manufacturing Extension Partnership	1,103	12.8	0.15
NM	New Mexico Manufacturing Extension Partnership	640	5.1	0.14
PA	IMC-PA	455	6.7	0.14
DE	Delaware Valley Industrial Resource Center	813	4.8	0.11
KY	Kentucky Manufacturing Assistance Center	331	10.3	0.08
ME	Maine Manufacturing Extension Partnership	794	4.8	0.08
CA	Corporation for Manufacturing Excellence (Manex)	513	5.0	0.08
AZ	Arizona Manufacturing Extension Partnership	602	4.0	0.07
WV	West Virginia Manufacturing Extension Partnership	230	4.5	0.06
SD	South Dakota Manufacturing Extension Partnership	230	1.7	0.06
HI	Hawaii HTDC- MEP	105	2.6	0.06
MD	University of Maryland Manufacturing Assistance Program	222	2.0	0.05
PA	MANTEC	845	1.9	0.05
WI	Northwest Wisconsin Manufacturing Outreach Center	363	5.3	0.04
NE	Nebraska Manufacturing Extension Partnership	71	1.7	0.04

State	Center Name	Sum of Job Impacts	Jobs per client	Jobs per SME
NV	Nevada Industry Excellence	324	2.4	0.04
VT	Vermont Manufacturing Extension Center	121	2.0	0.02
WY	Manufacturing-Works	151	0.9	0.02
ID	Idaho TechHelp	250	1.2	0.02
NH	New Hampshire Manufacturing Extension Partnership	218	2.4	0.02
AK	Alaska Manufacturing Extension Partnership	61	1.1	0.01
RI	Rhode Island Manufacturing Extension Services	212	0.5	0.01
CO	Colorado Association for Manufacturing and Technology	210	0.7	0.01
ND	North Dakota Manufacturing Extension Partnership	128	0.6	0.01
DE	Delaware Manufacturing Extension Partnership	292	0.5	0.01

Appendix C2

Open-ended Responses from Center Directors

SOURCE: Information request to Center Directors, June-August 2012.

What strategies and tools do you use to convert touches into clients?	How do you see the "new strategy" being funded?	What are the major challenges that you see over the next three years?
Find pain points. Do assessments. Use referrals to motivate action.	New pools of funding are limited, and we have to invest in new strategies before staff can deliver services to generate fees. So we have to tighten budgets and shift existing funds to invest in new strategies.	Reductions in base funding from NIST and state partner. Increased emphasis on reporting metrics to justify/counteract threats to federal and state funding. Slow economic growth limiting the availability of client fees.
Mostly face to face discovery meetings where we visit with the CEO and discuss their major business pains.		Transformation Center to an Innovation Practice with the adjustments in staffing and development of a balanced portfolio
Documented sales process and experienced sales staff		

We build relationships with C level executives and growth oriented companies. We follow a process we refer to as "Value-based relationships". It is a 10 step guide from first contact through project success and measurement		Maintaining our growth plans which require staff to be more productive.
open enrollment continuing education offerings		Shifting focus to technology deployment and product innovation
Sandler sales and management system. Sales Logix CRM package. Business Advisors.	Revenue from clients	Too many priorities. funding. Finding the right people to staff the Center. Increasing stick to revenue to offset un stable state and federal funding
Follow up personal visits		Increased market penetration will occur with smaller companies. This will require new sources of funding because smaller companies lack the capital to invest in growing their business.
Value selling, high level (CEO, COO) partnering of overall growth view.		Client fees C level appointments / meetings

-Incentive funding when available -Referrals to satisfied clients -Regular follow-up		-Economic Uncertainty -Maintaining federal commitment to importance of manufacturing -Changes in State priorities and funding (which matches MEP) -developing collaborations that avoid competition for limited funds
Intense Innovation workshops, marketing seminars, value-added sales seminars, Lean training, energy audits, Innovation Engineering Leadership Institutes.		Federal Funding Cuts State budget deficits continued slow recession recovery
Newsletters, invitation to events, follow up sales calls	Providing additional Growth Services	developing staff skill set to support growth services
Follow up contacts. IE Jumpstarts (new)	We see increased state and economic development support as sell as new fee generating opportunities.	Evolving to complete adoption of MEP. Innovation Practice throughout the state.
seminars focused on Critical Business Issues. Solution Selling to get into the plants. And then to go thru the discovery phase to find the "pain" of the company & offer a solution.	we need to develop a practice & then promote from inside where possible & hire from outside as needed.	1. Adjusting to NIST - CORE metrics, (specifically getting companies to answer the surveys). Sales to Smm's 3. Adjusting to mandate of Innovation Engineering as the lead product offering

Build a valued relationship with them by providing resources that help them become more competitive and help them to grow through innovation	If NIST MEP does not "fund" the opportunity, we will invest in the opportunity to the point we can cash flow the strategy.	Unquestionably it is 2:1 cost share. 1:1 cost share is reasonable
Our base strategy is based on relationships so "personal" invitations work best for conversion.		Eroding funding... no change in base MEP funding for a decade and it is becoming a threat. Also, heavy investment in staff training will be required to meet evolving client needs.
We start at the corner office when we can. We want to be a strategic partner first and then the opportunities will come out way.		The biggest is funding. My staff has gone from 23 to 12 in the last 3 years and the State could continue to cut more in the years to come. I need to do a better job in finding new sources of revenue.
We do not maintain a sales staff. Companies contact us via field staff and partner referrals.		Transitioning to meet new MEP directives.

Low-cost introductory assessments. MEP Success Stories that address the issue uncovered during an initial meeting.	I believe that MEP and other public funding will be needed to launch "new strategy" services until they become self-sufficient from client revenue. The initial "seed capital" for these new services will likely be from federal and state grants, but only until self-sufficiency is reached. This may take 2-4 years, but should be achievable. Federal and state support will then be used to support "new-new" strategies so that we continuously update our service set to meet changing manufacturer needs and evolving technologies, business models and successful methodologies. Some existing funding will be shifted from existing programs, as they have become (or should by now be) self-sustaining.	Transition to innovation. Adapting to reduction in state funding. Creating value proposition and calculation ROI for clients.

Person follow-up meetings, continuing to send them electronic newsletters and notices of events.		Maintaining talented staff with decreasing state funds. Managing to increasing requirements from MEP for a Cooperative Agreement that is not much money. As state funds are decreased, Centers may need to maintain funding that is not necessarily aligned with the MEP mission such as targeting larger companies that are willing to invest.
Counselor sales - relate, discover needs, advocate for solutions (even if the solution source is not within the MEP), and follow up, follow up, follow up. Stay in touch with the client on a regular basis. Provide information that the client might find useful. Invite the client to events. Make business connections for the client.	Lean will continue to be a solid base. Growth or innovation with enhance the C-level conversation. The vast majority of the current staff "get it." New hires need to already be ready to talk with the C-level.	Regaining state funding or obtaining other sources of funding. The small and mid sized manufacturers desperately need help and would be hard pressed to get it without the MEP.
on-site visits		Being appropriately staffed, collecting impacts. There seems to be building resistance to the survey, how

		many times we survey and an unwillingness to share. Also, with the D&B difficulties we are turning small businesses away.
CEO entry level, listen, address issues prohibiting growth/profitability		Entry to and buy-in by small to mid size mfg. company CEOs, resources (primarily $) to fund growth activities with smaller clients (less than 100 employees)
solution selling		ability to adopt NIST MEP services and new requirements that create add ional administrative or operational burden
partner organization, references		
Assessment tool(CR) In the past, we run about 9.2 proposals for every 10 CR's.	We have already made the switch to the new strategy. We hired three new agents to help us increase our market share. This is a new goal from the NIST metrics. We let go some positions that the skill set didn't match our current or future needs. We then hired the current skill set to help us grow.	Potential reduced State and Federal funds. Loss of Manufacturing in the United States.

Develop relationships, conduct discovery at companies, draft discovery agreements, reach consensus, draft proposals		New management and new field staff development, potential restructuring
Our Field Staff has face-to-face contact with clients.		Less funding and the single focus on Innovation Engineering
Personal visits, email	MEP has unfortunately not been increasing funding to Centers despite increased funding at the national level	Growing MEP reporting requirements. The new CORE system coming on-line October 1 is overkill and nothing more than a system behavior modification project. Too much emphasis on innovation vs. needed core services
Detailed , multiple conversations addressing all questions and explaining how and where we can help them improve or help them solve a specific problem. FOLLOW UP.	Centers are going to have to develop additional funding sources for some of the new initiatives, banks, state EDA funding , partnerships, etc. As the success grows. Additional MEP funding should be provided for initial efforts.	By far the largest challenge for us will be the transition to the NGS from the traditional service offerings. Our client base consist of the types of companies that still require the traditional services. So making that transition and maintaining some balance between the new and old will be a large challenge.

What are your plans for the medium and long term? How do you see your Center changing over three years and again further out, in the next 5-8 years?	Recommendations --Assuming that MEP funding levels remain constant, what changes would you recommend be made at the Federal and state levels to make your use of MEP funding more effective?	And if funding were to be increased, what would your priorities be for utilizing additional funding (perhaps also explain what level of additional funding would be required to implement the changes you recommend)
Less reliance on federal and state funding. Greater reliance on delivering professional services from contractors. AMEP will move up the hierarchy of companies to work with top tier oilfield services companies.	Maintain and refine balanced scorecard metrics. Apply leverage to solve the myriad problems at Dun & Bradstreet NAICS reporting system.	Apply leverage to solve the myriad problems at Dun & Bradstreet NAICS reporting system. Get funding directly into job creating technology startups. Expand NIST MEP client definitions to include technology & software clients.
Much more focus on export	Stronger state support and collaboration in business retention and expansion	$200,000 focus on technology grants for growth and export services to encourage the smaller companies to get involved.
We have invested heavily ($2M) in the past four years in new Growth services and consulting staff. In the coming 1-3 years it will be a focus on staff development to take advantage of our investment in new	Communication from MEP to Centers and back. Things are directed to Centers as if they are all alike...they are not.	We would hire more experienced staff with backgrounds in Business Growth consultative sales and delivery of services.

services.		
	Change cost share to 50/50	Technology Adoption and Scouting, and product innovation
Focusing more on innovation and long term engagements with clients. Longer term, spending more time on succession issues and possibly creating a family business Center.	refine focus. put as much funding out the Centers as possible. Increase base funding.	Increase market penetration and continue to focus on top line growth and innovation as a way to revitalize American manufacturing. Add staff in outreach and key delivery areas. Double or triple the base funding amount. THE ROI is extremely high on the program and the funding is a small amount and has been stagnant for a long while.
3 yrs - Transition from a traditional business model with Project Management generalists assigned to geographic territories to a mix of specialists with a focus on company growth opportunities in industry clusters. 5-8 yrs - Develop and train staff on a full suite of tools to address both top and	Predictability of funding levels at the Federal and state level would make the use of the funds we receive more effective.	Increased funding could be used to assist smaller companies that don't have the resources to do the improvement projects that they know they need to do.

bottom line growth for companies that is responsive to their needs in an evolving global market.		
Sorry- you're asking for a strategic plan in a tiny box- this is not realistic. Basically higher level long term engagement vs. point source fixes	Less time in surveys and delete NIST requirement for Duns registration- MAJOR issue for clients	More innovation, deeper assessments as lead in low cost activities
1. Moving to longer term engagements with clients as trusted advisor (in contrast to "point solutions" provider) 2. More focus on assisting companies with growth strategies based on innovation 3. Aggressive effort to broaden and diversify funding base.	Align MEP with other new programs such as NNMI and build MEP's into delivery model to provide access for small-to-mid-sized manufacturers	-More funding support for MEP's to use in leveraging resources from universities/colleges -Reinstitute TIP (or some variation thereof)
More collaboration with strategy More engagement at the highest level of leadership More innovation leadership communities formed	elimination of the match requirements that serve more of a hindrance than a catalyst for success.	More innovation measures and more tools to demonstrate change and success to potential clients.
Expanding Technology Acceleration Services Developing Innovation Services	Allow for a predictable NIST funding Cash Flow. Only receive 1/3 of funding within in the fiscal year	We would add addition field staff to cover our expanded new service offerings. Would be looking for a 15%

		increase
Medium--Have implemented MEP.ms Innovation Practice throughout five MEP.ms Centers and in use in all MEP.ms projects. Long term: MEP.ms is helping proactive companies in Mississippi to grow.	At the Federal Level, maintain at least current support to MEP.ms and lower the cost-share requirement to 50%. At the state level, continue increasing state support for MTA (host) and MEP.ms, particularly in the emerging innovation areas.	Could use additional funding to implement new growth and sustainability areas for MEP.ms such as in E3, Export, and implementation of IE Practice.
we are focused on helping companies "transform" working with their "C" management in a advisory roll. Setting Strategic direction and implementing change.	1. Get the money out to the Centers. The Centers are the working with companies. This would allow Centers to work more with companies and increase the penetration NIST/MEP wants. 2. Have Commerce & NIST change the "Match" to 1:1, a majority of the Centers are having difficulty making the match because they have lost state funding 3. have a 3rd party evaluation done of Innovation engineering to determine the ROI.	you could require Centers to increase penetration based on additional money. Every Center should receive - or be eligible to receive extra dollars. this way you can gain political support for the program on both sides of the aisle. As a Center you can say I benefited form the additional money. And as a result the MEP system is more effective and reaching more manufacturers. A review of all the Centers should be done on a per capita basis - those who are under funded should get a larger portion of

		funding.
Focus more toward leading companies with Growth Strategies. We will continue to assist companies with helping them to reduce operating costs...but will focus on ways to help them reduce operating costs as part of a strategy for growth.	Reduce Cost share to 1:1.	Increase Staff to serve more manufacturers. NOTE: As it stands now, since cost share is 2:1 and many states cannot get that much cost share, increasing federal funding would not get down to the Centers. Many MEP Centers cannot "draw" down all their program money now due to the cost share issue.
Engineers will have to learn to shorten the sales cycle from contact to contract. Both because it will make them more efficient and because clients will need faster engagement and delivery.	Relax match requirement so funds can be used to match other grants. In a similar fashion, if MEP would provide greater subsidies for the training conference and other training events it would help stretch Center funds greatly.	A total of $150,000 per year would allow us to add an engineer which would increase our capacity by 20%. In turn, the effort in client growth and resulting impact would increase more than 20%.

We are good in listening to the needs of out customers so the core work we are doing won't change too much. We have been doing Top line for 6 years and expect to implement Innovation Engineering very well over the next year. Things change very quickly these days but I have been in manufacturing for 43 years and most of our problems have been around for a long time. I see the basic work we are doing not changing that much because we have been working with the corner office for a long time. Having been a CEO of Manufacturing Companies for years, I know we are doing what they need. It is hard to look out 5 years but as long as we listen to out customers we will continue to be an important part of the growth of Manufacturing.	Finding talent for Manufacturing companies is the biggest problem our companies have these days and this will only get worse. MEP should lead the fight to help to improve the image of manufacturing in the US.	I wish I could work with more companies and with more money I could add the needed staff to get the job done.

Transition to "new strategy" services and efforts to serve "new clients" as defined by NIST MEP. No strategy beyond 5 years.	Change NIST MEP funding process to "block grant" format to reduce extensive overhead associated with current Cooperative Agreement and free up funds for additional deliverables to mfg clients.	50-50 block grant funding scenario.
Our plans are to identify the "legacy" services that have achieved self-sufficiency and retain the capability to deliver those services, but shed off those that are not self-sufficient UNLESS they are critical and directly tied to our public mission. Those that do not have a "public mission" fit AND are not self-sustaining financially need to be wound-down. This will free up resources for Next Generation services that are very important to our manufacturers. I suspect that our Center will see a complete transformation over the long term. We are currently known for LEAN, TRAINING and, to some extent, SUBSIDIES. In 4-6	I believe that national promotion of the MEP brand is critical to the improvement of all (collective) MEP effectiveness. We need to leverage that we are a network that rivals any "big" consulting firm, but right-sized for the small and mid-sized manufacturer. We need branding at a national level.	MEP program development (new service development). MEP national branding. MEP leadership development. MEP-funded research on manufacturing topics (industry reports).

years, we will likely be known for INNOVATION, ROI, CONSULTING and VALUE. This is already happening and will continue.		
Bolster our third party resources (we are a brokered model) for growth services. This has already begun. Further out we will be more integrated with the State's entities focused on economic development.	Much less reporting layers for the Centers. Fewer administrative staff at the Federal level. Centers have to produce the result but as Federal staff increases, it seems additional demands are placed on the Centers. Centers are struggling with decreased state funds and how to manage to increasing and changing metrics all the while trying	Marketing support from NIST-MEP National focused down to the State and Center Level. Packaged videos highlighting the national tools available to Centers that can be used with potential funding partners. Increasing Center base awards would sure be appreciated so we add some much needed administrative help to meet the current NIST-MEP

	to develop new growth services for companies. Something's got to give!	demands.
C-level conversations - a more strategic outlook. Building an MEP that will help small manufacturers grow and think strategically.	Reduce the cost share burden. Modify the definition of cost share. State changes are up to the Center to effect.	It dependents on the structure of the money. I don't see a use for additional restricted funds.
Double the size of our Center in terms of revenue. Become and integral part of the state and local economic development. Integrate further the resources of the University, faculty and students, with the NWMOC.	Reducing Match requirements	Hire additional staff, improve training for staff, and invest in new services. Unsure of the funding required,
Our 5 year strategic plan addresses: 1) Client revenue growth ROI and profitability ROI 2) Increasing stakeholder expectations (metrics) by ~3X 3) Attain a self-sustaining financial model of our Center.	1) Eliminate or reduce significantly the matching funds requirement (too much resource required to ensure making) 2) Allow the use of non-MEP funds for projects that support economic sectors that support manufacturing, e.g.	Use the funds to; 1) Create programs/initiatives that would provide subsidies to the mfg in the 10-100 employee range. They are hard pressed to identify resources (people & money) to address growth opportunities on a consistent basis. 2)

	utilities, distribution, transportation.	Create programs/initiatives that would address market penetration. Made in America & NGR are starts.
more partners and third party relationships	NIST MEP - implement the 1;1 match, as the law has intended it to be	Historically NIST MEP doesn't distribute additional funds to the Centers for the increases they have gotten so this seems unlikely
We are working on growing the company over the next four years. Looking at developing different revenue streams outside of the Federal and State money.	We would like to see several MEP's around the system help create the National NIST MEP Strategy.	Hire more agents to help with market share. Help develop a system to help Inventors grow into decent size companies. Increase the number of Road shows in the state to help educate the manufacturing community on new trends or services. Looking at additional $1,000,000 to do the above. It would help a great deal if we can get the Federal cost share reduced from 2 to 1 to 1 to 1. This would help a lot of MEP Centers.

Near: Increased focus on Food and Dairy, and recreation technology sectors with innovation. Medium: Restructuring team leadership and staff responsibilities. Long: More and more affiliations/partnerships on initiatives addressing key industry sectors, developing/improving state partnerships, sustainable programs with potential to be independent of grant funds.	Reduce match requirement at the federal level. Work to increase state funding of MEP program.	Additional MEP funding would be used to target the food and dairy sector with additional staff. Approximately 25% ($125,000) would allow us to add two staff to support this critical under-served sector
Develop a full innovation strategy which includes Innovation Engineering	Reduce match to 1X	2X current funding. We would do more partner funding
More emphasis on growth through innovation. More engagement at the C level of companies. Plan is to respond to the most pressing needs of Vermont SME's and the priorities of Vermont's stakeholder (State of VT) despite MEPs growing, heavy handed approach with all Centers (i.e. always asking for more and more from	Reduce staff at MEP HQ. Reduce number/frequency of costly Directors meeting. Eliminate MEP Regional Managers. They add little to no value. Fix the Cost Share problem so MEP Centers have enough match to draw down their federal funds.	Increase the base funding of all Centers to enable the addition of at least 2- 4 more FTEs ("feet on the street"). Do NOT increase staff levels at MEP HQ any further.

Centers without providing any increased funding - despite increases provided at the national level by Congress. MEP field staff has shrunk from about 1600 in 2008 to about 1200- 1300 today. MEP HQ has significantly increased its staff. What's wrong with that picture?		
The Center has to grow in terms of client fees and staff numbers. We should be adding additional staff over the next 3 years and beyond. We will be shifting expertise and skill sets towards energy work and the Next Generation strategies. Also a much tighter relationship with our university partners and a heavier involvement by our board and other partners in the state to promote the MEP expand the MEP.	Streamline the renewal and review process.	Increases in staff levels

Appendix D

Bibliography

American Small Manufacturers Coalition. 2009. Next Generation Manufacturing Study: Overview and Findings. Washington, DC: American Small Manufacturers Coalition. June.

Atkinson, R. D. 2011. *Explaining Anemic U.S. Job Growth: The Role of Faltering U.S. Competitiveness*. Washington, DC: The Information Technology and Innovation Foundation. December.

Autor, D., D. Dorn, and G. Hansen. 2011. "The China Syndrome: The Local Labor Market Effects of Import Competition" Cambridge, Massachusetts: MIT Working Paper.

Berger, S., et al. 2013. *A Preview of the Production in the Innovation Economy Report*. Cambridge: MIT Press.

Bureau of Economic Analysis. 2011. *Survey of Current Business 2006-2009*. Washington, DC: U.S. Department of Commerce. January.

Caputo, A., et al. 2002. "A methodological framework for innovation transfer to SMEs." *Industrial Management and Data Systems* 102(5):271-283.

Chapman, R. 1998. *Using Data Envelopment Analysis to Assess Performance of Manufacturing Extension Centers*. NISTIR 6198. Gaithersburg, Maryland: National Institute of Standards and Technology. July.

Cheney, D, C. Ordowich, J. Youtie, A. Fernández-Ribas, and P. Shapira. 2009. *Evaluating the Impact of MEP Services on Establishment Performance: A Preliminary Empirical Investigation*. Technical Report. Arlington, Virginia. November.

Cosmos Corporation. 1997. *MEP Successes: A Case Study Approach*. NIST Special Publication 916. Washington, DC: Government Printing Office.

Cosmos Corporation. 1998. *MEP Successes: Case Study Series Exemplary Projects Case Studies*. Washington, DC: Government Printing Office. November.

Cosmos Corporation. 1999. *MEP Successes: Case Study Series Transformed Firms Case Studies*. Washington, DC: Government Printing Office. April.

Davila, N. 2004. "Evaluating Manufacturing Extension: A Multidimensional Approach." *Economic Development Quarterly*. 18(3):286-302.

Deloitte. 2004. "Manufacturing Pennsylvania's Future: Regional Strategies that Build from Current Strengths and Address Competitive Challenges." Submitted to Industrial Resource Centers, PA Department of Community and Economic Development and Team PA Foundation.

Ellis, S. 1998. "Evaluation of Massachusetts Manufacturing Partnership: Selected Findings." In P. Shapira and J. Youtie, eds. *Manufacturing Modernization: Implications of Evaluation Results for Program Improvement and Policy Development. Proceedings of Fourth Workshop on the Evaluation of Industrial Modernization Programs*. Atlanta, Georgia: Georgia Institute of Technology.

Executive Office of the President. 2009. *A Framework for Revitalizing American Manufacturing*. Washington, DC: Executive Office of the President.

Ezell, S., and R. Atkinson. 2011. *International Benchmarking of Countries' Policies and Programs Supporting SME Manufacturers*. Washington, DC: Information Technology and Innovation Foundation. September.

Feller, I. A. Glasmeier, and M. Mark. 1996. "Issues and perspectives on evaluating manufacturing modernization programs." *Research Policy* 25(2):309-319.

Frazier, R. M. 2011. "The Imperatives of Successful Policy Implementation: An Evaluation of the Hollings National Institute of Standards and Technology-Manufacturing Extension Partnership (NIST-MEP) Program's Implementation in Arkansas." (Doctoral dissertation). University of Arkansas.

Glasmeier, A., K. Fuellhart, I. Feller, and M. Mark. 1998. "The Relevance of Firm-Learning Theories to the Design and Evaluation of Manufacturing Modernization Programs." *Economic Development Quarterly* 12(2):107-124. May.

Helper, S. 2008. *Renewing U.S. Manufacturing: Promoting a High-Road Strategy*. Washington, DC: Economic Policy Institute.

Helper, S., T. Krueger, and H. Wial. 2012. *Why Does Manufacturing Matter? Which Manufacturing Matters? A Policy Framework*. Washington, DC: The Brookings Institution. February.

Helper, S., and H. Wial. 2010. *Strengthening American Manufacturing: A New Federal Approach*. Washington, DC: The Brookings Institution.

Hollings Manufacturing Extension Partnership Advisory Board, with G. Yakimov and L. Woolsey. 2010. *Innovation and Product Development in the 21st Century*. Gaithersburg, Maryland. February.

Jarmin, R. S. 1998. "Manufacturing Extension and Productivity Dynamics: Preliminary Evidence." In P. Shapira and J. Youtie, eds. *Manufacturing*

Modernization: Implications of Evaluation Results for Program Improvement and Policy Development. Proceedings of Fourth Workshop on the Evaluation of Industrial Modernization Programs. Atlanta, Georgia: Georgia Institute of Technology.

Jarmin, R. S. 1999. "Evaluating the Impact of Manufacturing Extension on Productivity Growth." *Journal of Policy Analysis and Management* 18(1):99-119.

Kelly, M. 1997. "From Mission to Commercial Orientation: Perils and Possibilities for Federal Industrial Technology Policy." *Economic Development Quarterly* 11(4):313-328. November.

Kingsley, G., and H. Klein. 1998. "Interfirm Collaboration as a Modernization Strategy: A Survey of Case Studies." *The Journal of Technology Transfer* 23(1):65-74. Spring.

Luria, D. 1997. "Toward Lean or Rich? What Performance Benchmarking Tells Us About SME Performance, and Some Implications for Extension Center Services and Mission." In P. Shapira and J. Youtie, eds. *Manufacturing Modernization: Learning from Evaluation Practices and Results: Evaluation of Regionally-Based S&T Programs. Proceedings of Third Workshop on the Evaluation of Industrial Modernization Programs*. Atlanta, Georgia: Georgia Institute of Technology. Pp. 6-29.

Luria, D. 2011. "Evaluating the MEP Evaluation." Presentation at the National Academies workshop on "Strengthening American Manufacturing: The Role of the Manufacturing Extension Partnership." Washington, DC. November 14.

Luria, D., and E. Wiarda. 1996. "Performance benchmarking and measuring program impacts on customers: lessons from the Midwest Manufacturing Technology Center." *Research Policy* 25(2):233-246.

Manufacturing Extension Partnership. 1994. *Making a Difference for America's Manufacturers*. Gaithersburg, Maryland: National Institute of Standards and Technology.

Manufacturing Extension Partnership. 1997. *MEP National Data Highlights*. Gaithersburg, Maryland: National Institute of Standards and Technology. December.

Manufacturing Extension Partnership. 1998. *Review of Mission and Operations of Regional Centers of the Manufacturing Extension Partnership*. Gaithersburg, Maryland: National Institute of Standards and Technology. February.

Manufacturing Extension Partnership. 2008. *The Future of the Hollings Manufacturing Extension Partnership*. Gaithersburg, Maryland: National Institute of Standards and Technology. December.

Manufacturing Extension Partnership. 2011. *The Manufacturing Extension Partnership: Partnering for Manufacturing Innovation and Growth*. Gaithersburg, Maryland: National Institute of Standards and Technology. June.

Michigan Manufacturing Technology Center. 1996. *Fifth Year Review*. Ann Arbor, Michigan: Michigan Manufacturing Technology Center.

Modernization Forum and Nexus Associates. 1997. *Competition or Collaboration: The Role of Manufacturing Extension Centers in the Private Consulting Market. Working with Consultants: A Tool Kit for Manufacturing Extension Centers, Volume 3*. Washington, DC: National Institute of Standards and Technology.

National Academy of Public Administration. 2003. *The National Institute of Standards and Technology's Manufacturing Extension Partnership Program Report 1: Re-examining the Core Premise of the MEP Program*. Washington, DC. September.

National Academy of Public Administration. 2004. *The National Institute of Standards and Technology's Manufacturing Extension Partnership Report 2: Alternative Business Models*. Washington, DC. May.

National Commission on Fiscal Responsibility and Reform. 2010. *$200 Billion in Illustrative Savings*. A supplement to a draft proposal by the co-chairs. Washington, DC.

National Economic Council. 2011. *A Strategy for American Innovation: Driving Towards Sustainable Growth and Quality Jobs*. Washington, DC: The White House.

National Research Council. 1993. *Learning to Change: Opportunities to Improve the Performance of Small Manufacturers*. Washington, DC: National Academy Press.

National Research Council. 1996. *Conflict and Cooperation in National Competition for High-Technology Industry*. Washington, DC: National Academy Press.

National Research Council. 1999. *The Advanced Technology Program: Challenges and Opportunities.* C. W. Wessner, ed. Washington, DC: National Academy Press.

National Research Council. 1999. *Industry-Laboratory Partnerships: A Review of the Sandia Science and Technology Park Initiative.* C. W. Wessner, ed. Washington, DC: National Academy Press.

National Research Council. 1999. *New Vistas in Transatlantic Science and Technology Cooperation.* C. W. Wessner, ed. Washington, DC: National Academy Press.

National Research Council. 1999. *The Small Business Innovation Research Program: Challenges and Opportunities.* C. W. Wessner, ed. Washington, DC: National Academy Press.

National Research Council. 1999. *U.S. Industry in 2000: Studies in Competitive Performance*. D. C. Mowery, ed. Washington, DC: National Academy Press.

National Research Council. 2000. *The Small Business Innovation Research Program: A Review of the Department of Defense Fast Track Initiative.* C. W. Wessner, ed. Washington, DC: National Academy Press.

National Research Council. 2001. *A Review of the New Initiatives at the NASA Ames Research Center*. C. W. Wessner, ed. Washington, DC: National Academy Press.

National Research Council. 2001. *Building a Workforce for the Information Economy*. Washington, DC: National Academy Press.

National Research Council. 2001. *Capitalizing on New Needs and New Opportunities: Government-Industry Partnerships in Biotechnology and Information Technologies*. C. W. Wessner, ed. Washington, DC: National Academy Press.

National Research Council. 2001. *The Advanced Technology Program: Assessing Outcomes*. C. W. Wessner, ed. Washington, DC: National Academy Press.

National Research Council. 2001. *Trends in Federal Support of Research and Graduate Education*. S. A. Merrill, ed. Washington, DC: National Academy Press.

National Research Council. 2002. *Partnerships for Solid-State Lighting*. C. W. Wessner, ed. Washington, DC: National Academy Press.

National Research Council. 2003. *Government-Industry Partnerships for the Development of New Technologies: Summary Report*. C. W. Wessner, ed. Washington, DC: The National Academies Press.

National Research Council. 2003. *Securing the Future: Regional and National Programs to Support the Semiconductor Industry*. C. W. Wessner, ed. Washington, DC: The National Academies Press.

National Research Council. 2004. *Productivity and Cyclicality in Semiconductors: Trends, Implications, and Questions*. D. W. Jorgenson and C. W. Wessner, eds. Washington, DC: The National Academies Press.

National Research Council. 2004. *The Small Business Innovation Research Program: Program Diversity and Assessment Challenges*. C. W. Wessner, ed. Washington, DC: The National Academies Press.

National Research Council. 2005. *Deconstructing the Computer*. Dale W. Jorgenson and C. W. Wessner, eds. Washington, DC: The National Academies Press.

National Research Council. 2006. *Software, Growth, and the Future of the U.S. Economy*. D. W. Jorgenson and C. W. Wessner, eds. Washington, DC: The National Academies Press.

National Research Council. 2006. *The Telecommunications Challenge: Changing Technologies and Evolving Policies*. C. W. Wessner, ed. Washington, DC: The National Academies Press.

National Research Council. 2007. *Enhancing Productivity Growth in the Information Age: Measuring and Sustaining the New Economy*. D. W. Jorgenson and C. W. Wessner, eds. Washington, DC: The National Academies Press.

National Research Council. 2007. *Innovation Policies for the 21st Century*. C. W. Wessner, ed. Washington, DC: The National Academies Press.

National Research Council. 2007. *India's Changing Innovation System: Achievements, Challenges, and Opportunities for Cooperation*. C. W. Wessner and S. J. Shivakumar, eds. Washington, DC: The National Academies Press.

National Research Council. 2007. *Innovation Inducement Prizes at the National Science Foundation*. S. A. Merrill, ed. Washington, DC: The National Academies Press.

National Research Council. 2007. *SBIR and the Phase III Challenge of Commercialization*. C. W. Wessner, ed. Washington, DC: The National Academies Press.

National Research Council. 2008. *An Assessment of the SBIR Program*. C. W. Wessner, ed. Washington, DC: The National Academies Press.

National Research Council. 2008. *An Assessment of the SBIR Program at the Department of Energy*. C. W. Wessner, ed. Washington, DC: The National Academies Press.

National Research Council. 2008. *An Assessment of the SBIR Program at the National Science Foundation*. C. W. Wessner, ed. Washington, DC: The National Academies Press.

National Research Council. 2008. *Innovative Flanders: Innovation Policies for the 21st Century*. C. W. Wessner, ed. Washington, DC: The National Academies Press.

National Research Council. 2008. *Innovation in Global Industries: U.S. Firms Competing in a New World*. J. Macher and D. Mowery, eds. Washington, DC: The National Academies Press.

National Research Council. 2009. *21st Century Innovation Systems for Japan and the United States: Lessons from a Decade of Change*. S. Nagaoka, M. Kondo, K. Flamm, and C. Wessner, eds. Washington, DC: The National Academies Press.

National Research Council. 2009. *An Assessment of the SBIR Program at the Department of Defense*. C. W. Wessner, ed. Washington, DC: The National Academies Press.

National Research Council. 2009. *An Assessment of the SBIR Program at the National Aeronautics and Space Administration*. C. W. Wessner, ed. Washington, DC: The National Academies Press.

National Research Council. 2009. *An Assessment of the SBIR Program at the National Institutes of Health*. C. W. Wessner, ed. Washington, DC: The National Academies Press.

National Research Council. 2009. *Hidden Costs of Energy: Unpriced Consequences of Energy Production and Use*. Washington, DC: The National Academies Press.

National Research Council. 2009. *Revisiting the Department of Defense SBIR Fast Track Initiative*. C. W. Wessner, ed. Washington, DC: The National Academies Press.

National Research Council. 2009. *Understanding Research, Science and Technology Parks: Global Best Practices*. C. W. Wessner, ed. Washington, DC: The National Academies Press.

National Research Council. 2009. *Venture Funding and the NIH SBIR Program*. C. W. Wessner, ed. Washington, DC: The National Academies Press.

National Research Council. 2010. *Managing University Intellectual Property in the Public Interest*. Stephen Merrill and A. Mazza, eds., Washington, DC: The National Academies Press.

National Research Council. 2011. *Building the 21st Century: U.S.-China Cooperation on Science, Technology, and Innovation*. C. W. Wessner, rapporteur. Washington, DC: The National Academies Press.

National Research Council. 2011. *Growing Innovation Clusters for American Prosperity*. C. W. Wessner, rapporteur, Washington, DC: The National Academies Press.

National Research Council. 2011. *The Future of Photovoltaics Manufacturing in the United States*. C. W. Wessner, rapporteur. Washington, DC: The National Academies Press.

National Research Council. 2012. *Building Hawaii's Innovation Economy*. C. W. Wessner, rapporteur. Washington, DC: The National Academies Press.

National Research Council. 2012. *Building the Arkansas Innovation Economy*. C. W. Wessner, rapporteur. Washington, DC: The National Academies Press.

National Research Council. 2012. *Building the U.S. Battery Industry for Electric-Drive Vehicles: Progress, Challenges, and Opportuntities*. C. W. Wessner, rapporteur. Washington, DC: The National Academies Press.

National Research Council. 2012. *Clustering for 21st Century Prosperity*. C. W. Wessner, rapporteur. Washington, DC: The National Academies Press.

National Research Council. 2012. *Meeting Global Challenges: German-U.S. Innovation Policy*. C. W. Wessner, rapporteur. Washington, DC: The National Academies Press.

National Research Council. 2012. *Rising to the Challenge: U.S. Innovation Policy for the Global Economy*. C. W. Wessner and Alan Wm. Wolff, editors. Washington, DC: The National Academies Press.

National Research Council. 2013. *Building the Illinois Innovation Economy: Summary of a Symposium*. C. W. Wessner, rapporteur. Washington, DC: The National Academies Press.

National Research Council. 2013. *Building the Ohio Innovation Economy: Summary of a Symposium*. C. W. Wessner, rapporteur. Washington, DC: The National Academies Press.

National Research Council. 2013. *Strengthening American Manufacturing: The Role of the Manufacturing Extension Partnership—Summary of a Symposium*. C. W. Wessner, rapporteur. Washington, DC: The National Academies Press.

National Science Board. 2010. *Science and Engineering Indicators 2010.* Arlington, VA: National Science Foundation.

National Science Foundation, National Center for Science and Engineering Statistics. *Research and Development in Industry: 2006-07.* NSF 11-301. Arlington, Virginia: National Science Foundation.

Nexus Associates, Inc. 1996. *Evaluation of the New York Manufacturing Extension Program: Final Report.* Prepared for the New York Science and Technology Foundation/Empire State Development. Belmont, Massachusetts. March 18.

Nexus Associates, Inc. 1999. *The Pennsylvania Industrial Resource Centers: Assessing the Record and Charting the Future.* Report prepared the Ben Franklin/IRC Partnership Board. Belmont, Massachusetts. October.

Office of Management and Budget. 2002. *Program Assessment Rating Tool (PART), Detailed Information on the Manufacturing Extension Partnership Assessment.* Washington, DC: Government Printing Office.

Oldsman, E. 1996. "Does manufacturing extension matter? An evaluation of the Industrial Technology Extension Service in New York." *Research Policy* 25(2):215-232.

Oldsman, E. 1997. "Manufacturing extension centers and private consultants: collaboration or competition?" *Technovation* 17(5):237-244.

Oldsman, E. 2004. "Evaluating SME Programs: Learning from the NIST Manufacturing Extension Partnership." Paper presented at the SME Evaluation Workshop. Mexico City. September 23.

Oldsman, E., and C. Heye. 1998. "Waste not, want not—A look at the impact of manufacturing extension centers." *Journal of Technology Transfer* 23(1):37-41.

Pisano, G., and W. Shih. 2009. "*Restoring American Competitiveness.*" *Harvard Business Review* July.

President's Council of Advisors on Science and Technology. 2011. "Report to the President on Ensuring American Leadership in Advanced Manufacturing." Washington, DC: Executive Office of the President. June 2011.

Roper, S, J. Youtie, and A. Fernandez-Ribas 2010. "Knowledge, Capabilities and Manufacturing Innovation: A USA–Europe Comparison." *Regional Studies* 44(3):253–279.

Schact, W. 2011. *Manufacturing Extension Partnership Program: An Overview, 7-5700.* Washington, DC: Congressional Research Service.

Schrank, A., and J. Whitford. 2009. "Industrial Policy in the United States: A Neo-Polanyian Interpretation." *Politics and Policy* 37(4):521-553.

Sears, D., and P. Blackerby. 1998. "The MEP Evaluation Approach: Past, Present and Future." In P. Shapira and J. Youtie, eds. *Manufacturing Modernization: Implications of Evaluation Results for Program Improvement and Policy Development, Proceedings of Fourth*

Workshop on the Evaluation of Industrial Modernization Programs. Atlanta, Georgia: Georgia Institute of Technology. Pp. 161-170.

Shapira, P. 2001. "U.S. manufacturing extension partnership: Technology policy reinvented?" *Research Policy* 8(3):66-72.

Shapira, P. 2003. "Evaluating manufacturing extension services in the United States: experiences and insights." In P. Shapira and S. Kuhlmann, eds. *Learning from Science and Technology Policy Evaluation: Experiences from the United States and Europe*. Cheltenham, United Kingdom, and Northampton, Massachusetts: Edward Elgar.

Shapira, P. 2006. *Product and Service Innovation: Report to the Manufacturing Extension Partnership, National Institute of Standards and Technology,* Atlanta, Georgia, and Arlington, Virginia: Georgia Tech Program in Science, Technology, and Innovation Policy, and SRI International.

Shapira, P. 2009. "Innovation and small and midsize enterprises: innovation dynamics and policy strategies." In R. Smits, S. Kuhlmann and P. Shapira, eds., *Innovation Policy: Theory and Practice. An International Handbook*. Cheltanham, United Kingdom: Edward Elgar.

Shapira, P., and T. Rephann. 1996. "The adoption of new technology in West Virginia: Implications for manufacturing modernization policies." *Environment and Planning C: Government and Policy* 14:431-450.

Shapira, P., J. Roessner, and R. Barke. 1995. "New pubic infrastructures for small firm industrial modernization in the USA." *Entrepreneurship and Regional Development*. 7:63-84.

Shapira, P., and J. Youtie. 1995. *Assessing GMEA's Economic Impacts: Towards a Benefit-Cost Methodology*. Atlanta, Georgia: Georgia Institute of Technology. March. (Revised October 1996).

Shapira, P., and J. Youtie. 1997. "Coordinating Manufacturing Extension Services." *The Journal of Technology Transfer* 22(1):5-10. Spring.

Shapira, P., and J. Youtie. 1998. "Evaluation Industrial Modernization: Methods, Results, and Insights from the Georgia Manufacturing Extension Alliance." *The Journal of Technology Transfer* 23(1):17-28. Spring.

Shapira, P., J. Youtie, and L. Kay. 2011. "Building Capabilities for Innovation in SMEs: A Cross-Country Comparison of Technology Extension Policies and Programs." *International Journal of Innovation and Regional Development* 3-4:54-272.

Shapira, P., J. Youtie, and J. D. Roessner. 1996. "Current Practices in the Evaluation of U.S. Industrial Modernization Programs." *Research Policy* 25(2):185-214.

Shapira, P., J. Youtie, J. Wang, D. Hegde, D. Cheney, Q. Franco, and S. Mohapatra. 2004. *Re-assessing the Value of Information and its Impact on Productivity in Small and Midsize Manufacturers*. Atlanta, Georgia, and Arlington, Virginia: Georgia Tech Policy Project on Industrial Modernization and SRI International.

SRI and Georgia Tech. 2008. *Eureka! Winning Ways: Analysis of Early Client Experiences*. Arlington, Virginia. August.

SRI and Georgia Tech. 2009. *Making an Impact: Assessing the Benefits of Ohio's Investment in Technology-based Economic Development Programs*. Arlington, Virginia. September.

Stone & Associates and the Center for Regional Economic Competitiveness. 2010. *Re-examining the Manufacturing Extension Partnership Business Model: Alternatives for Increasing the Program's Impact on US Manufacturing Sector Performance*. Gaithersburg, Maryland. October.

Swamidass, P. 1994. *Technology on the Factory Floor II: Benchmarking Manufacturing Technology Use in the United States*. Washington, DC: The Manufacturing Institute.

Tassey, G. 2009. *The Technology Imperative*. Northampton, Massachusetts: Edward Elgar.

Tassey, G. 2010. "Rationales and Mechanisms for Revitalizing U.S. R&D Manufacturing Strategies." *Journal of Technology Transfer* 35:283-333.

Thompson, C. 1998. "Local Politics, National Policy, and the Taxpayer-Payback of Manufacturing Extension." *The Journal of Technology Transfer* 23(1):37-42. Spring.

U.S. Census Bureau. 2010. *Trade in Advanced Technology Products*. Washington, DC: U.S. Department of Commerce.

U.S. Department of Commerce. 2012. *The Competitiveness and Innovative Capacity of the United States*. Washington, DC: U.S. Department of Commerce. January.

U.S. Department of Commerce. 2012. *U.S. Competitiveness and Innovation Policy*. Washington, DC: U.S. Department of Commerce. January.

U.S. General Accounting Office. 1991. *Technology Transfer: Federal Efforts to Enhance the Competitiveness of Small Manufacturers*. GAO/RCED-92-30. Washington, DC: General Accounting Office.

U.S. General Accounting Office. 1995. *Manufacturing Extension Programs: Manufacturers' Views of Services*. GAO/GGD-950210216BR. Washington, DC: General Accounting Office.

U.S. Government Accountability Office. 2011. *NIST Manufacturing Extension Partnership Program Cost Share*. GAO-11-437R. Washington, DC: U.S. Government Accountability Office.

U.S. Government Accountability Office. 2013. *Corporate Income Tax: Effective Tax Rates Can Differ Significantly from the Statutory Rate*. GAO-13-520. Washington, DC: U.S. Government Accountability Office. May 30.

Voytek, K., K. Lellock, and M. Schmit. 2004. "Extension Partnership Program Developing Performance Metrics for Science and Technology Programs: The Case of the Manufacturing." *Economic Development Quarterly* 18(2):174-185. May.

Washington Post. 2010. "Made in Germany, Sold in China." September 17.

Welch, D., E. Oldsman, P. Shapira, J. Youtie, and J. Lee. 1997. *Net benefits: An assessment of manufacturing business networks and their impacts on member companies*. Chapel Hill, North Carolina: USNet and Regional Technology Strategies.

White, T. K., J. Reiter, and A. Petrin. 2012. *Plant-level Productivity and Imputation of Missing Data in U.S. Census Manufacturing Data*. National Bureau of Economic Research Working Paper No. 17816.

Wilkins, T. 1998. "Benchmarking Manufacturing Extension Centers." In P. Shapira and J. Youtie, editors. *Manufacturing Modernization: Implications of Evaluation Results for Program Improvement and Policy Development: Proceedings of Fourth Workshop on the Evaluation of Industrial Modernization Programs*. Atlanta, Georgia: Georgia Institute of Technology.

Yin, R., S. Merchlinsky, and K. Adams-Kennedy. 1998. *Evaluation of MEP SBDC Partnerships*. Bethesda, Maryland: COSMOS Corporation.

Youtie, J., and P. Shapira. 1997. *GMEA 1997: Review of Results*. Evaluation Working Paper E9701. Atlanta, Georgia. March.

Youtie J., and P. Shapira. 1997. "Tracking Customer Progress: A Follow-up Study of Customers of the Georgia Manufacturing Extension Alliance." *Journal of Technology Transfer* 22(2):43-52. Summer.

Youtie, J., and P. Shapira. 1998. *Summary of Manufacturing Extension Impact Studies*. Georgia Tech Policy Project on Industrial Modernization. Atlanta, Georgia: Georgia Institute of Technology.

Youtie, J., and P. Shapira. 2005. *The Challenge of Manufacturing Innovation: Industry, Rurality, and Competitiveness in the State of Georgia*. Evaluation Working Paper E2501. Atlanta, Georgia, June.

Youtie, J., P. Shapira, L. Kay, D. Dodonova, D. Sabbarese, and C. Morales. 2010. *Innovation in Manufacturing: Needs, Practices, and Performance in Georgia 2010-2012*. Atlanta, Georgia: Georgia Tech Program in Science, Technology, and Innovation Policy.

Youtie, J., P. Shapira, L. Kay, A. Rivera, B. Lynch, G. Cutler, and A. Fernandez-Ribas. 2008. *Innovation in Manufacturing: Needs, Practices, and Performance in Georgia 2008-2010*. Atlanta, Georgia: Georgia Tech Program in Science, Technology, and Innovation Policy.